Responsible Use of AI in Military Systems

Artificial Intelligence (AI) is widely used in society today. The (mis)use of biased data sets in machine learning applications is well-known, resulting in discrimination and exclusion of citizens. Another example is the use of non-transparent algorithms that can't explain themselves to users, resulting in the AI not being trusted and therefore not being used when it might be beneficial to use it.

Responsible Use of AI in Military Systems lays out what is required to develop and use AI in military systems in a responsible manner. Current developments in the emerging field of Responsible AI as applied to military systems in general (not merely weapons systems) are discussed. The book takes a broad and transdisciplinary scope by including contributions from the fields of philosophy, law, human factors, AI, systems engineering, and policy development.

Divided into five sections, Section I covers various practical models and approaches to implementing military AI responsibly; Section II focuses on liability and accountability of individuals and states; Section III deals with human control in human-AI military teams; Section IV addresses policy aspects such as multilateral security negotiations; and Section V focuses on 'autonomy' and 'meaningful human control' in weapons systems.

Key Features:

- Takes a broad transdisciplinary approach to responsible AI
- Examines military systems in the broad sense of the word
- Focuses on the practical development and use of responsible AI
- Presents a coherent set of chapters, as all authors spent two days discussing each other's work

This book provides the reader with a broad overview of all relevant aspects involved with the responsible development, deployment and use of AI in military systems. It stresses both the advantages of AI as well as the potential downsides of including AI in military systems.

Chapman & Hall/CRC Artificial Intelligence and Robotics Series
Series Editor: Roman Yampolskiy

Digital Afterlife and the Spiritual Realm
Maggi Savin-Baden

A First Course in Aerial Robots and Drones
Yasmina Bestaoui Sebbane

AI by Design: A Plan for Living with Artificial Intelligence
Catriona Campbell

The Global Politics of Artificial Intelligence
Edited by Maurizio Tinnirello

Unity in Embedded System Design and Robotics: A Step-by-Step Guide
Ata Jahangir Moshayedi, Amin Kolahdooz, and Liao Liefa

Meaningful Futures with Robots: Designing a New Coexistence
Edited by Judith Dörrenbächer, Marc Hassenzahl, Robin Neuhaus,
and Ronda Ringfort-Felner

*Topological Dynamics in Metamodel Discovery with Artificial Intelligence:
From Biomedical to Cosmological Technologies*
Ariel Fernández

A Robotic Framework for the Mobile Manipulator: Theory and Application
Nguyen Van Toan and Phan Bui Khoi

AI in and for Africa: A Humanist Perspective
Susan Brokensha, Eduan Kotzé, and Burgert A. Senekal

*Artificial Intelligence on Dark Matter and Dark Energy: Reverse Engineering of
the Big Bang*
Ariel Fernández

Explainable Agency in Artificial Intelligence: Research and Practice
Silvia Tulli and David W. Aha

An Introduction to Universal Artificial Intelligence
Marcus Hutter, Elliot Catt, and David Quarel

AI: Unpredictable, Unexplainable, Uncontrollable
Roman V. Yampolskiy

Transcending Imagination: Artificial Intelligence and the Future of Creativity
Alexander Manu

Responsible Use of AI in Military Systems
Jan Maarten Schraagen

For more information about this series please visit: https://www.routledge.com/
Chapman--HallCRC-Artificial-Intelligence-and-Robotics-Series/book-series/
ARTILRO

Responsible Use of AI in Military Systems

Edited by
Jan Maarten Schraagen

CRC Press
Taylor & Francis Group
Boca Raton London New York

CRC Press is an imprint of the
Taylor & Francis Group, an **informa** business
A CHAPMAN & HALL BOOK

First edition published 2024
by CRC Press
2385 NW Executive Center Drive, Suite 320, Boca Raton FL 33431

and by CRC Press
4 Park Square, Milton Park, Abingdon, Oxon, OX14 4RN

CRC Press is an imprint of Taylor & Francis Group, LLC

Funded by Ministerie van Defensie Financieel Administratie en Beheerkantoor FABK

Library of Congress Cataloging-in-Publication Data
Names: Schraagen, Jan Maarten, editor.
Title: Responsible use of AI in military systems / edited by Jan Maarten Schraagen.
Other titles: Responsible use of artificial intelligence in military systems
Description: First edition. | Boca Raton : CRC Press, 2024. |
Series: Chapman & hall/crc artificial intelligence and robotics series |
Includes bibliographical references and index.
Identifiers: LCCN 2023056193 (print) | LCCN 2023056194 (ebook) | ISBN 9781032524306 (hardback) |
ISBN 9781032531168 (paperback) | ISBN 9781003410379 (ebook)
Subjects: LCSH: Artificial intelligence—Military applications. |
Artificial intelligence—Moral and ethical aspects.
Classification: LCC UG479 .R47 2024 (print) | LCC UG479 (ebook) |
DDC 355.40285/63—dc23/eng/20240229
LC record available at https://lccn.loc.gov/2023056193
LC ebook record available at https://lccn.loc.gov/2023056194

ISBN: 978-1-032-52430-6 (hbk)
ISBN: 978-1-032-53116-8 (pbk)
ISBN: 978-1-003-41037-9 (ebk)

DOI: 10.1201/9781003410379

Typeset in Times
by codeMantra

Contents

Preface

The Netherlands organized an international summit (REAIM) on the topic of responsible AI in military systems on February 15-16, 2023. Leading up to this summit, various activities were organized. On behalf of the Dutch Ministry of Defence, an Expert Workshop on the Responsible Use of AI in Military Systems was organized on October 31 and November 1, 2022. The venue was the Naval Establishment in Amsterdam. Thirty academic experts and twenty-five representatives from government, industry, and research institutes came together for two days of highly interactive discussions on how to move the responsible development, deployment, and use of AI in the military domain forward. This Expert Workshop was followed up by an invitation to the academic experts to write a chapter addressing the issues in their field of expertise regarding the responsible use of AI in military systems. The chapters were written in the first half of 2023, taking into account both the outcomes of the Expert Workshop and the REAIM Summit, as well as the latest developments in AI, such as the rise of ChatGTP, which was released in November 2022, just after the Expert Workshop.

The resulting chapters in this book therefore represent the state of the art in the quickly developing field of Responsible AI. They are written from various perspectives and academic disciplines, including, law, ethics, computer science, human factors engineering, and policy making.

Acknowledgements

First and foremost, I would like to thank the Dutch Ministry of Defence for providing me with the opportunity to organize the Expert Workshop on the Responsible Use of AI in Military Systems and to contribute to the organization of the REAIM Summit. Without these activities, and without the Dutch MoD's continued funding and support, this book would not have been possible. In particular, I would like to thank Auke Venema of the Dutch MoD for his unwavering belief in me. Auke was the one responsible for hiring me during my secondment to the Dutch MoD and he stimulated me to take up the many activities of which this book is but one of the results. He also guided me through the intricacies of the MoD's bureaucracy and politics, making it much less of a maze than it otherwise would have been. Many thanks also to Geert Kuiper, Merle Zwiers, Josanne van Gorkum, Merel Roolvink, Lotte Kerkkamp (now with TNO), Joël Postma, Michiel van Dusseldorp, Tom van Hout, and Mirke Beckers, all, with the Dutch MoD. I have thoroughly enjoyed all of your support in the run-up to the REAIM Summit.

Thanks also to the team at the Ministry of Foreign Affairs with whom I collaborated for almost a year while organizing the REAIM Summit: Tessa de Haan, Maarten Wammes, Eline Bötger, Ingeborg Denissen, Veerle Sonneveld, Maud Duit, Maaike Stroeks, Thomas Kist, Pieter Ton, Robin Middel, and Maarit Wittebrood.

This book would also not have been possible without the many stimulating discussions I had with my colleagues at TNO: Jurriaan van Diggelen, Antoine Smallegange, Adelbert Bronkhorst, Marlijn Heijnen, Liisa Janssens, Leon Kester, Pieter Elands, Jasper van der Waa, Tjeerd Schoonderwoerd, Mark Neerincx, and Freek Bomhof. In particular, I would like to thank Jurriaan van Diggelen en Liisa Janssens for their detailed and constructive criticism of my concluding chapter. They cannot be held responsible for any remaining errors in the chapters I wrote for this book. I would like to especially single out Chris Jansen for his continued support in his capacity as my research manager.

Jan Maarten Schraagen

Editor

Jan Maarten Schraagen is a Principal Scientist at The Netherlands Organization for Applied Scientific Research (TNO). His research interests include human-autonomy teaming and responsible AI. He is the main editor of *Cognitive Task Analysis* (2000) and *Naturalistic Decision Making and Macrocognition* (2008) and co-editor of the *Oxford Handzbook of Expertise* (2020). He is editor-in-chief of the *Journal of Cognitive Engineering and Decision Making*. Dr. Schraagen holds a PhD in Cognitive Psychology from the University of Amsterdam, The Netherlands.

Contributors

Sandro Bjelogrlic
Data Science & Artificial Intelligence
    Centre, Chief Technology Office
NATO Communications and Information
    Agency
The Netherlands

Shannon Cooper
Department of Defence
Australia Defence Force
Australia

Damian Copeland
Department of Defence
University of Queensland
Australia

S. Kate Devitt
Human-centred computing, ITEE
University of Queensland
Australia

Linda Eggert
Institute for Ethics in AI, and Balliol
    College
University of Oxford
UK

Pieter Elands
Intelligent Imaging
TNO
The Netherlands

Richard G. Freedman
Smart Information Flow Technologies
USA

Guillaume Gadek
Engineering – Connected Intelligence
Airbus Defence and Space
France

Marlijn Heijnen
Human Machine Teaming
TNO
The Netherlands

Florian Keisinger
Airbus Defence and Space
Germany

Leon Kester
Autonomous Systems & Robotics
TNO
The Netherlands

Wolfgang Koch
Dept. Sensor Data & Information Fusion
Fraunhofer FKIE/University of Bonn
Germany

Jon R. Lindsay
School of Cybersecurity and Privacy, and
    the Sam Nunn School of International
    Affairs
Georgia Institute of Technology
USA

Diego Mauri
Department of Law
University of Florence
Italy

Christopher A. Miller
Smart Information Flow Technologies
USA

Mark Neerincx
Human Machine Teaming
TNO
The Netherlands

Ravi Panwar
United Services Institution of India
India

Lauren Sanders
School of Law
University of Queensland
Australia

Dan Saxon
Leiden University College
The Netherlands

Tjeerd Schoonderwoerd
Human Machine Teaming
TNO
The Netherlands

Afonso Seixas Nunes, SJ
School of Law
St. Louis University
USA

Thomas W. Simpson
Blavatnik School of Government
University of Oxford
UK

Michael Street
Data Science & Artificial Intelligence
   Centre, Chief Technology Office
NATO Communications and Information
   Agency
The Netherlands

Jasper van der Waa
Human Machine Teaming
TNO
The Netherlands

Jurriaan van Diggelen
Human Machine Teaming
TNO
The Netherlands

Kerstin Vignard
National Security Analysis Department
   (APL) and Johns Hopkins Institute for
   Assured Autonomy
Johns Hopkins Applied Physics Lab
USA

Introduction to Responsible Use of AI in Military Systems

1

Jan Maarten Schraagen

Technological developments in Artificial Intelligence (AI) continue to add new dimensions and complexities to world security and future conflict scenarios at an increasing pace. While the application of AI holds great potential for progress and economic growth as well as significant opportunities in the fields of security and defense, its potential misuse in international crises and conflicts may undermine the world's security interests and create risks for international peace and stability. The international community is now faced with the central question of how military application of AI can – and should – be dealt with responsibly while at the same time creating an effective deterrent.

This Introduction will set the stage for the chapters that follow, by providing a brief overview of relevant developments in AI, the military, as well as systems engineering practices. This will be followed by a brief introduction to each chapter, providing the reader with an overview of the contents of this volume.

ARTIFICIAL INTELLIGENCE: A BRIEF HISTORY

AI has a long and varied history, with periods of scientific and commercial successes followed by periods of disillusionment, instigated by scientific challenges as well as unrealistically high expectations (Nilsson, 2009). In the early days of AI (1956–1974),

DOI: 10.1201/9781003410379-1

the objective of making machines intelligent was primarily conceived as implementing general search strategies that could reason over symbolic task representations. However, it gradually became apparent that these general search strategies were insufficient for attaining high levels of performance. Researchers subsequently turned to ways of incorporating large amounts of domain knowledge into systems. AI moved from a search paradigm to a knowledge-based paradigm (Goldstein & Papert, 1977), culminating in the heyday of highly domain-specific expert systems in the 1980s (Feigenbaum et al., 1988). However, expert systems were brittle, meaning they only performed well on the limited scope they were designed for, and with the assistance of human experts who were required to close the gap between the designers' intentions and the real-world application (Woods, 2016). In a particular study on fault diagnosis with an expert system, technicians were required to follow underspecified instructions by the expert system, to infer machine intentions, and to recover from errors that led the expert system off-track (Roth et al., 1987). It should come as no surprise that expert systems did not live up to their expectations and rarely made it out of the lab to real-life usage (Leith, 2016). For a long time (roughly from 1990 until 2010), several alternative approaches (e.g., multiagent systems and the Semantic Web) were explored, with little to no success. Then, big data and machine learning entered the scene (Russell & Norvig, 2021). Deep learning turned out to be very successful, leading to unprecedented outcomes such as superhuman performance on image classification tasks, game-playing (Go, chess), and major breakthroughs in voice recognition and automatic language translation. Deep Neural Networks (DNNs) seem to bypass the problem of manual knowledge elicitation and modeling common-sense knowledge that haunted expert systems in the 1980s. However, manual labeling work is still required, for deep learning image classifiers still require labels in order to be able to learn. To obtain a label (for instance, that a certain image qualifies as a 'cat' and another as a 'dog'), a dataset usually requires humans to point out the area and indicate which type of object resides there. As deep learning requires a lot of data, this burden of manual labeling work is often too large or simply not feasible. A second problem with DNNs is that they are no longer understandable by humans. Performing calculations with tens of millions of parameters, the functioning of a deep learning network is inherently incomprehensible to humans (the problem of so-called 'black-box AI models'). Finally, DNNs may turn out to be brittle after all, as small perturbations in the input image may easily fool a neural image classifier (Moosavi-Dezfooli et al., 2016). In conclusion, AI is still very much in development and a future AI era may well go beyond deep learning and evolve into a hybrid of multiple connectionist AI techniques, symbolic approaches, and humans handling unexpected situations that inevitably arise (Peeters et al., 2021).

ARTIFICIAL INTELLIGENCE IN MILITARY SYSTEMS

In the past, AI was funded largely by defense-related funds. This changed around 2010 when AI became a huge commercial success, giving rise to billion-dollar civilian industries in highly automated driving and data analytics. Still, the recent developments in AI have not gone unnoticed by the defense sector. AI is generally viewed as having large promises in a number of defense areas. AI is expected to speed up and improve decision-making processes, as it is able to process large amounts of data at speeds that are not matched by humans. AI may also be able to select the right information out of large amounts of data, thereby enhancing decision-making processes. AI may also be used to control robots and information agents that can perform dull, dirty, and dangerous tasks without a human operator, thereby freeing up already scarce personnel to focus on more demanding cognitive tasks. Instead of a single tele-operated robot, such as a drone, AI may be used not only to free up personnel, but also to scale up to numerous drones. Also, in communication-denied environments (e.g., underwater or through jamming), where tele-operation is impossible, AI can enable autonomy. The applications of AI lie in several military domains, such as unmanned autonomous systems, decision-making support and intelligence, cyber security, logistics and maintenance, business processes (HR, training, medical, automating work processes), and safety (own personnel as well as civilians).

AI can enhance power on the battlefield, as well as efficiency and effectiveness in the use of unmanned autonomous systems. It can also make work more attractive by delegating particular dull, dangerous, and dirty tasks to AI. If AI takes over certain dangerous tasks, it may make the work of military personnel safer. In military decision support and intelligence, AI can perform automated analysis, combination, and selection of huge amounts of data. This may enhance situation awareness and sensemaking on the battlefield, as well as speed up and qualitatively improve the intelligence-gathering process. AI may also play a role in the automated detection of attacks and vulnerabilities. AI may do this orders of magnitude faster than humans. Also, AI may assist in the automated analysis of the condition of systems, enabling better and faster predictive maintenance and proactive logistics. AI may assist in the automation of work processes, recruitment of personnel, training and education of personnel, as well as in health monitoring and diagnosis.

In conclusion, there are many potential applications of AI in military systems, going beyond merely weapons systems. It is also important to stress that AI will be used to enhance current systems rather than act as a stand-alone 'AI system'. This implies that AI will be used as an add-on to existing systems in the domains mentioned above.

CHALLENGES OF USING AI IN THE MILITARY DOMAIN

Apart from the perceived benefits, there are also challenges associated with the use of AI. First, if AI-based solutions are to be used, they need to be trusted. This is achieved with sound development and validation methods at different phases of a system's life cycle. This in turn requires explainability, so that the developers and certification authorities can scrutinize the solution. Explainability is defined here as the capability of an AI agent to "produce details or reasons to make its functioning clear or easy to understand" (Arrieta et al., 2020, p. 85). Moreover, in some cases also the user or regulator could scrutinize the results of an AI-based solution if it were explainable. As mentioned above, DNNs are no longer understandable by humans and currently have a hard time explaining themselves. The field of 'Explainable AI' is rapidly developing and has grown exponentially over the past few years (Arrieta et al., 2020). Hence, the challenges associated with this topic will remain with us for the foreseeable future. One particular research challenge, to be discussed in more detail below, is what trust repair strategies should be adopted by intelligent teammates working in human-agent teams.

Second, to the extent that large data sets are used by the AI, there is a risk that the data sets are biased. For example, they may work for white males but not for black females, thus leading to discrimination of particular groups in society. There is also the related risk that, as the world constantly changes, there will be 'distributional drift' or 'prediction drift' in the data. In settings with significant changes/distribution shifts, the model based on the past data may not survive contact with the world as it currently is (a state of affairs that has long been recognized in the military, as witnessed by the saying that 'no plan survives first contact with the enemy'). Therefore, the model needs to be monitored and the data need to be as unbiased as possible. This is important not only from an ethical point of view, but also from a performance point of view (biased AI may simply not be effective in particular situations). On the other hand, to the extent that the military is bound by legal obligations on data gathering, as well as dealing with inherently complex situations with a lot of contextual factors, there may in many cases actually be a shortage of data, while the demand for data may be much higher than in civilian settings (e.g., in e-commerce). This may also negatively impact the quality of the models developed in AI.

Third, to the extent that AI takes over certain tasks from humans, there is a fear of humans not being in control anymore over what AI does. This plays a role in the discussion on the use of AI in autonomous weapons systems (AWS). Given the difficulties associated with clearly defining 'meaningful human control', and the fact that 'control' is not a requirement, whereas compliance with the law of war is, the U.S. Department of Defense (2023) prefers the term 'appropriate levels of human judgment' instead of 'meaningful human control'. In response to this, Human Rights Watch (2023) claims that it is not clear what constitutes an "appropriate level" of human judgment. Human Rights Watch also claims that human "control" is an appropriate word to use because it encompasses both the mental judgment and physical act needed to prevent AWS from posing moral, ethical, legal, and other threats. Hence, the debate on the use of the word 'control' is far from over. To make matters more complicated, Ekelhof (2019) has rightfully pointed out that control is distributed over multiple persons at various junctions in the decision-making cycle involved in the target selection and engagement process. Therefore, different forms of control are exercised even before weapons are activated. And even after an AWS has been activated, there may be a human 'in the loop' or 'on the loop', leading to disengagement of the weapon system prior to impact (this is not the case for all AWS; moreover, this discussion largely depends on one's definition of what an AWS is). This leads us, finally, to the issue of the definition of 'autonomous weapons systems' or 'autonomy' in particular. The arguments surrounding this definition are highly contested as well. The UN Convention on Certain Weapons (CCW) established a Governmental Group of Experts (GGE) to discuss emerging technologies in the area of lethal autonomous weapon systems (LAWS). Over the period 2014–2019, the CCW/GGE has not arrived at a shared definition of AWS. Indeed, in a recent review, Taddeo and Blanchard (2022) identified 12 definitions of AWS proposed by States or key international actors. Clearly, this approach is detrimental in facilitating agreement around conditions of deployment and regulation of their use. However, for the purposes of this article, the discussions surrounding LAWS should not be confused with discussions on the use of AI in military systems. Autonomy in military systems may be enabled by AI, but there are also other technologies to enable autonomy.

RESPONSIBLE USE OF AI

While automation based on AI holds great potential for the military domain, it can also have unintended adverse effects due to various imperfections introduced throughout the life cycle. This can be due to biased data, wrong modeling assumptions, etc. In order to advance the trustworthiness of AI-enabled systems, and hence their ultimate use, an iterative approach to the design, development, deployment, and use of AI in military systems is required. This approach, when incorporating ethical principles such as lawfulness, traceability, reliability, and bias mitigation, is called 'Responsible AI' (U.S. Department of Defense, 2022). This implies that the military use of AI will be conducted in a recognized, responsible fashion across the enterprise, mission support, and operational levels in accordance with international law. The normative statements below constitute a first step toward the responsible use of AI in military systems. It is important to recognize that Responsible AI is not identical to 'explainability' or 'transparency', and therefore should not be confused with the field of Explainable AI. An AI model is considered to be transparent if by itself it is understandable (Arrieta et al., 2020), hence without the need for further explanations. Responsible AI involves other ethical principles besides explainability or transparency, such as lawfulness, bias mitigation, and reliability. In that sense, it encompasses explainable AI but cannot be reduced to it.

In terms of incorporating ethical principles such as data protection and bias mitigation, safe and secure AI will be enabled by the development of sustainable, privacy-protective data access frameworks that foster better training and validation of AI models utilizing quality data. Proactive steps should be taken to minimize any unintended bias in the development and use of AI applications. Adequate data protection frameworks and governance mechanisms should be established first within Defense and next with industry at the national or international level, protected by judicial systems, and ensured throughout the life cycle of AI systems.

AI applications should be appropriately understandable and transparent, including through the use of review methodologies, sources, and procedures. To this end, AI applications should provide meaningful information, appropriate to the context and user, and consistent with the state of art. Transparency and explainability are factors that can improve human trust in AI systems. The level of transparency and explainability should always be appropriate to the context and impact, as there may be a need to balance between transparency and explainability and other principles such as privacy, safety, and security.

An iterative socio-technical systems engineering and risk management approach should be adopted to ensure potential AI risks (including privacy, digital security, safety, and bias) are considered from the outset of an AI project. Efforts should be taken to mitigate or ameliorate such risks and reduce the likelihood of unintended consequences. A robust testing process should be developed, allowing for the assessment of AI applications in explicit, well-defined use cases. This includes continuous identification, evaluation, and mitigation of risks across the entire product lifecycle and well beyond initial deployment.

Appropriate oversight, impact assessment, audit, and due diligence mechanisms should be developed to ensure accountability for AI systems and their impact throughout their life cycle. Both technical and institutional designs should ensure auditability and traceability of (the working of) AI, in particular to address any conflicts with human rights norms and standards and threats to environmental and ecosystem well-being.

AI actors should ensure traceability, including in relation to datasets, processes, and decisions made during the AI system lifecycle, to enable analysis of the AI system's outcomes and responses to inquiry, appropriate to the context and consistent with the state of art.

CONCEPTUAL DISTINCTIONS AND CLARIFICATIONS

It is important to make a number of conceptual distinctions and clarifications, particularly when talking about the responsible use of AI in military systems.

First, in response to recent fast developments in AI, many organizations, agencies, and companies have published AI ethics principles and guidelines. In a meta-analysis, Jobin, Ienca, and Vayena (2019) included 84 documents containing ethics principles and guidelines. The most frequently mentioned principles were: transparency, justice and fairness, non-maleficence, responsibility, and privacy. The principles and guidelines have been criticized by some for being (i) too abstract to be practical, (ii) reflecting mainly the values of the experts chosen to create them (hence, not being inclusive), and (iii) serving the priorities of the private entities which funded some of this work ('ethics washing') (Hagendorff, 2020; Hickok, 2021). Although some of these criticisms are justified, one should realize that the principles are a starting point. There is great value in all of these documents being publicly accessible (several websites track them and make them available for analysis purposes, e.g., aiethicslab.com and algorithmwatch.org). Some of these principles are useful for structuring the discussion regarding the challenges for human use, for instance, bias mitigation, explainability, traceability, governability, and reliability (taken from the North Atlantic Treaty Organization (NATO) Principles of Responsible Use of AI, 2021).

Second, 'military systems' are much broader than just weapons systems. AI may be of use in a broad array of systems and applications, including business process applications, predictive maintenance, and highly automated responses to cyber-attacks. This does not in any way diminish the importance of discussing the use of AI in (offensive) weapons systems.

Third, 'autonomy' and 'AI' are not identical. AI may be used to achieve the goal of system autonomy, in the general (and, admittedly, vague) sense of achieving tasks with little or no human intervention (Endsley, 2017). However, there are other ways of achieving this goal, including the use of logic-based programming as used in classical automation. An example of the latter would be close-in weapon systems, such as the Goalkeeper or the Phalanx, which are completely automatic weapon systems for short-range defense of ships. These weapon systems may be called 'autonomous' as defined in the U.S. DoD Directive 3000.09 (2023): "A weapon system that, once activated, can select and engage targets without further intervention by an operator". Yet, these close-in AWS do not need AI to function as intended. This is not to deny that data and AI may be key enablers of autonomy.

Fourth, the definition of the concept of 'autonomy' is driven by political and strategic motivations, as briefly discussed above, and is not value-neutral. It is beyond the scope of the current chapter to arrive at a value-neutral definition of 'autonomy' (see Taddeo and Blanchard, 2022, for such an attempt). I will take up the issue of how to define autonomy in the final concluding chapter of this volume.

OVERVIEW OF THE CHAPTERS IN THIS BOOK

The chapters in this book are organized into four major sections. Section I presents models and approaches for implementing military AI responsibly. Section II is an overview of legal aspects regarding the liability and accountability of individuals and states when using AI in the military domain. Section III addresses the shifting role of human control in military teams in which humans and AI have to work together. This section includes both philosophical and human factors contributions. Section IV broadens the scope to include political and economic aspects of using AI in the military domain. Section V contains a concluding chapter in which the issues addressed in the previous sections are critically evaluated. Below, I will briefly summarize the contents of each chapter.

Section I: Implementing Military AI Responsibly: Models and Approaches

This section starts with the chapter by Heijnen et al. who present a Socio-Technical Feedback loop (SOTEF) methodology to establish and maintain the required value alignment at the levels of governance, design, development, and operation of military AI throughout its life cycle. Value alignment is important as the use of military AI forces us to think about what values are at stake and how we want to ensure these values are accounted for. SOTEF takes an iterative, transdisciplinary, and multistakeholder approach, tailored to the prevailing objectives, context, and AI technology. Ethical, legal, and societal aspects as well as objectives for the human-AI system in high-risk situations are made explicit, commensurable, and auditable (including the attribution of responsibility and accountability). An illustrative scenario and an example set of methods and functions for value alignment exemplify the methodology.

In the second chapter of this section, Koch and Keisinger argue that democracies must be able to defend themselves "at machine speed" if necessary, to protect their common heritage of culture, personal freedom, and the rule of law in an increasingly fragile world. The use of AI in defense in their view comprises responsible weapons engagement as well as military use cases such as logistics, predictive maintenance, intelligence, surveillance, or reconnaissance. Responsibility as a notion poses a time-less question: How to decide 'well' according to what is recognized as 'true'? To arrive at an answer, responsible controllability needs to be turned into three tasks of systems engineering: (i) Design artificially intelligent automation in a way that human beings are mentally and emotionally able to master each situation; (ii) Identify technical design principles to facilitate the responsible use of AI in defense; and (iii) Guarantee that human decision-makers always have full superiority of information, decision-making, and options of action over an opponent. Koch and Keisinger discuss The Ethical AI Demonstrator (E-AID) for air defense as paving the way by letting soldiers experience the use of AI in the targeting cycle along with associated aspects of stress as realistically as possible.

The third chapter by Panwar takes a risk management approach to the responsible use of AI in military systems. Risks posed by different military systems which lever-age AI technologies may vary widely and applying common risk-mitigation measures across all systems will likely be suboptimal. Therefore, a risk-based approach holds great promise. Panwar presents a qualitative model for such an approach, termed as the Risk Hierarchy, which could be adopted for evaluating and mitigating risks posed by AI-powered military systems. The model evaluates risks based on parameters that adequately reflect the key apprehensions emerging from AI-empowerment of military applications, namely, violation of International Humanitarian Law (IHL) and unreliable performance on the battlefield. These parameters form the basis for mapping the wide spectrum of military applications to different risk levels. Finally, in order to mitigate the risks, modalities are outlined for evolving a differentiated risk-mitigation mechanism. Factoring in military ethos and analyzing risks against the backdrop of realistic con-flict scenarios can meaningfully influence risk evaluation and mitigation mechanisms. The rigor that underpins the Risk Hierarchy would facilitate international consensus by providing a basis for focused discussions. The chapter suggests that mitigating risks in AI-enabled military systems need not always be a zero-sum game, and there are com-pelling reasons for states and militaries to adopt self-regulatory measures.

Street and Bjelorglic, of the NATO Communications and Information Agency, have written a chapter from the perspective of those developing AI solutions for military users. Their chapter addresses some practical steps to ensure that military AI is devel-oped and deployed responsibly. Specifically, several high-level principles relating to the responsible use of military AI are considered, together with the steps which developers can take to demonstrate that these areas have been addressed responsibly when developing effective AI solutions for military use. A framework is presented that allows a pragmatic balance between the risks involved in any given AI solution and the tests, checks, and mitigations to be applied during its development.

Section I concludes with a chapter by Gadek, who promotes an existing EU-supported AI application assessment framework, ALTAI, by reviewing three military use cases and highlighting its relevance and shortcomings. Gadek claims that ethics assessments do bring an added value to AI development and that potential solutions such as "explainable AI" or "exhaustive tests", even if desirable, are neither sufficient nor necessary to decide to use AI systems.

Section II: Liability and Accountability of Individuals and States

Cooper, Copeland, and Sanders argue that while AI promises more rapid decision-making, great efficiencies, and enhanced lethality, it also presents a range of risks. States developing new AI capabilities for use in the military domain must establish national processes that allow them to identify and mitigate the risks across the entire life cycle of the AI capability. Their chapter canvases existing military regulatory and governance frameworks designed to address these challenges, particularly during the acquisition and use of highly technical, military capabilities. To mitigate such risks, the chapter identifies and explains the national weapon review process and proposes how such a process may be modified to enable a broader risk-based approach to address legal, ethical, human control, and operational risks associated with the military use of AI technologies.

In his chapter, Mauri argues that the increasing use of AI techniques in the military raises multiple questions, related not only to the ability of AWS to operate within the rules that international law provides for the use of force, but also to issues of international responsibility. In the event that, on the battlefield, AWS (e.g., a drone equipped with systems to select and engage targets without the need for human intervention) are directed to employ force, even lethal force, against an impermissible target (e.g., an unarmed civilian), who is to be held responsible? Numerous authors have begun to speak of possible 'responsibility gaps'. This chapter addresses the issue of the international responsibility of the State and its alleged limitations in regulating AWS.

The chapter by Saxon addresses the use of military AI in unlawful attacks in the ongoing armed conflict in Ukraine and challenges to hold individuals and States accountable for those crimes. The analysis focuses on the more limited context of Russia's 2022–2023 aerial campaign to destroy Ukrainian energy infrastructure. First, the chapter reviews the facts known about these attacks and the technology operating one of the primary weapons used by the Russian armed forces to carry them out – the Iranian-made Shahed drone. Next, it explains the basic principles of IHL, in particular the rules of targeting, which are particularly relevant to the use of military AI. The remainder of the chapter examines how Russia's operation of the Shahed weapon system in the context of repeated targeting of Ukraine energy installations likely constitutes war crimes, and the possibilities of holding persons and States (e.g. Russia and Iran) accountable for these offenses. It concludes that Russia's use of military AI technology that increases the accuracy of its long-running attacks illustrates the greater likelihood that breaches of IHL occurred, as well as Russia's responsibility for those crimes.

Seixas Nunes, in chapter 10, argues that AWS have thrown into question the traditional framework for assessing accountability in war. Some scholars 'scapegoat' military commanders while others 'scapegoat' AWS for violations of IHL caused by those systems. Seixas Nunes offers a different approach. Specifically, he posits that designers and programmers should be considered as potentially liable for violations of war crimes committed by their systems.

Section III: Human Control in Human-AI Military Teams

The first chapter in this section, by Eggert, examines the normative limits of 'meaningful human control' (MHC). That AWS must, like other weapons, remain under MHC is a popular demand in response to various worries about AWS. These include (i) that AWS may not be able to comply with the laws of war; (ii) that delegating life-and-death decisions to algorithms presents a grave affront to human dignity; and (iii) that it may become impossible to ascribe responsibility for harms caused by AWS. Eggert probes the relationship between the moral significance of human control on the one hand and autonomy in weapon systems, conceived as a certain degree of independence from human agency, on the other. In challenging the justificatory force of MHC in mainstream discussions, Eggert offers a starting point for rethinking what role the notion should play in debates about the ethics of AWS.

Simpson, in his chapter, starts by focusing on a move played by DeepMind's AI programme AlphaGo In a match against Lee Sedol, one of the greatest contemporary Go players. AlphaGo played a move which stunned commentators at the time, who described it as 'unthinkable', 'surprising', 'a big shock', and 'bad'. Move 37 turned out to be key to AlphaGo's victory in that game, and it displays what Simpson describes as the property of 'unpredictable brilliance'. Unpredictable brilliance also poses a challenge for a central use case for AI in the military, namely in AI-enabled decision-support systems. Advanced versions of these systems can be expected to display unpredictable brilliance, while also posing risks, both to the safety of blue force personnel and to a military's likelihood of success in its campaign objectives. This chapter shows how the management of these risks will result in the redistribution of responsibility for performance in combat away from commanders, and toward the institutions that design, build, authorize, and regulate these AI-enabled systems. Surprisingly, this redistribution of responsibility is structurally akin to systems in which humans are 'in the loop' as it is for those in which humans are 'out' of it.

The chapter by Devitt explores moral responsibility for civilian harms by human-AI teams. Devitt argues that although militaries may have some bad apples responsible for war crimes and some mad apples unable to be responsible for their actions during a conflict, increasingly militaries may 'cook' their good apples by putting them in untenable decision-making environments through the processes of replacing human decision-making with AI determinations in war-making. Responsibility for civilian harm in human-AI military teams may be contested, risking operators becoming detached, being extreme moral witnesses, becoming moral crumple zones, or suffering moral injury from being part of larger human-AI systems authorized by the state. Acknowledging military ethics, human factors, and AI work to date as well as critical case studies, this chapter offers new mechanisms to map out conditions for moral responsibility in human-AI teams. These include: (i) new decision responsibility prompts for critical decision method in a cognitive task analysis, and (ii) applying an AI workplace health and safety framework for identifying cognitive and psychological risks relevant to attributions of moral responsibility in targeting decisions. Mechanisms such as these enable militaries to design human-centered AI systems for responsible deployment.

Miller and Freedman, in the final chapter of this Section, define Responsible Human Delegation as the making of a "responsible" decision (i.e., a technically and ethically sound one) to "delegate" a task or function to automation. Delegation implies that there will be at least periods of no human oversight, after some initial period of the human operator's learning about and perhaps configuring the automation's behavior and performance. Neglect Tolerance is a concept from research on human-robotic interaction which, roughly, uses the amount of time a robot can be "neglected" (i.e., have a function delegated to it for autonomous performance) in context while still maintaining an acceptable level of performance. In this chapter, Miller and Freedman show how Neglect Tolerance can be adapted to a set of moral or ethical hazards and thereby used to provide a quantitative test of whether or not, in a specified set of conditions with a specified set of automation behaviors, a delegation decision can be "responsible". They provide a sample analysis using a hypothetical delegation decision and a Bayesian modeling approach, though alternatives are also discussed.

Section IV: Policy Aspects

Visions of the future of military AI are evergreen, but the reality of military automation is more complicated, Lindsay claims in his chapter. Information system performance is often more about the quality of people and organizations than the sophistication of technology. This is especially true of machine learning, which lowers the costs of prediction but increases the value of data and judgment. For commercial AI, economic institutions help to provide quality data and clear judgment. These enabling complements are likely to be missing or less effective in the contested environment of war. In other words, the economic conditions that enable AI performance are in tension with the political context of violent conflict. This strategic tension is likely to lead to several unintended consequences. These include unmanageable organizational complexity, as militaries and governments struggle to provide quality data and clear judgment, and strategic controversy, as adversaries target the data and judgment that become sources of strength for an AI-enabled organization. The irony, according to Lindsay, is that increasing military automation will make the human dimension of war even more important.

In the final chapter of this Section, Vignard poses the important question of what can be learned from how the international community has approached the development of norms of responsible State behavior in the absence of appetite for new treaties? Would a similar approach focusing on reaffirming existing international law, agreement on norms, identification of confidence-building measures, and the development of capacity-building initiatives suffice in the field of military applications of AI? Or have these approaches proven too slow to keep pace with the speed of innovation while excluding key stakeholders, such as technologists and the private sector? This chapter identifies key lessons from the UN negotiations on cyber in the context of international security (from 2004 to 2021) and those on lethal AWS (2014-present) applicable to the objectives of developing a shared understanding of Responsible AI (RAI) and accelerating international operationalization of RAI practices.

Section V: Bounded Autonomy

In the final Section, Schraagen critically evaluates the issues addressed in the previous chapters. The aim of this concluding chapter is to reflect on some common themes that run throughout this book, as well as to highlight some issues and research challenges that were not sufficiently highlighted by the contributors. The first issue critically discussed is the debate on 'killer robots'. Three arguments are advanced against the Stop Autonomous Weapons campaign. Secondly, a critical discussion on the various definitions of the concept of 'autonomy' is carried out, resulting in an argument for the concept of 'bounded autonomy'. This concept basically states that the capacity of a system to display autonomous behavior is very limited compared with the variety of the environments in which adaptation is required for objectively autonomous behavior in the real world. This leads to a discussion of the concept of 'meaningful human control'. The argument is that more attention to the testing, evaluation, and certification process of weapon systems is required, rather than to the control exercised by individual commanders or operators. Finally, research challenges in the field of Human Factors and Ergonomics are formulated, in the context of Responsible AI for military systems.

REFERENCES

Arrieta, A. B., Díaz-Rodríguez, N., Del Ser, J., Bennetot, A., Tabik, S., Barbado, A., García, S., Gil-López, S., Molina, D., Benjamins, R., Chatila, R., & Herrera, F. (2020). Explainable Artificial Intelligence (XAI): Concepts, taxonomies, opportunities and challenges toward responsible AI. *Information Fusion, 58*, 82–115.

Ekelhof, M. (2019). Moving beyond semantics on autonomous weapons: Meaningful human control in operation. *Global Policy, 10*(3), 343–348.

Endsley, M. R. (2017). From here to autonomy: Lessons learned from human-automation research. *Human Factors, 59*(1), 5–27.

Feigenbaum, E. A., McCorduck, P., & Nii, H. P. (1988). *The rise of the expert company.* New York: Times Books.

Goldstein, I., & Papert, S. (1977). Artificial intelligence, language, and the study of knowledge. *Cognitive Science, 1*(1), 84–123.

Hagendorff, T. (2020). The ethics of AI ethics: An evaluation of guidelines. *Minds and Machines, 30*, 99–120.

Hickok, M. (2021). Lessons learned from AI ethics principles for future actions. *AI and Ethics, 1*, 41–47.

Human Rights Watch (2023). Review of the 2023 US policy on autonomy in weapons systems. Review of the 2023 US Policy on Autonomy in Weapons Systems I Human Rights Watch (hrw.org)

Jobin, A., Ienca, M., & Vayena, E. (2019). The global landscape of AI ethics guidelines. *Nature Machine Intelligence, 1*, 389–399.

Leith, P. (2016). The rise and fall of the legal expert system. *International Review of Law, Computers & Technology, 30*(3), 94–106.

Moosavi-Dezfooli, S. M., Fawzi, A., & Frossard, P. (2016). Deepfool: A simple and accurate method to fool deep neural networks. In *Proceedings of the IEEE conference on computer vision and pattern recognition* (pp. 2574–2582).

Nilsson, N. J. (2009). *The quest for artificial intelligence*. Cambridge: Cambridge University Press.

North Atlantic Treaty Organization (NATO) Principles of Responsible Use of AI (2021). Summary of the NATO Artificial Intelligence Strategy.

Peeters, M. M. M., Van Diggelen, J., Van den Bosch, K., Bronkhorst, A., Neerincx, M. A., Schraagen, J. M., & Raaijmakers, S. (2021). Hybrid collective intelligence in a human-AI society. *AI & Society, 36*, 217–238.

Roth, E. M., Bennett, K. B., & Woods, D. D. (1987). Human interaction with an "intelligent" machine. *International Journal of Man-Machine Studies, 27*, 479–525.

Russell, S., & Norvig, P. (2021). *Artificial Intelligence: A modern approach* (4th ed.). Pearson Education.

Taddeo, M., & Blanchard, A. (2022). A comparative analysis of the definitions of autonomous weapons systems. *Science and Engineering Ethics, 28*, 37–59.

U.S. Department of Defense (2022). *Responsible artificial intelligence strategy and implementation pathway*. Washington, DC: Department of Defense.

U.S. Department of Defense (2023). *DoD Directive 3000.09. Autonomy in weapon systems*. Washington, DC: Department of Defense.

Woods, D. D. (2016). The risks of autonomy: Doyle's catch. *Journal of Cognitive Engineering and Decision Making, 10*(2), 131–133.

SECTION I

Implementing Military AI Responsibly

Models and Approaches

A Socio-Technical Feedback Loop for Responsible Military AI Life-Cycles from Governance to Operation

2

Marlijn Heijnen, Tjeerd Schoonderwoerd, Mark Neerincx, Jasper van der Waa, Leon Kester, Jurriaan van Diggelen, and Pieter Elands

INTRODUCTION

Major investments in AI technologies are extending AI capabilities substantially, leading to new applications in the military domain. Such applications often contain embedded learning algorithms and some form of autonomous information processing and decision-making. These applications are highly needed in complex and time-critical military operations, such as AI-based cybersecurity countermeasures in response to adversarial (AI-based) attacks, or deployments of autonomous weapons for a ship's

DOI: 10.1201/9781003410379-3 **CC BY-ND - Attribution-NoDerivs** self-defense. In general, AI can substantially reduce the risks and improve the precision of military operations, due to its capacity to process large amounts of data quickly and by reacting quickly based on the learned and pre-programmed models.

However, the military application of AI is under debate. For example, scientists and non-governmental organizations have warned against the emergence of "killer robots",[1] that is, autonomous weapon systems that select and attack targets without meaningful human control (MHC). Generally MHC refers to the requirement that not AI but humans should ultimately remain in control of and morally responsible for (AI-driven) military operations. However, in its advice to the Dutch government, the Advisory Council on International Affairs/Advisory Council on Peace and Security (AIV/CAVV) concluded that there is not yet agreement on an international definition of MHC, except that human judgment should always be preserved. In a philosophical account of MHC, Santoni de Sio and Van den Hoven (2018) identified two general necessary conditions for MHC: tracking (a system should be able to respond to the moral reasons of humans deploying the system) and tracing (a system should always allow to trace back the outcome of its operations to at least one human along the chain of design and operations).

The concern about MHC is not without reason: AI technology might bring along unintended (side-) effects that must be prevented, such as deterioration of human control, an unbalanced ("biased") situation awareness (i.e., caused by the frequency distribution of objects or phenomena in the training dataset), or the opponent or enemy anticipating predictable behavior of AI technology. A multitude of, often interdependent, factors can bring about such unforeseen negative effects, such as shortcomings in the technology (e.g., biased training data, incomplete world models, poorly designed user interfaces), and performance changes of humans who work with the AI systems (neglect due to loss of oversight or over-reliance). The challenge is to establish MHC: enabling humans to have oversight and take responsibility in decision-making (Aliman, 2020; Amoroso & Tamburrini, 2020; Scharre, 2018; Van Diggelen et al., 2023) throughout the AI lifecycle: MHC is not only to be achieved during the operation of AI-based systems, but also during governance, design and development activities.

To study responsible AI, we must set it in a realistic context. This can be done using scenarios that are operationally relevant, capture the complexity of defense operations, and express a clear need to use AI. In this chapter, we use a short scenario described in Box 2.1 to illustrate that moral decisions regarding the deployment and use of AI systems are made at several stages. The combination of these decisions determines how the use of the AI system is embedded and controlled in the operation, how the uncertainties are taken into account, how the risk assessments are made, and how the responsibilities are allocated (cf. Ekelhof, 2019). In this example, the military organization decided to use AI-enabled and remotely-activated, rapid-fire guns, and the commander authorized the installation and use of these guns to guard against light vehicles with explosives within a demarcated defense zone. When the risk of "vehicle attacks" is deemed high, a human guard can command the guns to fire at identified hostile vehicles. In this scenario, there are risks of collateral damage and incorrect target identifications.

Note that Box 2.1 presents a small and simplified scenario, in which, for example, changes of the human-machine capabilities are not addressed. Incidents and problems can appear and accumulate at the individual entity (e.g., malfunctioning actuators), team organization (e.g., inappropriate allocation of responsibilities), and societal level (e.g., discrimination against specific population groups). As the technology is new and adaptive, and is operating in a dynamic environment, there will be uncertainties in the predictions of outcomes. The military decision-making processes, embedding advanced AI technologies, should incorporate careful consideration and weighing of the relevant ethical, legal, and societal aspects (ELSA), taking into account the uncertainties, risks, and unintended side effects. For example, a diminished value awareness can be a risk; we might think that an AI system is aligned with our values, while in reality they are only partially aligned (such circumstances set high requirements for AI technology's explanation capability).

In the example, objectives include the prevention of further violent escalations and supporting the government's administration, rule of law, and law enforcement. It represents a non-international armed conflict, to which Common Article 3 of the Geneva Conventions applies.[2]

What are the challenges concerning the design, development, and maintenance of human-AI systems, and how to identify the moral consequences of the deployment of

---

BOX 2.1   SIMPLIFIED MILITARY SCENARIO TO ILLUSTRATE MORAL DECISION-MAKING AT THE LEVEL OF AVAILABILITY, DEPLOYMENT, AND USE OF A (SEMI-) AUTONOMOUS AI SYSTEM[3]

RAPID DEFENCE SCENARIO

An urban operating base has been under attack by terrorists for several months. These attacks largely involve automobiles, disguised as civilian traffic but equipped with large quantities of explosives, driven by suicidal adversaries who accelerate when nearing the entry gate to the base. Since buildings and base personnel are located near the gate, there is very limited time for gate guards to target and respond to such attacks even when they can identify them, and much destruction and death have resulted over the past months.

To improve reaction times, the base commander previously authorized the installation and use of *RivalReveal*, that is, remotely activated, rapid-fire guns capable of stopping a light vehicle as it heads toward the gates. These guns can fire at various levels of autonomy. They can fire "automatically" within a previously defined target zone, after receiving prior orders from a human guard. Due to space limitations in the urban environment, the presence of base personnel in the target zone cannot be completely avoided. This means that there is a possibility that guns cause collateral damage to innocent bystanders. This risk of collateral damage to innocent bystanders is the primary hazard. Another risk is that the human guard will incorrectly identify a target, either positively or negatively. This scenario is suitable for following the Socio-Technical Feedback (SOTEF) loop, to be described below, because the violent nature makes it highly morally sensitive, and the high-speed decision-making justifies the choice of considering AI-based techniques.

new AI technologies in future operations? This is challenging, in particular, because these questions need to be addressed continuously during the complete lifecycle of the socio-technical systems[4] (STS), while the operational circumstances and conditions are continuously changing, affecting the decision-making processes and outcomes. Furthermore, the values concerning certain military operations can change over time. This means that the decision-making processes and outcomes should continuously be evaluated to ensure alignment with values. Thus, *value alignment* is an important continuous process for the identification of the moral values that are at stake. However, it remains difficult to identify relevant moral values (especially considering that they may change over time), and to ensure that the human-AI system continues to operate in accordance with these moral values and their context-dependencies. Another difficulty lies in ensuring that the human-AI system can notice in time when an outcome is in violation of one or more moral values. This is especially relevant for military AI systems, as a violation of values might have a severe impact.

To date, there is no consensus on how MHC must be operationalized (AIV/CAVV, 2021)[5] and how to achieve value alignment in the development and deployment of AI. There is agreement on guiding principles, such as formulated by the UN Group of Governmental Experts,[6] NATO,[7] and the TAILOR consortium.[8] The NATO Principles of Responsible Use for AI in Defense will help steer efforts in accordance with moral values, norms, and international law, but a comprehensive prescriptive approach is lacking for building and implementing AI technology in such a way that it is under MHC, during its complete lifecycle. In this chapter we propose the SOTEF loop: a methodology to establish MHC at the levels of society, organization, and operation, addressing regulation, design, development, maintenance, and modification processes of a specific human-AI system in a specific context (Aliman et al., 2019; Aliman & Kester, 2022; Peeters et al., 2021).

Known approaches such as value-sensitive design (Friedman & Kahn, 2003; Friedman & Hendry, 2019; Van Den Hoven, 2013), participatory multistakeholder analyses of the ELSA[9] (Van Veenstra et al., 2021), and responsible research and innovation (Stilgoe et al., 2013; Von Schomberg & Hankins 2019),[10] share important aspects with the SOTEF loop such as stakeholder involvement, value analyses, and multidisciplinary design. Other guidelines (e.g., Dunnmon et al., 2021) also have become more concrete on how to implement ethical principles. The SOTEF loop incorporates these approaches and applies them not only in the design phase of an AI system, but ensures value alignment throughout the STS lifecycle. The SOTEF loop differs from these existing approaches and guidelines as it takes a comprehensive (socio-technical) engineering perspective: including all stakeholders (in addition to the defense organization, for example, regulators, subject-matter experts, and AI manufacturers). Additionally, the SOTEF loop focuses on the iterative nature of the human-AI system where continuous feedback, adaptation, and improvement throughout its lifecycle are essential. As such, it connects current approaches with each other and allows the functionality of the human-AI system to develop over time in a responsible way.

The SOTEF loop describes a process to (i) identify the ELSA to which the behavior of the human-AI system should adhere (including assigning responsibilities that apply during operation, (ii) ensure that the human-AI system can operate according to those aspects, and (iii) enable stakeholders on different levels to regularly reflect and give feedback on the system's behavior and propose appropriate value-based adjustments. There is no single solution that achieves these three goals in all possible applications of AI technologies and this means that solutions are situation-dependent, that is, they are affected by the specific AI system deployed and the specific context of the governance, design, configuration, and operation. And because context, as well as values, change over time, the involvement and feedback from different stakeholders is needed during the complete lifecycle of the STS, from redesigns between iterations in order to realign to such new values to human support capabilities in order to intervene during operation when misalignment occurs. Instead of aiming for a one-size-fits-all solution, the SOTEF loop introduces a set of methods to identify and operationalize the relevant ELSA (given the mission goals) in order to establish MHC of a specific AI system in a specific context. The applicability of each method should be carefully considered in each specific context and might range from setting rules to which the human-AI system should adhere, predefining the behavior of the AI system, learning from human-selected data, to using goal functions and augmented utilitarianism (Aliman & Kester, 2022) (see definitions in Appendix 2.A).

We believe that the SOTEF loop is especially useful when dealing with high-risk military AI applications. Risk can be defined as the likelihood that unintentional harm of any kind (e.g., social, psychological, physical, or technical) can be done, with high risk implying that is it very likely that such harm will be done in the context of operation. Therefore, high-risk AI can be considered as unintentional harm being highly likely, as a result of the context in which the AI system is applied, the capabilities of the AI system, and/or the way in which it is applied (e.g., the human-AI interactions that take place). This makes MHC over AI systems in such contexts highly relevant, as the behavior that results from assessing situations and weighing values will have a major impact. Our assumption is that the higher the risk of the human-AI system application, the more extensive these moral considerations need to be for that risk to be mitigated. As a result, higher (ethical) demands are placed on the behavior of the human-AI system. If the design process results in a requirement that the AI system must base its behavior on moral values, then the AI system's implemented internal processes need to explicitly incorporate these values.

THE SOTEF LOOP

Santoni de Sio and Van den Hoven (2018) identified tracking and tracing requirements for meaningful control of autonomous AI systems, being: (i) responsiveness to the environment and moral considerations of the humans designing and deploying the AI systems, and (ii) providing the possibility to trace back the outcomes to a human during the design and operation process. We propose the SOTEF methodology as a way to explicate and embed these requirements in the design process for a human-AI system in a specific context. The SOTEF methodology operationalizes these requirements at different control levels in four feedback loops: governance, design, development, and operation. The SOTEF methodology aims to structure the process for achieving MHC in an iterative fashion and offers validated methods for operationalization. Furthermore, the SOTEF methodology recognizes that this process and the methods used will differ per application, as each will be unique with respect to ELSA.

The SOTEF methodology prescribes how to set up the governance, design, development, and operation of a human-AI system. Each of these topics forms four distinct feedback loops at various timescales required in iteratively constructing and improving a human-AI system to behave according to ELSA. These intertwined feedback loops are based, respectively, on standardization efforts (Zielke, 2020), design processes such as the Double Diamond (Kunneman et al., 2022), system engineering processes such as the V-Model (Clark, 2009), and human-machine teaming interaction principles (Van der Waa et al., 2020). However, these processes are not explicitly intertwined with governance (Coeckelbergh, 2019). For this reason, the SOTEF loop includes a governance feedback loop to include regulation, policy, and laws in the construction and maintenance of human-AI systems.

A single method toward responsible military AI does not exist, as the possible applications of human-AI systems differ too greatly in terms of their goals, required tasks and capabilities, and ethical, legal, and societal context. A context that varies per society and the experiences that society acquired, and will continue to change over time as more experiences are gathered (Winston & Edelbach, 2013). This requires the development of multiple methods that can be applied in an iterative fashion for the human-AI system to adapt to a changing context. The feedback loops of the SOTEF methodology thus require varying methods, as the application and context demand. Methods that need to be developed and evaluated based on potential applications of human-AI systems and their ethical, societal, and legal contexts.

The feedback loops of governance, design, development, and operation are visualized in Figure 2.1. All are depicted as persisting feedback loops on various timescales. Each larger feedback loop governs the feedback loops on its lower timescale, while iteratively improving itself as it obtains feedback from those smaller loops. For example, the governance loop dictates the design process (e.g., NATO Principles signifying what should be considered during design), whereas the design process dictates what should be developed (e.g., how humans and AI systems should be interacting) and how the human-AI system should operate (e.g., as supervisory control). In turn, the design loop ideates on novel applications of human-AI systems that influence the governance loop. Below, we describe each loop and state its current challenges.

The Governance Feedback Loop

The governance loop dictates the principles and regulations for military applications of AI to provide the necessary ethical, legal, and societal context in which human-AI systems need to be designed, developed, and operated. This can include applying existing laws regulations, policies, and permissible opportunities for the military application of AI (cf. case law), or developing new ones. Examples are the international laws, regulations, standards, and guidelines established in the international human rights law,

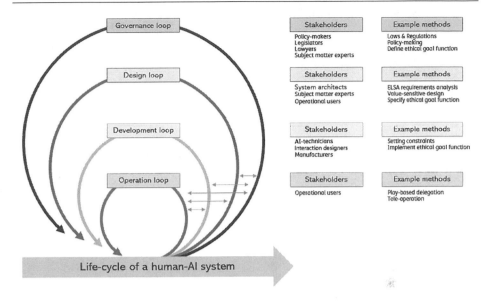

FIGURE 2.1   An illustration of the Socio-Technical Feedback (SOTEF) loop.

International humanitarian law, the EU AI Act, the high-level expert group on AI, ISO/ IEC, and the NATO principles of responsible use of AI. The timescale of the governance loop is the largest of the four, as it attempts to be the most encompassing. It will provide the context that defines the inner, smaller loops. In turn, the smaller loops will provide critical reflection on whether the governance is sufficient or lacking in any way. Many efforts currently focus on the governance loop through methods such as committees and debates. However, these are often one-time exercises that lack the required iterative nature to match the progress and societal changes that occur over longer periods of time. Furthermore, such exercises only incorporate the feedback of specific AI applications in an ad-hoc fashion depending on societal events. Hence, the governance loop has various proven methods in place, but these methods need to be applied in a structured and iterative fashion.

The Design Loop

The design loop within the SOTEF methodology consists of an interdisciplinary design process that starts with determining the context, problem space, and the envisioned application of AI. Within this loop, the human-AI system should be defined, including a specification of the respective roles, tasks, goals, competencies, responsibilities, functions, and interactions of the humans and AI systems. The design loop should conclude with specified requirements on the human-AI system in the specific context that is considered, and in particular explicitly state the involved ELSA and how these should be addressed or implemented during development. Throughout this phase, the application of relevant laws, and the identification and specification of moral values and societal issues should come into play (Van de Poel, 2009). Numerous methods exist to support specific steps: from methods to involve potential or future users, for example,

Participatory Design (Schuler & Namioka, 1993; Ten Holter, 2022), to methods to integrate values in design and engineering, for example, Value Sensitive Design (Friedman & Kahn, 2003; Friedman & Hendry, 2019) or Rapid Ethical Deliberation (Aliman & Kester, 2022; Steen et al., 2021).

There are two main challenges in the design loop. The first challenge is the appropriate selection of the stakeholder(s) that need to be involved in selecting and carrying out subsequent design methods. Part of this selection involves determining the distribution of responsibility for the design across various stakeholders. The second challenge is the translation and specification of the identified values, ethics, and laws into concrete requirements that will guide the development process. The focus here is on specifying the desired functioning and behavior of the human-AI system. For example, what behaviors adhere to principles such as transparency and human agency, and values such as safety and integrity? What is required from the human-AI system to be able to show this behavior? For example, an AI system might be required to assess the safety of a given situation in order to warn a human operator when it encounters a dangerous situation. In order to be able to do so, it needs to be specified in the design loop what *safety* entails and how it can be assessed by the AI system. The design loop requires interaction with governance stakeholders to assess whether the system design warrants operation within existing laws and regulations. They are involved as a stakeholder in the design of the high-level capabilities and in the preparation of software specifications in order to indicate what developers must strictly adhere to and where they are allowed some flexibility in the implementation. This includes the design of processes for developers to reflect and communicate when the boundaries that have been set are in conflict with the given requirements (e.g., because of technical limitations). Thus a necessary discussion in any design loop is on who to involve with what responsibility to derive requirements from identified relevant moral values, ethics, and laws given the application that is considered. Methods are required that can facilitate this, on top of (existing) methods that shape the more general design process.

The Development Loop

The development loop should translate an agreed-upon design into a human-AI system while adhering to set translation restrictions to prevent diverging from the principles underlying the design. This translation effort should encompass all aspects of creating and maintaining a human-AI system. This loop includes all technological development efforts. In addition, it includes developing an organization to enable and support the designed human-AI system. Finally, it encompasses the validation and verification of the developed human-AI system as a whole. For example, a technical effort might include the development of a formal model that facilitates morally correct behavior for the AI system. Similarly, a new technical education and training regime can be developed that will impact human behavior. Finally, the resulting human-AI system – with trained humans and an AI system with a moral model – should be verified to assess whether it adheres to the specifications and validated whether it behaves as intended.

Two main challenges are part of the development loop: (i) the translation of design specifications into an implementation that adheres to the underlying ethical and legal aspects on which these specifications were based; and (ii) the reliable validation and verification of such an implementation in light of these specifications. Methods are required to address these challenges, as it is essential to prevent any design choices from being made in the development loop (e.g., a developer making decisions on how a moral value should impact the human-AI system's behavior). If this occurs, the development loop would dictate (parts of) the human-AI system's behavior in an ill-supported manner.

The Operation Loop

Finally, the operation loop is the most inner feedback loop in the SOTEF methodology. This is where the actual human-AI interaction takes place, for example, to prepare or carry out a mission task. Here, the human-AI system is being instantiated, that is, it is decided when and how the human and AI-agents act (according to the predefined policies, rules, and requirements of the other loops) and where their behaviors can be observed. If the decision is not to deploy the AI, a reflection is required on the human-AI system's design and implementation. If the decision is in favor of deployment, the human-AI system must first be configured to tailor it to the foreseen deployment. This can include, for example, setting constraints for operation, providing specific training exercises, or specifying the task-specific goals of the system. There is a time gap between the development and operation of a human-AI system. A gap that is currently often neglected, as the discussion is about either how to design responsibly or how a human-AI system should operate. This ignores the fact that there is often time – and a necessity – to tailor the human-AI system to a particular deployment in a specific context. At times the moment for configuration is clear, for instance in the case of configuring autonomous AI systems just before a mission. At other times, this is much more diffuse, for instance in a classification algorithm running on always-on sensor systems. At those times, defining what constitutes configuration should be defined as part of the design process. Currently, methods are lacking to support this configuration component in the operational loop, so effort should be put into developing them.

The debate on MHC often focuses on either the governance, design, or operation loop. However, in practice, any control is likely to be a mix of control methods from governance, design, and operation. In the end, however, the final control mechanism resides in the operation. However, in some applications of AI, operational control is limited because *direct* human intervention in an AI agent's behavior is limited or even impossible. For instance, when communication is difficult or decisions need to be made in a short amount of time. However, within the SOTEF methodology, the operation loop is defined more broadly than mere direct human intervention. It also encompasses less direct interventions through reviews and feedback, which can be provided before, during, or after a specific instance of use (e.g., a specific military operation). Such methods (e.g., run time verification) can provide feedback to the other loops, potentially triggering a new design, development, or even new governance.

The goal of the SOTEF methodology is to ensure that the entire human-AI system behaves in a morally acceptable manner and abides by relevant governance while effectively achieving the set goals. The methodology takes a high-level perspective, giving opportunities to develop new control mechanisms where governance, design, and development control mechanisms are intertwined with operational control. This is paramount for the responsible use of AI technologies. It broadens the discussion about how humans can control an AI system during operation, toward the discussion of how humans can control the behavior of a human-AI system in a combination of governance, design, development, and operational control.

Adaptation through Iteration

These four loops involve different human actors and act on their own timescales and should continue throughout the entire life cycle of the human-AI system. This means that the design of the system could always be reconsidered, a developed human-AI system should be open for change, and even during operations, the human-AI system should be able to adapt when needed. For example, humans might decide to not use the AI system during a specific operation as the context changes. To make this decision they might receive explanations from the AI system that convey the risks of using it, thus giving the AI system a role in the decision. A role that is defined in the design and development loop as the explanations are designed and implemented.

The SOTEF methodology recognizes that such complex and intertwined control mechanisms require iterations and places them in an overarching process during the lifetime of a human-AI system. These feedback loops are required as no matter the used methods, it cannot be guaranteed that a human-AI system always behaves responsibly according to ELSA in every possible situation. The SOTEF methodology addresses this by connecting governance, design, development, and operation activities over iterations to ensure the best possible human-AI system that is improved as experience is gathered – from sandbox environments to real operations.

The feedback loops allow for adaptation to changing circumstances, new insights, and changing values. Without these, the resulting human-AI system would be static in an ever-changing context which would eventually result in its failure to comply with the then-current moral values. As such, the SOTEF methodology recognizes not only that feedback occurs within each loop but also across loops, in both a top-down and bottom-up fashion. For example, the governance loop can change as new laws and regulations are implemented that set new requirements for human-AI systems. Similarly, these new laws and regulations can arise due to gained experience by already deployed human-AI systems as they pass through multiple operation loops.

The Involved Stakeholders, Their Responsibility and Accountability

Following the SOTEF methodology requires the involvement of many stakeholder groups. These include, but are not limited to, lawyers, subject matter experts, policy-makers, legislators, operational users, technicians, AI experts, manufacturers, and designers. The SOTEF methodology does not dictate that generic responsibilities should be assigned to each stakeholder group. Rather, it dictates that for each of the four feedback loops methods should be available and responsibility should be assigned based on the selected methods and who should be involved in them. This selection is expected to occur according to governance and during the design, where governance would likely dictate relevant stakeholder groups, and in the design of a human-AI system specific representatives are selected.

The methods in the SOTEF methodology should be developed and evaluated on their contribution to the responsible use of AI for particular application domains. Such methods will dictate the participation and role of stakeholder representatives from which responsibility and accountability can be derived. It should be noted that there are dependencies between the loops and that there needs to be shared awareness of the outcomes across the loops (e.g., to assure that performance at the operations addresses the regulations of the governance). In addition, the overarching responsibility of developing and evaluating such methods lies with those involved in developing the SOTEF methodology further That is, those involved in the development of a method are *responsible* for communicating its strengths and limitations, while those who choose to apply a method are *accountable* for the implications of this method in the context of the application.

BOX 2.2   ILLUSTRATION OF SOTEF LOOP USING THE RAPID DEFENSE SCENARIO

SOTEF IN THE SCENARIO OF RAPID DEFENSE

Within the Rapid defense scenario, the iterative, transdisciplinary, value-sensitive approach of the SOTEF methodology can be illustrated as follows. Before the *RivalReveal* (RR) system is put into operation, weapon reviews of RR are performed (*governance*) using existing regulations. This informs the *design* of interaction and autonomy, for example, by stating requirements for operator interfaces and implementing automatic failure mode responses. Furthermore, training programs are developed for users to *configure* and *operate* the system. In this phase, collaboration between the various stakeholders is essential to manage the interdependencies between the cycles, for example, governance dictates design choices, and design possibilities inform compliance. After the initial system is evaluated, the system is taken into use. As the SOTEF methodology is a lifecycle approach, it does not end there. Compliance with ethical guidelines such as appropriate levels of judgment and care are continuously monitored against their effectiveness. On all longer timescales, this can lead to changing the regulatory framework, for example, changing the ethical principles, or sharpening them to be more precise. At the levels of design and operation, incidents and practical experiences with RR are continuously monitored and are assessed in relation to public values. These insights could be used at all levels: to train operators, sharpen system design, and even to reconsider ethical guidelines based on the way they play out in practice. A functioning SOTEF methodology does not arise naturally: it requires careful consideration of all stakeholders, providing them with the right information at the right moment and empowering them to act.

METHODS AND FUNCTIONS
IN THE SOTEF LOOP

Responsible military AI must be achieved by empowering humans to align the behavior of the STS with human values. The SOTEF loop facilitates this process by offering various control methods and functions for identifying, implementing, monitoring, and adjusting relevant values and goals in a STS. Methods are concrete processes and procedures that can be used to implement part of a control loop. For example, utility elicitation and value-sensitive design are methods to identify relevant moral values within a domain, which is one of the goals in the Design loop. Functions are the prescribed capabilities of the STS that should be implemented in a specific component (e.g., AI system, human, environment) of the STS. For example, explainability of system behavior might be a function that is required for the Governance loop, as those who govern need to understand the relationship between the applicable regulation and the behavior of the system in the operational context for which they might be held accountable. Team Design Patterns can be used to explore the allocation of functions in the STS (Van der Waa et al., 2020).

Table 2.1 lists methods and functions that can be used for the instantiation of the SOTEF loop methodology. Note that methods and functions can be applied in combination and might entail interdependencies. For example, an ethical goal function (method) can prescribe the selection of training data (function). As Table 2.1 shows, the methods and functions that are being used in a SOTEF-loop may take a variety of forms, differing with respect to:

- The **component** of the STS for which the method is implemented (e.g., the AI system, the human, the environment, the interaction, etc.)
- The **response time** between the act of controlling and the moral behavior (outcome of moral decision-making) of the STS that is being controlled. *Long* means more than one day, *medium* means between ten seconds and one day, *immediate* means less than ten seconds
- The **human actor** that executes control over the STS
- The **feedback mechanism** by which the STS's behavior is monitored and used as input to further control the STS (e.g., to realign the STS with relevant moral values)

The repertoire of methods and functions that can be applied to instantiate the feedback loops will evolve over time. The sections above already mentioned value sensitive design (Friedman & Kahn, 2003; Friedman & Hendry, 2019), rapid ethical deliberation (Steen et al., 2021), participatory design (Bratteteig & Verne, 2018), and moral programming (Aliman & Kester, 2022). Table 2.1 shows other relevant methods and functions as a further illustration of the repertoire of methods. Current research of the ELSA labs in the Netherlands[11] will extend this list and provide a comprehensive state-of-the-art overview.

TABLE 2.1 Methods and functions to instantiate (part of) the feedback loops of the SOTEF-methodology

NAME	LOOP	COMPONENT	RESPONSE TIME	HUMAN ACTOR	FEEDBACK MECHANISM	EXAMPLE
Restricting use context	Governance	Legal Context	Long	Legislators and policymakers	Law enforcement	Prohibition stating that AI system must not be used in urban environments
Value identification	Governance/ Design	Human, Ethical context	Long	Various stakeholders	Value deliberation and validation	Identifying ethical considerations for AI healthcare applications (Char et al., 2020)
Requirement analysis	Design	Human, AI, inter-action	Long	Military authorities, Human-AI interaction experts	Requirement validation	Scenario-based requirements engineering (Sutcliffe, 2003)
Algorithm auditing	Governance Design Development	AI	Long	Various stakeholders	Explainable AI	Risk rating, surrogate explanations, post-processing, etc. (Koshiyama et al., 2022)
Defining ethical decision framework	Design	Human, AI, inter-action	Long	Military authorities, Human-AI interaction experts	Decision quality validation	Allocating ethical decision-making in a human–AI team (van der Waa, 2020)
Shaping infrastructure	Design	Environmental context	Long	Military planners	Incident management system	Placing a fence around the AI's workplace
Selecting training data	Development	AI	Long	AI engineers	Explainable AI	Engineers compose image datasets of representative "hostile vehicles"
Ethical goal function	Governance Design Development	AI	Long	Various stakeholders	Explainable morality	Value and harm model in an autonomous car (Reed et al., 2021)

(Continued)

TABLE 2.1 (Continued) Methods and functions to instantiate (part of) the feedback loops of the SOTEF-methodology

NAME	LOOP	COMPONENT	RESPONSE TIME	HUMAN ACTOR	FEEDBACK MECHANISM	EXAMPLE
Norm engineering	Development	AI	Long	AI engineers	Explainable AI	Privacy-enhancing technologies (e.g., Liu et al., 2021)
Human task training	Operation	Human	Long	Military trainers, doctrine developers	Incident management system	Training a soldier to work with a particular AI system
Human resilience training	Operation	Human	Long	Military trainers	Simulation-based training	Appraisal training to decrease the effects of traumatic experiences (Beer et al., 2020)
Play-based delegation	Operation	Inter-action	Medium	Human teammate	Progress appraisal	Human calls predefined play for doing an area surveillance during a mission (Miller & Parasuraman, 2007; van Diggelen et al., 2021)
Collaborative planning	Operation	Inter-action	Medium	Human teammate	Progress appraisal, AI-assisted feedback	Human and AI formulate mission plan together
Tele-presence	Operation	Inter-action	Immediate	Tele-operator	Visual, sound & other senses	AI autonomously performs surveillance, but is taken over by the human in unexpected situations
Adaptive automation	Operation	Inter-action	Immediate	Operator	Adjustable work agreements, explaining displays	Attuning the level of automation to the momentary situation and operator workload (De Tjerk et al., 2010)

Note: Response time is a relative concept; the context and momentary risk level of the (planned) operation determine its actual value.

Note that each of these methods and functions can be implemented in the SOTEF loop. The outcome of a specific method does not guarantee ethically aligned behavior of the STS over time. Under the assumption that a combination of methods will lead to better ethically aligned behavior, verification and validation should entail the comprehensive sum of these outcomes, that is, the results of all loop levels. As the human, AI, and environment change over time due their "inner" feedback loops *and* their adaptations to each other, regular reviews and adjustments should be made to the STS during its complete life cycle. This process should be directed by those who have an overview of – and insight in – the STS in the context in which it is deployed. Since different stakeholders are involved in each loop, the review and adjustment process has to take place in interaction with each other.

DISCUSSION AND CONCLUSION

In this chapter, we provided an overview of the SOTEF methodology: a comprehensive, iterative STS-engineering approach that distinguishes a governance, design, development, and operation loop for responsible military AI life cycles. The implementation of these feedback loops will be done within a specific context for a specific set of objectives, affecting (1) the scope and types of moral considerations and (2) the choice and modes of AI applications. Table 2.1 presents a set of methods and functions that can be applied to instantiate the loops (with their distinctive features and some examples). Such instantiations involve the combination of the most appropriate methods and functions to establish the desired situated value alignment and MHC. An illustrative scenario exemplified the proposed value-alignment process for MHC (i.e., how the SOTEF implementation can be achieved). There will be some challenges to fully implement the SOTEF methodology. We will briefly discuss these below.

One challenge is to select relevant stakeholders at an early stage and to provide them with the needed resources. The involvement of stakeholders is key in the SOTEF loop, and should already be arranged at the start of an exploration or a design of AI functionality for military operations. However, stakeholder involvement (e.g., legal experts, legislators, ethicists, military users, system engineers, AI developers, and NGOs) has its challenges in all feedback loops. For example, stakeholders might want to protect themselves from co-optation by other stakeholders, safeguarding freedom of speech and maintaining independence (confidentiality). Another practical example concerns resources. Time and money may constrain relevant stakeholders to participate in value dialogues (Krabbenborg, 2020).

Another challenge is that the engagement of stakeholders also raises questions about communication and empowerment. Although the SOTEF loop relies on ideals of willingness to cooperate, openness, and harmony, it is known that these ideals are rarely realized in practice (Blok, 2014). The SOTEF methodology will provide the arguments and tools to establish the required engagement, referring to the applicable standards, methods, and functions. Furthermore, we aim to build up and share experiences on "who to involve how" (e.g., the implementation of the stakeholder roles and involvement of representatives of "unaccustomed" stakeholder groups such as citizens in a mission area), and how different values can be conceptualized, expressed, reported, and balanced.

The third challenge is that humans may find it hard to acknowledge, explicate and verbalize their values, because their primary assessment of right and wrong is often *implicit*, based more on emotional responses and less on rational (conscious) considerations (Haidt, 2001; Van Diggelen, Metcalfe, Van den Bosch, Neerincx, & Kerstholt, 2023). Furthermore, people's moral assessments are not unitary but *multi-dimensional* and *context-dependent* (i.e., related to the specific situation, work, and social roles; cf., Hannah, Thompson & Herbst, 2020; Aliman & Kester, 2022). The provision of scenarios, vignettes, and simulations in a virtual reality environment might help to systematically reflect on the moral aspects at stake, making the implicit explicit (cf., Parsons, 2015).

In conclusion, the SOTEF loop methodology comprises the assessment of a specific human-AI system operating in a specific context through an iterative, transdisciplinary, and multistakeholder approach. Although military AI creates new challenges and concerns for moral decision-making, it can also provide part of the solution. The use of military AI forces us to think about what values are at stake and how we want to ensure these values. SOTEF supports making ELSA of human-AI system deployment explicit, comparable, and auditable. It provides a way to better explicate attribution of responsibility and accountability; as such it is a way forward to operationalize MHC of military AI-based systems. It challenges stakeholders to make explicit and validate their goals and moral values, for the specific context the human-AI system is to operate in. Currently, we are operationalizing this methodology for realistic use cases, in order to refine and test the applicability of various methods and functions.

ACKNOWLEDGMENTS

We would like to express our gratitude to Dr. A.W. Bronkhorst and J.A.P. Smallegange for their invaluable contributions to this work. Their insightful comments and feedback during discussions helped us to refine our ideas and approaches. We also thank them for taking the time to review the chapter several times, and for providing constructive criticism that has helped us to clarify our arguments and conclusions.

APPENDIX 2.A: GLOSSARY

The table below provides working definitions of core concepts in this chapter. The TAILOR Handbook of Trustworthy AI provides a more generic overview of relevant definitions of trustworthy AI in the form of a publicly accessible Wiki: http://tailor.isti.cnr.it/handbookTAI/TAILOR.html.

CONCEPT	WORKING DEFINITION
Morality and ethics	Both morality and ethics pertain what is right ("good") and wrong ("bad"). The word morality is more used in relation to the personal normative aspects, whereas ethics more in relation to the normative standards within a certain community or social setting.
High-risk	The likelihood that unintentional socio-psycho-techno-physical perceived serious harm can be done.
Moral model	A formal model that represents what is right and what is wrong (e.g., in terms of action's benefits and harms) and, as such, univocally governs the behavior of the socio-technical system (STS). An AI agent's moral model is a formal model of how it should behave such that it contributes to a morally acceptable behavior of the STS as a whole.
Socio-technical system	A holistic perspective of a system containing an interconnection between humans (society as a whole) and (AI-based) technologies, including both social and technical aspects.
Socio-technical feedback loop	The human-centered methodology that addresses the context of all stakeholders of an AI application comprehensively, and prescribes a life-cycle enduring review and refinement process to enhance the models, reasoning, and adaptations to changing circumstances.
Value-alignment	The continuous process including the identification of the moral values that are at stake, and how they are addressed in a military operation.
Human-AI system	All of the humans and AI agents combined that collaborate to achieve a shared goal during operation.
Methodology	A methodology is a set of methods employed by a discipline. In the context of this chapter, it is the discipline of arriving at a responsible application of AI in the military domain.
Moral value	Something held to be right/wrong or desirable/undesirable at a certain moment in time by a certain group of people. Moral values describe what people value in terms of what they believe is morally acceptable. Fundamental examples include honesty and respect. Pragmatic examples include being fair and respecting another's privacy.
Goal function	A model of what the AI application should pursue such that it can be used to steer the AI application's behavior. Examples include optimization functions (loss, reward, fitness, utility functions) and logic inference rules (drawing conclusions and rule resolution).
Augmented utilitarianism	A non-normative meta-ethical framework that builds upon the foundational principles of deontological ethics, consequentialist ethics, and virtue ethics and combines them in one framework. Augmented utilitarianism tries to capture a more nuanced and comprehensive understanding of human harm perception from the perspective of moral psychology, for example, "dyadic morality". AU functions as a scaffold to encode human ethical and legal conceptions in a machine-readable form (e.g., ethical goal functions).
Ethical goal function	A goal function that also models moral values and thus governs an AI application's behavior in terms of what should be pursued in terms of how those values are modeled. Examples include multiobjective functions, utility functions, or multicriteria optimization functions whose attributes approximate observable moral values, and inference engines whose inference rules incorporate deontic logic.

NOTES

1 https://www.stopkillerrobots.org/
2 IHL Treaties – Geneva Convention (III) on Prisoners of War, 1949 – Article 3 (icrc.org)
3 TER report HFM-RWS 322: meaningful human control (MHC) of artificial intelligence (AI)-based systems, https://scienceconnect.sto.nato.int/tap/activities/11639
4 STS can refer to either socio-technical or socio-technological system. They both relate to the interaction between social and technical elements in a human-AI system, but might emphasize slightly different aspects. The former is more often used to emphasize the need for a holistic approach focusing both on technical and human factors, while the latter is used to highlight the role of technology in shaping interactions, behaviors and outcomes of a human-AI system. We use the term socio-technical as this is the more established and commonly used term in scientific literature.
5 AIV/CAVV advice 2021 and cabinet response 2022: https://www.adviesraadinternationalevraagstukken.nl/documenten/publicaties/2021/12/03/autonome-wapensystemen
6 Guiding Principles affirmed by the Group of Governmental Experts on Emerging Technologies in the Area of Lethal Autonomous Weapons System: https://www.ccdcoe.org/uploads/2020/02/UN-191213_CCW-MSP-Final-report-Annex-III_Guiding-Principles-affirmed-by-GGE.pdf
7 NATO Principles of Responsible Use: https://www.nato.int/docu/review/articles/2021/10/25/an-artificial-intelligence-strategy-for-nato/index.html
8 The TAILOR Handbook of Trustworthy AI (http://tailor.isti.cnr.it/handbook-TAI/TAILOR.html).
9 https://elsalabdefence.nl/
10 https://rri-tools.eu/
11 https://nlaic.com/en/category/building-blocks/human-centric-ai/elsa-labs-en/

REFERENCES

Aliman, N.-M. (2020). Hybrid cognitive-affective strategies for AI safety [Doctoral dissertation, Utrecht University]. https://doi.org/10.33540/203

Aliman, N. M., & Kester, L. (2022). Moral programming: Crafting a flexible heuristic moral meta-model for meaningful AI control in pluralistic societies. In *Moral design and technology* (pp. 494–503). Wageningen Academic Publishers.

Aliman, N.-M., Kester, L., Werkhoven, P., & Yampolskiy, R. (2019). Orthogonality-based disentanglement of responsibilities for ethical intelligent systems. *In International conference on artificial general intelligence (AGI)*, Shenzhen.

Amoroso, D., & Tamburrini, G. (2020). Autonomous weapons systems and meaningful human control: Ethical and legal issues. *Current Robotics Reports*, *1*(4), 187–194.

Beer, U. M., Neerincx, M. A., Morina, N., & Brinkman, W. P. (2020). Computer-based perspective broadening support for appraisal training: Acceptance and effects. *International Journal of Technology and Human Interaction (IJTHI)*, *16*(3), 86–108.

Blok, V. (2014). Look who's talking: Responsible innovation, the paradox of dialogue and the voice of the other in communication and negotiation processes. *Journal of Responsible Innovation*, *1*(2), 171–190.

Bratteteig, T., & Verne, G. (2018). Does AI make PD obsolete? Exploring challenges from artificial intelligence to participatory design. *In Proceedings of the 15th participatory design conference: Short papers, situated actions, workshops and tutorial-volume 2* (pp. 1–5).

Clark, J. O. (2009, March). System of systems engineering and family of systems engineering from a standards, V-model, and dual-V model perspective. In *2009 3rd annual IEEE systems conference* (pp. 381–387). IEEE.

Coeckelbergh, M. (2019). Artificial intelligence: Some ethical issues and regulatory challenges. *Technology and Regulation*, 2019, 31–34.

De Greef, T. E., Arciszewski, H.F.R., Neerincx, M. A. (2010). Adaptive automation based on an object-oriented task model: Implementation and evaluation in a realistic c2 environment. *Journal of Cognitive Engineering and Decision Making*, *4*(2), 152–182.

De Greef, T. E., Arciszewski, H.F.R., Neerincx, M. A. (2020). Identifying ethical considerations for machine learning healthcare applications. *The American Journal of Bioethics*, *20*(11), 7–17.

Dunnmon, J., Goodman, B., Kirechu, P., Smith, C., & Van Deusen, A. (2021). *Responsible AI guidelines in practice: Lessons learned from the DIU portfolio*. Washington, DC: Defense Innovation Unit.

Ekelhof, M. (2019). Moving beyond semantics on autonomous weapons: Meaningful human control in operation. *Global Policy*, *10*(3), 343–348.

Friedman, B., & Hendry, D. G. (2019). *Value sensitive design: Shaping technology with moral imagination*. Cambridge, MA: MIT Press.

Friedman, B., & Kahn, P. (2003). Human values, ethics, and design. In J. Jacko & A. Sears (Eds.), *The human-computer interaction handbook: Fundamentals, evolving technologies and emerging applications* (pp. 1177–1201). Mahwah, NJ: Lawrence Erlbaum Associates.

Haidt, J. (2001). The emotional dog and its rational tail: A social intuitionist approach to moral judgment. *Psychological Review*, 108, 814–834.

Hannah, S. T., Thompson, R. L., & Herbst, K. C. (2020). Moral identity complexity: Situated morality within and across work and social roles. *Journal of Management*, *46*(5), 726–757.

Koshiyama, A., Kazim, E., & Treleaven, P. (2022). Algorithm auditing: Managing the legal, ethical, and technological risks of artificial intelligence, machine learning, and associated algorithms. *Computer*, *55*(4), 40–50.

Krabbenborg, L. (2020). Deliberation on the risks of nanoscale materials: Learning from the partnership between environmental NGO EDF and chemical company DuPont. *Policy Studies*, *41*, 372–391

Kunneman, Y., Alves da Motta-Filho, M., & van der Waa, J. (2022). Data science for service design: An introductory overview of methods and opportunities. *The Design Journal*, *25*(2), 186–204.

Liu, X., Li, H., Xu, G., Chen, Z., Huang, X., & Lu, R. (2021). Privacy-enhanced federated learning against poisoning adversaries. *IEEE Transactions on Information Forensics and Security*, *16*, 4574–4588.

Miller, C. A., & Parasuraman, R. (2007). Designing for flexible interaction between humans and automation: Delegation interfaces for supervisory control. *Human Factors*, *49*(1), 57–75.

Parsons, T. D. (2015). Virtual reality for enhanced ecological validity and experimental control in the clinical, affective and social neurosciences. *Frontiers in Human Neuroscience, 9*, 660.

Peeters, M. M., van Diggelen, J., Van Den Bosch, K., Bronkhorst, A., Neerincx, M. A., Schraagen, J. M., & Raaijmakers, S. (2021). Hybrid collective intelligence in a human-AI society. *AI & Society, 36*, 217–238.

Reed, N., Leiman, T., Palade, P., Martens, M., & Kester, L. (2021). Ethics of automated vehicles: Breaking traffic rules for road safety. *Ethics and Information Technology, 23*(4), 777–789.

Santoni de Sio, F., & Van den Hoven, J. (2018). Meaningful human control over autonomous systems: A philosophical account. *Frontiers in Robotics and AI, 5*, 15.

Scharre, P. (2018). *Army of none: Autonomous weapons and the future of war.* WW Norton & Company.

Schuler, D., & Namioka, A. (1993). *Participatory design: Principles and practices.* Hillsdale, NJ: Lawrence Erlbaum Associates.

Steen, M., Neef, M., & Schaap, T. (2021). A method for rapid ethical deliberation in research and innovation projects. *International Journal of Technoethics (IJT), 12*(2), 72–85.

Stilgoe, J., Owen, R., & Macnaghten, P. (2013). Developing a framework for responsible innovation. *Research Policy, 42*, 1568–1580

Sutcliffe, A. (2003). Scenario-based requirements engineering. In Proceedings 11th IEEE *international requirements engineering conference, 2003* (pp. 320–329). IEEE.

Ten Holter, C. (2022). Participatory design: lessons and directions for responsible research and innovation. *Journal of Responsible Innovation, 9*(2), 275–290.

Van de Poel, I. (2009). Values in engineering design. In A. Meijers (Ed.), *Handbook of the philosophy of science. Volume 9: Philosophy of technology and engineering sciences* (pp. 973–1006). Elsevier.

Van den Hoven, J. (2013). Value sensitive design and responsible innovation. In R. Owen, J. Bessant, and M. Heintz (Eds.) *Responsible innovation* (pp. 75–83). (Chichester, UK: John Wiley & Sons, Ltd).

van der Waa, J., van Diggelen, J., Cavalcante Siebert, L., Neerincx, M., & Jonker, C. (2020). Allocation of moral decision-making in human-agent teams: A pattern approach. In *Engineering psychology and cognitive ergonomics. Cognition and design: 17th international conference, EPCE 2020, held as part of the 22nd HCI international conference, HCII 2020, Copenhagen, Denmark,* July 19–24, 2020, Proceedings, Part II 22 (pp. 203–220). Springer International Publishing.

van Diggelen, J., Barnhoorn, J., Post, R., Sijs, J., van der Stap, N., & van der Waa, J. (2021). Delegation in human- machine teaming: Progress, challenges and prospects. In *Intelligent human systems integration 2021: Proceedings of the 4th international conference on intelligent human systems integration (IHSI 2021): Integrating people and intelligent systems, February 22–24, 2021, Palermo, Italy* (pp. 10–16). Springer International Publishing.

van Diggelen, J., Metcalfe, J. S., Van den Bosch, K., Neerincx, M, & Kerstholt, J. (2023). Role of emotions in responsible military AI. *Ethics and Information Technology, 25*, 17. https://doi.org/10.1007/s10676-023-09695-w

Van Veenstra, A. F., Van Zoonen, L., & Helberger, N. (2021). *ELSA Labs for human centric innovation in AI.* Netherlands AI Coalition.

Von Schomberg, R., & Hankins, J. (Eds.). (2019). *International handbook on responsible innovation: A global resource.* Edward Elgar.

Winston, M., & Edelbach, R. (2013). *Society, ethics, and technology.* Cengage Learning.

Zielke, T. (2020). Is artificial intelligence ready for standardization? In *Systems, software and services process improvement: 27th European conference, EuroSPI 2020, Düsseldorf, Germany, September 9–11, 2020, proceedings 27* (pp. 259–274). Springer International Publishing.

How Can Responsible AI Be Implemented?

3

Wolfgang Koch and Florian Keisinger

SOME POLITICAL PRELIMINARIES ON MILITARY AI

"All kinds of instruments are turned into weapons. […] We love the world of Kant but must prepare to live in the world of Hobbes. Whether you like it or not" (Gutschker, 2021). This statement of Josep Borrell, High Representative of the European Union for Foreign Affairs and Security Policy, in November 2021 was prophetic and marks a new epoch we live in. The engineering communities have also woken up to this world. As the news from the Russian war in Ukraine proves, external security is the prerequisite to achieve all other individual, social, political, or ecological goals. Previously, Islamist terror, organized crime, and political radicalization taught us to value domestic security. Approximately 80 years after the end of WW II, at least Europeans must learn again what a truly "sustainable" and precious commodity "security" actually is, without which personal freedom and cultural fruits perish. Without external and internal security, there can be no calculable economic processes, no steady inflow of raw materials, no robust supply chains for export-dependent nations, no services of general interest, and no social balance. Even without secure technology, modern societies would be unstable, as they also depend on intrinsically risky technology and processes.

Also apart from the Russian aggression against Ukraine, the democratic world is facing major challenges in terms of foreign policy and security. Part of the new reality is that armaments activities are increasing around the globe, with the focus not only on the rather symbolic pursuit of new and more nuclear weapons, but, above all, on the application of new technologies. Computer science in particular has to deliver its own contribution to defend human culture, the freedom of nations, the legal order,

DOI: 10.1201/9781003410379-4

and world peace. In "the Age of AI," i.e., of mathematical algorithms, this discipline is crucial since there can be no effective armed forces, without information superiority and decision dominance in all military domains. We have long since found AI not only in the military dimensions of Land, Air, Sea, and Space, but also in Cyber and Information space. Hypersonic missiles, underwater warfare, or military aviation are just a few examples where AI already is or soon will be applied.

Critical Discussions are Necessary

The use of artificial intelligence in the defense sector is accompanied by critical discussions, which we appreciate as a necessary part of the democratic discourse. As always, the primary factor in all technology in general, and in Artificial Intelligence in particular, is the responsible use of it, i.e., its concrete application. The politically charged buzzword "autonomous weapon systems," weapons that supposedly define and attack targets without any human intervention, has taken root in this context. However, this terminology is misleading since the use of artificial intelligence in weapons systems currently takes place at best in the context of semi-automated applications – and is not beyond human control.

Germany's position on this point is clear: "We reject lethal autonomous weapon systems that are completely removed from human control. We are actively promoting their international outlawing" reads the coalition agreement of the current German government. The first military AI applications can be found in assistance systems, in which certain processes, especially in reconnaissance and data generation, are carried out semi-automatically. What is done with this information, or what consequences it entails, is up to the decision of the armed forces, i.e., of the human being. For future weapon systems – even with increasingly automated functions – the principle of human control always applies. The notions "AI in national and alliance defense" and "autonomous weapon systems" must therefore not be equated. The former serves as a support and decision-making tool to comprehensively evaluate the ever-increasing volumes of data – and is thus an important decision-making aid. The latter is a military horror scenario, which the German government and the Bundeswehr, together with numerous European and non-European partners, are right to oppose – specifically, for example, in the context of the UN negotiations in Geneva to ban lethal autonomous weapons systems.

Many citizens are understandably afraid of intelligent machines that develop a supposed "life of their own" beyond human control. Therefore, it is all the more important to critically reflect on Artificial Intelligence – just like on all new technologies. However, we should refrain from evoking emotions by using the wrong terminology and contexts, which lead to wrong associations among people outside computer science or the respective field of application. This is especially important given the public response to large-scale language models and other Generative AI that add a new dimension to the simulation of human intelligence performance. Quantum Computers, on the other hand, also have comparable transformative potential and can be used in morally questionable ways. Nevertheless, Quantum Computing seems less threatening and uncontrollable than Artificial Intelligence, Machine Learning, or Technical Automation in defense. The responsibility aspect in dealing with these and other technologies is the same.

Critical Understanding is Inevitable

The call for a general outlawing of AI in defense sounds reasonable on the surface and is unlikely to go unheard in Western societies. What is not mentioned are the resulting consequences for European and transatlantic defense capabilities. Nevertheless, disarmament and regulatory negotiations, for example within the framework of the United Nations, must continue to be pursued with vigor. In this context, it must always be borne in mind, however, that the Western countries can engage in disarmament and regulatory negotiations far more effectively from a position of participation and inclusion of modern technologies than – except in the case of autonomy, which must be categorically denied – by relying on strict rejection. Moreover, it would be little more than a symbolic gesture on a global scale.

In addition, if AI-based technologies in offensive weapons systems are outlawed, their use in defense against AI-based weapons would also have to be outlawed. Only those who understand the threats can counter them confidently and effectively. This is not just a question of becoming militarily and technologically disengaged; those who help shape the future also have a say in which direction the technological development progresses. In other words, they can bring their own values into the discourse.

How should Western countries position themselves when it comes to artificial intelligence in defense? A differentiated approach considers not only technological and security policy aspects, but also legal, ethical, and operational issues. A normative framework should be developed that ensures human control in the application of AI – including in weapon systems – as well as adherence to ethical standards in the sense of Western values, while at the same time meeting the security policy and operational realities of the 21st century. Among the latter is the need for Western countries to be technologically capable of defense against AI-enabled weapons systems. The general exclusion of AI would result in irresponsible structural military inferiority and would operationally increase the risk of collateral damage, for example due to a lack of precision in target acquisition. Against this backdrop, the military use of AI is not only ethically justified, but even required.

What such a process for the responsible use of AI in defense could look like in concrete terms can currently be seen in potentially the largest European defense project of the 21st century, the "Future Combat Air System" (FCAS). To accompany the development of the technology, a multistakeholder committee was set up in 2019 by Airbus Defense and Space and Fraunhofer FKIE with the involvement of players from the fields of security and defense, research, and science, think tanks as well as churches and society. The ambition is to define a path in order to guarantee overall human control of the system.

Within this framework, and to experience the use of AI in an FCAS in the context of potential use cases as realistically as possible, a special tool is currently being developed under the auspices of this body in the German part of the FCAS project, which will concretely allow the application of AI to be experienced in various scenarios. This is being done based on AI-based assistance and in the context of simulated combat decisions. The scenarios considered are realistically designed and include military expertise from the Bundeswehr. The aim is to provide a realistic picture of the opportunities, limitations, and ethical implications of Artificial Intelligence in defense in a specific use case. The resulting findings can provide input for the technical FCAS design.

This approach, which is sketched in this chapter, is a novelty in a major defense project, at least in Germany, if not worldwide, and could become a model for other major international defense projects as well – and point the way in answering the morally difficult question of guidelines for the application of AI in defense from a security policy and operational perspective without putting German, European and transatlantic defense capability and security policy responsibility at risk.

COGNITIVE AND VOLITIVE MACHINES IN DEFENSE

Before any scientific reflection or technical realization, intelligence and autonomy are ubiquitous as natural abilities. All living creatures fuse different sensory impressions with already learned knowledge and messages of other living beings. In this way they create an image of their environment, the prerequisite for situation-appropriate action in the biosphere, for processes to avoid dangers and to achieve goals. What is meant is instrumental intelligence, which serves a particular end. In characteristic gradations natural intelligence (NI) and autonomy are bound to the corporeality of creatures, which is not fully described by such instrumentally understood processes. The ethical and legal questions of human–animal interaction with the respective intelligence and autonomy seem to be much closer to a solution.

Artificially intelligent automation, on the other hand, provides new types of machines that greatly enhance the perceptive mind and the active will of persons, who alone are capable to perceive intelligently and to act autonomously in a proper sense.

1. *Cognitive machines* fuse massive streams of sensor, observer, context, and mission data for producing comprehensive situation pictures, the basis for conscious human cognition to plan, perceive, act, and assess effects appropriately
2. *Volitive machines* transform deliberately taken overall decisions of responsible human volition into complex chains of automatically executed commands for data acquisition, sub-system control, and achieving effects on objects of interest

Processes triggered by such machines and running automatically are therefore to be distinguished from NI and autonomy. Nevertheless, certain processes that underlie conscious perception and causal action and that were previously reserved for humans are, so to speak, "excarnated," i.e., transferred to machines. This understanding is in line with the US AI Strategy, which defines AI as "the ability of machines to perform tasks" that "normally require human intelligence (Allen, 2020)." This would include long-established technologies. It also includes physical assistance through AI-controlled exoskeletons or robots. The immediate physical presence of humans is becoming increasingly dispensable for their perception and action.

Such machines will become key elements of the FCAS, the largest European armament effort ever for protecting European sovereignty. In this program, manned jets are only elements of a larger networked system of systems, where unmanned "remote carriers" protect the pilots as "loyal wingmen" and support them on reconnaissance and combat missions. By technically assisting their minds and wills cognitively and volitively, air commanders and staffs will remain capable of appropriately acting even on short time scales in the complex "technosphere" of modern warfare with spatially distributed and highly agile assets. This is particularly true when targeting cycles are vastly accelerated and to be executed "at machine speed" in a network-centric and collaborative way.

This book chapter is harvesting fruits of ongoing discussions in the working group *Responsible Technology for an FCAS* (Keisinger & Koch, 2023) and evolves insights published earlier (Koch, 2020, 2021a, b, 2022). Our considerations correspond to some extent to the *IEEE 7000 Model Process for Addressing Ethical Concerns During System Design* (Spiekermann, 2021). A large community of engineers and technologists is addressing ethical problems and technical realizations to mitigate them throughout the various stages of system initiation, analysis, and design for particular use cases, for example, Fair, Accountable, or Transparent AI.

After concluding this section with remarks on foundational documents of the German Armed Forces as examples for military views on the topic and on possible metaphors for understanding the interaction of human beings with AI-driven cognitive and volitive machines, we introduce in the next section the notions of reflective and normative assistance in military decision-making with a focus on "combat clouds." Here, ethically relevant implications demanded by official documents are considered, shaping the ethics, ethos, and morality of dealing with AI-based weaponry. Based on the fundamental notion of "responsibility" and its relation to systems engineering aspects, we next discuss the design principles of the FCAS Ethical AI Demonstrator (E-AID), the core contribution of this chapter. Considerations toward normative assistance close this section. The problem of transparent criteria development is addressed in the following section, which has implications for acquiring "digital virtues" in dealing with AI in defense and might establish an analogy between the Hippocratic Oath and soldierly ethos. Finally, we try to draw conclusions in a more generalizing sense in the final section.

Looking into Foundational Documents

"The more lethal and far-reaching the effect of weapons are, the more necessary it is that people behind the weapons know what they are doing," observes Wolf von Baudissin (1907–1993), the visionary architect of Germany's post-WW II armed forces, the *Bundeswehr*. "Without the commitment to the moral realms, the soldier threatens to become a mere functionary of violence and a manager" he continues and thoughtfully adds: "If this is only seen from a functional point of view, i.e., if the goal to be achieved is in any case put above human beings, armed forces will become a danger" (von Baudissin, 1969). It is in this sense that we consider aspects of AI-driven targeting cycles and their responsible design.

The more general key question this chapter is intending to help answer therefore reads: How can the information fusion community *technically* support responsible use of the great power we are harvesting from artificially intelligent automation? While facing soberly the risks of digitalization in defense, we nevertheless beware of exaggerating them, which may become a risk in itself and prevent innovation in defense. Despite our clear military focus, we hope that our considerations below might enjoy a broader consent also in civil decision-making.

As will become visible, the use of AI in defense systems of systems such as FCAS is intended to unburden military decision-makers from routine or mass tasks. We in particular need to tame technical complexity in such a way that commanders, staffs, and soldiers will be able to focus on doing what only persons can do, i.e., to consciously perceive a situation intelligently and act responsibly. The importance of automation for armed forces was recognized as early as 1957 when von Baudissin wrote that thanks to automation, "human intelligence and manpower will once again be able to be deployed in the area that is appropriate to human beings" (von Baudissin, 1969). Seen from this perspective, armed forces do not face fundamentally new challenges as users of artificially intelligent automation, since the technological development has long extended the range of perception and action.

In order to be able to argue in a more focused way in the sense of a use case, we will examine conceptual documents of the German *Bundeswehr* in our approach that span the period from its founding in the 1950s, when the term "AI" was actually coined, to its most recent statements on the matter. Since these armed forces have learned lessons from the totalitarian tyranny in Germany from 1933 to 1945 and the horrors of "total war,"[1] characterized by the high technology of this time, they are presumably in a conceptual way well prepared for mastering the digital challenge we are confronted with today. This is even more the case since the *Bundeswehr* is a parliamentary army enshrined in the German Constitution, *Grundgesetz*, which acts exclusively in accordance with specific mandates from the *Bundestag* and therefore on behalf of the German people.

Remarks on Human–Machine Interaction

From a technology-agnostic perspective, the "well-intentioned" but technically unrealizable transparency demand that has been made in various places is: "AI systems must be fundamentally explainable." On the one hand, model-based AI systems that in principle are "fundamentally explainable" turn out to be so complex in practice that they become "gray boxes" as well. On the other hand, despite all the research into Explainable AI, it will at least be possible to turn the "black boxes" of data-driven algorithms into gray boxes. Due to their NI, however, humans are very successful in dealing responsibly with other NIs, also gray boxes, for example with animals, even if with hunting dogs or riding horses better than with cats, but also with people, for example with colleagues when solving tasks together.

Human interaction with artificially intelligent machines can therefore be designed according to a hunting dog or rider metaphor (Flemisch et al., 2014) or in the sense of collegiality. The prerequisite is a specific education of the human being. This will enable naturally intelligent people to deal with artificial gray boxes by means of suitably designed human–machine interaction systems just as responsibly and successfully as with natural gray boxes. Likely, AI research will purposefully develop "dog-like AI" that is more suitable for certain tasks and less risky than "predatory AI."

Chatbots will be revolutionizing the interfaces between humans and cognitive or volitive machines also in the military domain. Their "eloquence" not only facilitates human–machine communication. Rather, "dialogs" with chatbots stimulate processes of reflection, categorization, or speculation in the users. They clarify their question or make the users aware of what they "actually" wanted to know. AI programs such as DALL-E, which generate photorealistic images from text descriptions, perform comparable tasks in information visualization and take the idea of Chernoff Faces further. The American statistician Herman Chernoff (*1923) had represented multivariate data as human-like "faces," whose "content" is much easier to grasp by humans than, for example, tables, since they perceive even tiny changes in facial expressions (Chernoff, 1973).

This kind of AI-driven human–machine interaction will influence our mental and interpersonal habits. Chatbots are far superior to conventional keyword searches in vast "knowledge repositories." How do humans remain mentally resilient to the new power? How do they prove ultimately responsible authority?

Without an image of a human as a conscious and free person, any AI-based assistance becomes questionable, especially in military use. Especially in the military technosphere, soldiers use personal judgment and decisiveness to act on their own responsibility: "Due to the scientization and mechanization of the military craft [the superior] is largely dependent on the judgment of his specialists in his assessment of the situation and his decisions," Wolf von Baudissin (1907–1993) underlines the finality principle of military action. "His subordinates, down to the lowest level of the hierarchy, must solve their small situation on their own initiative in a complicated interplay of technique and tactics - always within the framework of the overall intention, of course" (von Baudissin, 1969).

ARTIFICIALLY INTELLIGENT COMBAT CLOUDS

From a digitalization perspective, the core infrastructure for future air defense and combat systems is *air combat clouds*. While sensors are collecting data, combat clouds distribute, verify, validate, organize, evaluate, process, and fuse data to enable adaptive management of sensors, platforms, communication links, and effectors such as weapons "at machine speed." In the digital age, information superiority in complex situations and decision dominance even at very short time scales decide between success and failure of a mission. According to the introductory remarks, the architecture of a combat cloud, i.e. of the informational backbone for military air operations, has to facilitate the responsible use of weapon systems by human decision-makers. Artificially intelligent automation is crucial here since it enables complexity management and responsible action by providing cognitive and volitive assistance. In parallel, "digital twins" accompanying the technological development from the very beginning have to ensure that comprehensive ethical and legal compliance is not at the expense of effectiveness in air defense and combat.

We here use the term "Artificial Intelligence" in a sense that does not only comprise machine or deep learning (DL), for example, but a whole "cloud" of data-driven and model-based algorithms, including approaches to Bayesian learning, game theory, and adaptive resources management. It seems worthwhile to consider "Artificial Instinct" as a more appropriate of the acronym "AI" that was proposed by the Polish science fiction author, philosopher, and futurologist Stanisław Lem (1921–2006) nearly 40 years ago (Lem, 1983).[2]

A "cloud of algorithms," realized by the art and craft of programming and enabled by qualitatively and quantitatively appropriate testing and training data, drives a data processing cycle that starts from elementary signals, measurements, and observer reports collected from multiple and heterogeneous sources. For us, "AI" denotes the process that fuses such streams of mass data and context knowledge, which provide pieces of mission-relevant information at several levels, for producing comprehensive and near real-time situation pictures. On their basis, air commanders and their staff become aware of the current situation in a challenging environment and the status of the mission. Human decision-making for acting according to the ends of the mission to be achieved is carried out at different levels of abstraction and degrees of detail. Technical Automation transforms deliberate acts of will into complex command sequences to control networking platforms, multifunctional sensors, and effectors.

Algorithms for comprehensively harvesting information by data fusion and adaptively managing the various processes of data collection as well as weapon engagement and effect assessment belong to the methodological core of cognitive and volitive machines that assist the intelligent mind and autonomous will of decision-makers. They exploit sophisticated methods of applied mathematics and run on powerful computing devices, where quantum computing may become a game changer (Govaers et al., 2021; Stooß et al., 2021). The concepts of mind and will and therefore of consciousness and responsibility bring human beings as persons into view that are "somebody" and not "something."

Reflective and Normative Assistance

While artificially intelligent automation is indispensable for achieving situational awareness, a prerequisite of reducing collateral damage, for example, as well as of commanding resources, it also implies specific vulnerabilities such as:

1. *Loss of data integrity* causing invalid situation pictures and improper decisions due to unintended malfunction of sensors, programming errors, misuse of training data, or data incest.
2. *Artifacts generated by AI* algorithms from sensor and context data that do not exist in reality, or blind spots, that are disabling situation pictures to show what is actually present in reality.

3. *Hostile intervention* at various levels to be taken into account, where adversaries take over sensors or subsystems, which then produce deceptive data or initiate unwanted action.
4. *More general issues* of automated systems such as misuse, disuse, abuse, non-use, and blind or overly trust, which are not specifically AI-related, but must be taken into account as well.

In consequence, resilient cognitive and volitive machines for defense systems of systems have to comprise the detection and compensation of such deficits in the sense of "Artificial Self-criticism."

Any ethically and legally acceptable use of cognitive and volitive machines relies on "truth," defined as "equivalence between awareness and the actual situation" and "goodness," defined as "equivalence between the choices made and norms." Their proper use, however, needs to be supported by "reflective" and "normative" assistance functions, seen as part of ethically aligned cognitive and volitive machines as discussed below. The fusion of sensor data and non-sensor information provides mission-relevant insights. Apparently, comprehensive information fusion is the key to seamlessly integrating also formalized ethical or legal constraints, seen as a particular type of context knowledge, into reconnaissance or combat missions. For the sake of simplicity, we confine the discussion of the normative framework to the Rules of Engagements (RoEs) that have to mirror the risks of artificially intelligent automation and must permeate the technical system design.

Ethical Implications of Basic Documents

According to the foundational document of the German *Bundeswehr* (German MoD, 2018), updated in 2018, artificially intelligent automation is expanding its capability profile by providing:

1. Perception of a military situation as reliably as possible by "obtaining, processing, and distributing information on and between all command levels, units and services with minimum delay and without interruption or media disruption."
2. Support of "targeted deployment of forces and means according to space, time and information, [...] where characteristics of military leadership are the personal responsibility of decision-makers and the implementation of their will at any time."

Readiness to defend ourselves against highly armed opponents must not only be technologically credible, but also correspond to the consciously accepted "responsibility before God and man, inspired by the determination to promote world peace as an equal partner in a united Europe," as the very first sentence of the German Constitution, the *Grundgesetz,* proclaims (Federal Republic of Germany, 1949). Guided by this spirit and for the first time in Germany, an intellectual struggle over the *technical* implementation of ethical and legal principles accompanies a major defense project from the outset. The goal of the working group on *Responsible Use of New Technologies in an FCAS* is to operationalize ethically aligned engineering.

Official documents released by the German Ministry of Defence implicitly define elementary requirements that are relevant for ethically-aligned FCAS systems design and have direct implications for the E-AID to be discussed below. With a focus on ethically critical tasks within the targeting cycle to be executed by FCAS commanders, E-AID demonstrates, in which way cognitive, volitive, reflective, and normative assistance systems should be developed and how they interact with each other. Also in view of the international law, considerations on the ethical implications are encouraged, since Article 36 of the Additional Protocol I of the 1949 Geneva Conventions requires states to conduct legal reviews of all new weapons, means, and methods of warfare in order to determine whether their use is prohibited (Von Baudissin, 2015). To be mentioned are the 11 guiding principles affirmed by a group of governmental experts within the framework of the UN Convention on Certain Conventional Weapons (CCW, 2019).

On Ethics, Ethos, and Morality

For properly designing cognitive and volitive machines in the context of FCAS, ethical implications need to be clarified while avoiding moralizing. The following distinction proves to be helpful in designing reflective and normative assistance:

1. *Digital ethics* denotes theoretical reflections about right decisions in using artificially intelligent automation. Required is a philosophically founded conception of what characterizes human being "by nature," i.e. an Image of Man that makes notions such as *mind*, *will*, and, therefore, *consciousness* and *responsibility* conceptually possible.
2. *Digital ethos* addresses the attitude of decision-makers on all levels. "The more momentous the decisions and actions of individual soldiers are, the more their ethos must be determined by responsibility," as von Baudissin observed. For developing an ethical attitude, the notion of *virtue* seems helpful which is understood here as perfection of mind and will toward the good of reason.
3. *Digital morality*, finally, comprises the formulation of concrete guidelines for dealing with artificially intelligent automation, not only on the battlefield, but also in research, development, procurement, planning, and mission preparation.

Along with such considerations, the German Ministry of Defence underlines that "the importance of AI does not lie in the choice between human or artificial intelligence, but in an effective and scalable combination of human and artificial intelligence to ensure the best possible performance" (German MoD, 2019b). Comprising ergonomic as well as ethical and legal dimensions of AI, this statement implicitly demands responsible systems engineering and, as such, aims as well at fulfilling the military requirements previously mentioned. In particular, numerous research questions for systems engineering result that aim at a fundamental military requirement: "Characteristic features of military leadership are the personal responsibility of decision-makers and the implementation of their will in every situation," according to the "Concept of the Bundeswehr" (German MoD, 2018).

DECISION-MAKING FOR WEAPON ENGAGEMENT

A challenge for valid situational awareness and responsible decision-making for weapon engagement in the FCAS domain is the ever-decreasing time available for human involvement in the decision-making process. Further problems are limited explainability and deceivability of both, algorithmically generated information and automated execution of complex command chains. The following issues need to be addressed:

1. While in certain applications, occasional malfunctioning of AI-enabled automation may have no consequences, rigorous safety requirements must be guaranteed for FCAS with all legal consequences. The military use of technically uncontrollable technology is immoral *per se*.
2. The notion of *meaningful human control* needs to be interpreted more broadly than the concept of *human-in/on-the-loop* suggests.[3] A more fundamental notion is "accountable responsibility." Since the use of fully automated effectors on unmanned platforms may well be justifiable, even necessary in certain situations, the overall system design must guarantee that always a distinct "somebody" is responsible.

In view of these considerations, artificially intelligent automation for FCAS poses a timeless question: Which design principles facilitate "good" decisions according to what is recognized as "true" according to the previous definitions? Turned into systems engineering, this implied two tasks:

1. Design cognitive assistance in a way that human beings are not only mentally, but also emotionally able to master each situation.
2. Design volitive assistance to guarantee that human decision-makers always have full superiority of information and the options of action.

In consequence, digital ethics as well as a corresponding ethos and morality are essential soft skills to be built up systematically in parallel to technical excellence. Personality development plans should encourage ethical competence for responsibly designing *and* using AI-based cognitive and volitive assistance.

On the Notion of "Responsibility"

Literally, the very word "responsibility" is rooted in the language at courts of justice designating the obligation of being called upon to "respond" to questions about one's own actions by a judge, a primal situation of human existence as a person. This overall concept has far-reaching implications:

1. To speak of responsibility is only reasonable if it is assumed voluntarily. Responsibility, thus, presupposes the notion of a "free will" and an Image of Man as a free and "autonomous" person. Here, "autonomy" is understood as a moral right and the capability to think for oneself and decide in a way that achieves a freely set effect.
2. The concept of free will as the decisive cause of decisions to action implies the idea of an accountable person, which is legally relevant and an essential criterion in the International Law.
3. Responsibility, as considered here, implies in addition to the legal notion of accountability the ability of a person to act freely and the willingness to act well even in case of absent or contradicting rules. Casuistry, the formalization of human action by just following well-defined rules, seems impossible.
4. The will, responsible in freedom, is not absolute, but depends on the understanding mind. In a philosophical sense, the "True" as the formal object of the mind and the "Good" as the formal object of the will thus form the intellectual basis of responsible action. Admittedly, it is not trivial to actually achieve what is true or good.

Figure 3.1 illustrates the core elements of the concept of responsibility as a triangle relationship, insofar as it is relevant to the technical systems design. It implies the

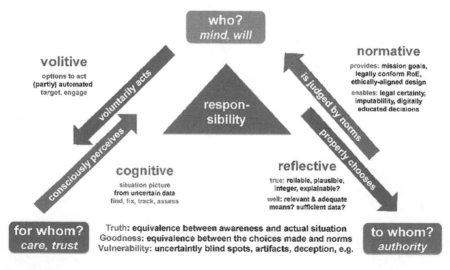

FIGURE 3.1 Artificially intelligent automated assistance enabling responsible action for FCAS. (© Fraunhofer FKIE.)

notion of persons or groups of persons as preciously sketched and establishes characteristic relationships between them. Responsible systems design is, thus, by definition "anthropocentric."

1. *Who bears responsibility?* Military capability development takes place at various levels and requires responsible action in research, development, certification, and qualification of military Command & Control, ISR, and weapon systems as well as in the preparation and execution of military operations.
2. *For whom is responsibility borne?* The relationship between responsible persons and those for whom they are responsible is characterized by "care" and "trust" and therefore determined by prospective action and reaction. In a proper sense, responsibility can only be assumed by persons for persons. Indirectly, one might speak of responsibility toward animals, cultural heritage, or the environment, for example, insofar as these are related to persons.
3. *To whom is responsibility assumed?* Responsibility implies the notion of a personal authority exercising his or her authority by judgment. The responsible person recognizes this authority by his or her justification. The relationship between responsible persons and a personal authority is retrospective in nature.

Voluntarily assumed responsibility, which shows itself in "care" and "trust," as well as in the readiness to justify itself and to choose properly in obedience to norms, keeps military forces stable in combat. It can and should be supported by normative and reflective assistance systems to be specified below. Purely legal constructs, however, such as liability for damage caused by one's actions, are insufficient, especially in military operations.

According to these considerations, no machine can act responsibly or irresponsibly, i.e. in a "good" or "evil" way by responding to moral challenges in one way or another, but persons only. In a figurative sense, it would be possible to speak of "Good" technical systems to encourage the morally acceptable and efficient use of them to achieve military objectives. "Evil" systems facilitate their irresponsible use.

FCAS E-AID

By the FCAS E-AID discussed here we wish to clarify on which technically realizable basis human operators are enabled to make balanced and conscious decisions regarding the use of weaponry based on artificially intelligent automation. One might speak of "meaningful authorization." This is particularly pressing in cases where AI algorithms such as DL are under consideration, which have the character of a "black box" for the user.

For approaching a viable solution, it is important to make AI-based findings comprehensible, plausible, or "explainable" to human decision-makers. On the other hand, soldiers should not confirm recommendations for action without weighing them up themselves, simply based on some kind of "trust" in the AI-based system. To this end, we introduce the concept of "reflective" assistance as indicated in Figure 3.1.

Especially for FCAS, engineers must aim at developing comprehensible, plausible, or "explainable" methods. With the help of E-AID air commanders and staff experience the use of AI in militarily relevant and close-to-reality scenarios by displaying all associated aspects of psychological stress including ethical conflicts as realistically as possible. Selected features of the E-AID, such as automated target recognition for decision-making in air combat enable interaction with an actual AI developed for military use in order to enable a realistic view of the possibilities, limitations, ethical implications, and engineering demands of this technology in military practice.

Discussions with officers of the German Air Force have clarified the scenarios to be considered. One of the missions envisaged for FCAS is the elimination of enemy air defense using remote carriers with electro-optical and signal intelligence sensors that collect data on positions of equipment that is supporting enemy air defense. The (much)-simplified steps in such a use case proceed as follows:

- The user will detect, identify, and track enemy vehicles in different scenarios with and without AI support for comparison, by exploiting control of multiple sensor systems on a remote carrier.
- The output of the AI system is used to graphically highlight relevant objects accordingly and enrich them with basic context information (e.g., type of detected vehicle, certainty level).
- The user, who is in the role of a virtual payload operator of the remote carrier flying ahead, has the task of recognizing and identifying all relevant objects.
- Manual target designation needs to be analyzed here as well and compared with those done by AI.
- To facilitate the user's ability to perform this task, optional confirmation dialogues provide information for all individual objects recognized or preselected by the AI system at a much greater level of detail.

This dialogue will enable the following:

1. To request a magnified image of the object in question to confirm the target by visual address, and to understand in the magnified section by means of appropriate highlighting of Explainable AI (XAI) which has recognized elements of the tracked object.
2. To enhance sensor data fusion with additional data sources, to understand which sensor technology, if any, has "tipped the scales" for classification as a hostile object, and to visualize corresponding levels of confidence for the respective sensor category.
3. To check compliance with the ROE for the object in question, insofar as deterministic algorithms can provide support here; to confirm compliance with the ROE as checked.

Ostensibly, such a dialogue should provide a more unambiguous identification of an object as "hostile." In other words, the design needs to allow the operator in critical situations to selectively query all technical information from the system that is relevant to rationalize the targeting process. Evidently, there are limitations of this approach if there are too many objects or if time is too short to query all technical information. The demonstrator is expected to clarify what "critical" means in dense target situations.

Elements of Normative Assistance

As indicated before, the RoEs provide the underlying normative framework considered here. In designing a technical system for normative assistance, the possibilities and limitations of implementing legal principles need to be addressed. The following discussion was inspired by comments by the German lawyer Tassilo Singer (personal communication, September 12, 2022, see also Singer, 2019).

1. According to current understanding of the legal state of the art, certain legal principles formulated in the RoEs cannot be translated into an algorithmic form or in such a way that they can make human-type, evaluative decisions (for example, moral or ethical opinions, weighing decisions). An example in the context of the international humanitarian law is the principle of "proportionality," i.e. prohibition of excess. It will be part of the work with E-AID to identify those legal principles.
2. If it is possible to translate a legal principle, such as "An attack may not be directed against a civilian population. A distinction must be made between civilians and combatants," into an algorithm or an AI-model, certain criteria, threshold values or parameters are decisive prerequisites for the legally compliant behavior of an AI-controlled system, i.e. the effective restrictability (with probability bordering on certainty) of the actual behavior. At least on the level of mission execution, a large portion of rule type RoEs should be translatable in algorithmic form. The thresholds mentioned are already present, at least verbally, in military documents such as the procedures of military reporting, and are even assigned to numerical values: "possibly" (<30%), "likely" (30%–90%), and "probable" (>90%).
3. This leads to a key thesis: Provided a legal principle can be translated into an AI-model with quantitative criteria being integrated in the previous sense, a legally compliant implementation of legal principles can be achieved through technical system design, supplemented by sandboxing, testing, auditing.
   a. If this is the case, a control mechanism needs to be integrated, either additionally in the AI model or as part of the training, for example, a definable "no fly zone." A threshold value in connection with a rule could be: Only from a certain probability on may a target be classified as a combatant. Below this threshold, the system cannot automatically attack. Nevertheless, the use of such parameterizations is limited since it might imply attacking two civilians in 100 attacks is acceptable, for example (see the discussion of non-translatable legal statements).
   b. Further elements are appropriate safety and security as well as anti-tampering systems that automatically block all automated engagement of effectors in the event of any tampering with the system control and only allow them to be unlocked using special keys, for example.

    c.   Those legal principles that are translatable in this sense with additional parameters that enable a certain "fine-tuning," i.e. an individual or subsequent application-related adjustment and the consideration of special reservations, could make a legally compliant autonomous system possible in the respect previously sketched.

    d.   In order to achieve operational readiness, test simulations, comprehensive sandboxing with digital twins, real-life tests and objective, third-party audits (possibly by certification authorities) would be necessary in addition to the fulfilment of information and IT security standards yet to be defined. In addition, appropriate operator training and familiarization with the system and its capabilities (trust by understanding the system) is inevitable.

4. Overall, however, it should be pointed out that for the development of a comprehensively legally compliant system, the combination of several individual solutions (legal rates+parameters/thresholds) and the systemic combinability must be given and, thus, building a certain "box" around artificially intelligent automation for weapon engagement.

5. A hurdle that cannot be crossed from today's point of view will remain in the area of decisions on proper values, as a technological solution for support is currently not apparent.

TRANSPARENT CRITERIA DEVELOPMENT

In consequence, systems engineering for designing responsible assistance by cognitive and volitive machines, which technically support ethically and legally compliant behavior, has to fulfill four major requirements:

1. Situational awareness to enable responsible action
2. Identification of responsible options to act
3. Comprehensive plausibility of propositions
4. Resilience against failure or hostile intervention

These are basic for ensuring responsible decisions before, during, and after the mission in order to successfully achieve clearly defined ends and intermediate purposes in a given operating theatre. To what extent collateral effects can be tolerated, is part of this decision-making.

Realization in the Life Cycle

The following, requirements should be met in the research, development, procurement, deployment, and use phases of assistance systems for responsible action.

1. Transparent criteria for development must accompany military capability development from the very outset. Philosophers, lawyers, and the military pastoral care bring in basic insights. Legal standards that apply to defense research, development, and procurement are indispensable. Finally, yet importantly, the experience of commanders and soldiers must be taken into account. Analogous to industrial quality assurance and certification processes, these considerations support responsible action not only in battle, but also on all levels of responsibility well before.

2. Evolutionary innovation, on the one hand, replaces outdated technology while letting procedures and processes largely unchanged, whereas disruptive innovation, on the other hand, opens up fundamentally new applications, which require both conceptual and organizational changes. Ultimately, the innovative potential of defense digitization is only realizable if it takes into account the mindset and *esprit de corps* of the armed forces and, last but not least, the maxims of military licensing, certifying, and qualification bodies.

3. Mission-relevant decisions can be evaluated and correspond to the mission-specific RoEs that define the framework for action in a legally binding manner. RoEs, thus, have to have a direct impact on the technical systems design, but can be so complex that computer-aided "synthetic legal advisors" are indispensable for identifying RoE-compliant options for action in battle. This is particularly true in the spatially delimited and accelerated operations "at machine speed," which FCAS is designed for, where ethically relevant knowledge itself must be made electronically accessible. Synthetic legal advisors may operate at different levels of automation.

4. In a first step, RoE assistants would be helpful that at least mechanize the simple part of the rules, accompanied with the capability to query underlying information in order to validate the underlying rationales. In this phase, the complex part can still remain with the human controller. Over time, more and more aspects might be taken over by the system, alongside growing operator trust by understanding the capabilities of the novel AI-enabled supporting functions and, more generally, trust in responsible systems design.

Remarks on Soldierly Virtues

The notion of virtues as part of the soldierly ethos has been reflected in military philosophy. Since their ethos must be determined by responsibility, "the more momentous the decisions and actions of individual soldiers are," as von Baudissin had observed in the context of "scientization and mechanization of the military craft," it seems worthwhile to reflect the relevance of virtues even in the Age of AI. Carl von Clausewitz (1780–1831), for example, the Prussian general and military theorist who stressed the moral, psychological, and political aspects of war, speaks of "the courage of responsibility, be it before the judgment seat of some external power or the inner one, namely conscience" (von Clausewitz, 1832). It is a "disposition of the mind," which he equates with "courage against personal danger." The Clausewitzian philosophy is rooted in the notion of "virtues," habits of "good" behavior, which are acquired by some sort of "supervised" moral "training" over time and appear under different names in most cultures.

The so-called four "cardinal virtues," prudence, justice, bravery, and temperance, fundamental to Western ethics, are examples with a potential for wider consent.

The willingness to "accept wounds in the struggle for the realisation of the good" (Pieper, 1996) characterizes bravery as a particularly soldierly virtue, which is closely related to the Clausewitzian "courage of responsibility" previously mentioned. The virtue of justice, on the other hand, is to be seen as the perfection of prudence, which perceives reality, such as a military situation, as it actually is. Bravery can only indirectly complement justice, since it is not directly aiming at the "good," but rather at the obstacles that arise in the realization of the "good." "Only the prudent can be brave. Bravery without prudence is not bravery." The proper meaning on "temperance," which is also an essential element of the soldierly ethos, "makes a unified whole out of disparate parts," remarks the philosopher Josef Pieper (1904–1997). "This is the first and proper sense of the Latin verb *temperare*; and only based on this broader meaning can *temperare* – negatively – mean 'to restrain'. [...] 'Temperance' means: to realize order in oneself." (Pieper, 1996).

Beyond mere "functioning," but in the sense of acquiring soldierly virtues, i.e. ethical attitudes that prepare the mind and will toward the good of reason, that are adapted to the requirements of the digital age in combat, E-AID may serve as a simulator for training the responsible execution of the targeting cycles of FCAS.

Hippocratic Oath – An Analogy?

Only if based on an Image of Man that is compatible with the responsible use of technology along the lines previously discussed, can digital assistance systems support morally acceptable decisions.

> It is the responsibility of our generation, possibly the last to look back to a pre-digital age and into a world driven by artificial intelligence, to answer the question of whether we continue to recognize the integrity of the human person as a normative basis,

thoughtfully observes the German political theologian Ellen Ueberschär (2019).

It sheds some light from a perhaps unexpected perspective on the problem of digital ethics, ethos, and morality, that the conceptual architect of Germany's post-WW II armed forces sees this task assigned to the military pastoral care. It pronounces the necessity of such an Image of Man, especially in the military service, and to provide educational offer toward a realization of this conception. It would be worth considering in this context, whether the swearing-in ceremony, which was considered indispensable when the *Bundeswehr* was founded, shouldn't be reviewed with a fresh eye in the spirit of the Hippocratic Oath, generally regarded as a symbol of another professional ethos that is committed to responsibility for life and death. For von Baudissin it is

> one of the essential tasks of the military clergy to point out the sanctity of the oath, as well as of the vow, to show the recruit the seriousness of the assumption of his official duties on his own conscience, but at the same time also the limits, set by God for everyone, and therefore for this obligation as well.

(von Baudissin, 1969)

AN ATTEMPT OF A SUMMARY

Only alert NI is able to assess plausibility, develop understanding, and ensure control. "The uncontrolled pleasure in functioning, which today is almost synonymous with resignation to technical automatism, is no less alarming than the dashing, pre-technical feudal traditions because it suggests the unscrupulous, maximum use of power and force," von Baudissin observed in the 1950s (von Baudissin, 1969). These words ring true not only for shaping the soldierly ethos in the digital age. There is a more general need for a new enlightenment in dealing with AI maturely, ethically, and intelligently, i.e., "man's release from his self-imposed immaturity. *Sapere aude*—Have the courage to use your own intellect!" (Kant, 1784). Anthropocentrism in this sense underlines the ethical and legal dimensions of artificially intelligent automation, which characterize the use of AI in defense systems.

Since we feel encouraged to assume that a broader consent among the information fusion community might be achieved, we are closing with some recommendations that address certain blind spots, at least according to the observations of the authors.

1. Digital ethics and a corresponding ethos and morality should be built up systematically for responsibly using artificially intelligent automation in the military domains. In particular, such skills enable commanders "to assess the potential and impact of digital technologies and to manage and to lead in a digitized environment," as an official German document states (German MoD, 2019a). In particular, leadership philosophies and personality development instruments should encourage such competences.

2. In addition to the operational benefit of artificially intelligent automation in closing capability gaps, expanding the range of capabilities, and developing corresponding concepts, operational procedures, and other organizational measures, ethical and legal compliance needs to be achieved. Only then, cognitive and volitive assistance will become acceptable before the conscience of the individual commanders, but also in the broader view of the Common Good of the society as such. Success in both aspects will indicate a real innovation

3. Defense projects should be accompanied from their very beginning by comprehensive analyses of technical controllability and personal accountability in a visible, transparent, and verifiable manner. Otherwise, the paradigm shifts and large material efforts associated with artificially intelligent automation would hardly be politically, societally, and financially enforceable. Of course, there will be more and less problematic projects, implying that an exemplary approach according to these lines would be appropriate.

"Firmly confident in his better inner knowledge, the military leader must stand like the rock where the wave breaks," observed Carl von Clausewitz (I.6, p. 96). Artificially intelligent automation therefore requires the ethos of digitally educated commanders and staffs. They do not need to know how to design and program AI-based defense systems, but to assess their strengths and weaknesses, risks, and opportunities. The associated digital morality and competence are teachable. It addresses a key question of the soldierly ethos, which is aggravated by artificially intelligent automation but not fundamentally new.

NOTES

1 "I ask you: Do you want total war? If necessary, do you want a war more total and radical than anything that we can even imagine today?" (Sportpalast speech of Nazi propaganda minister Joseph Goebbels (1897–1945) on February 18, 1943).
2 Lem anticipated that the metaphor "instinct control" seems to be appropriate for what we call today "autonomous driving," for example. "The wasp probably possesses a sufficient number of nerve cells that it could just as well steer a truck [...] or control a transcontinental missile."
3 Aspects discussed in this context are: (i) Context Control: controlling the space, duration, time and conditions, (ii) Understanding the System: functioning, capabilities and limitations in given operational circumstances, (iii) Understanding the Environment: situational awareness and understanding of the environment, proper training, (iv) Predictability and Reliability: knowledge of the consequences of use and reliability as the likelihood of failure, both in realistic operational environments against adaptive adversaries, (v) Human Supervision and Ability to Intervene, (vi) Accountability: certain standard of authority and accountability framework of human operators, teammates and commanders, and (vii) Ethics and Human Dignity: preserve human agency and uphold moral responsibility in decisions to use force.

REFERENCES

Allen, G. (2020). *Understanding AI technology.* US Department of Defense, Joint Artificial Intelligence Center. https://apps.dtic.mil/sti/pdfs/AD1099286.pdf

CCW (2019). Annex III. *Meeting of the high contracting parties to the convention on prohibitions or restrictions on the use of certain conventional weapons which may be deemed to be excessively injurious or to have indiscriminate effects.* Geneva, Switzerland, 13 December 2019. https://documents-dds-ny.un.org/doc/UNDOC/GEN/G19/343/64/PDF/G1934364.pdf?OpenElement

Chernoff, H. (1973). The use of faces to represent points in K-dimensional space graphically. *Journal of the American Statistical Association, 68*(342), 361–368. https://doi.org/10.1080/01621459.1973.10482434

Federal Republic of German. (1949, May 23). *Basic Law for the Federal Republic of Germany.* https://www.gesetze-im-internet.de/englisch_gg/.

Flemisch, F. O., Bengler, K., Bubb, H., Winner, H., & Bruder, R. (2014). Towards cooperative guidance and control of highly automated vehicles: H-mode and conduct-by-wire. *Ergonomics, 57*(3), 343–360.

German MoD. (2018). *Konzeption der Bundeswehr [Concept of the Bundeswehr]*. Berlin: Ministry of Defense (MoD). https://www.bmvg.de/resource/blob/26544/9ceddf6df2f48ca 87aa0e3ce2826348d/20180731-konzep-tion-der-bundeswehr-data.pdf

German MoD. (2019a). *Umsetzungsstrategie, Digitale Bundeswehr' [Implementation Strategy 'Digital Bundeswehr']*. Berlin: Ministry of Defense (MoD). https://www.bmvg.de/de/ themen/ruestung/digitalisierung/umsetzungsstrategie-digitale-bundeswehr.

German MoD. (2019b). Erster Bericht zur Digitalen Transformation *[First Report on Digital Transformation]*. Berlin: Ministry of Defense (MoD). https://www.bmvg.de/resource/blo b/143248/7add8013a0617d0c6a8f4ff969dc0184/20191029-down-load-erster-digitalbericht-data.pdf

Govaers, F., Stooß, V., & Ulmke, M. (2021). Adiabatic quantum computing for solving the multi-target data association problem. In *2021 IEEE international conference on multisensor fusion and integration for intelligent systems (MFI)* (pp. 1–7). https://doi.org/10.1109/ MFI52462.2021.9591187

Gutschker, T. (2021, November 11). Europa ist in Gefahr [Europe is in Danger]. Frankfurter Allgemeine Zeitung. https://www.faz.net/redaktion/thomas-gutschker-11132364.html

Kant, I. (1784). *An Answer to the Question: What is Enlightenment?* https://donelan.faculty. writing.ucsb.edu/enlight.html.

Keisinger, F., & Koch, W. (2023). The responsible use of new technologies in a future combat air system. https://www.fcas-forum.eu/

Koch, W. (2020). On ethically aligned information fusion for defense and security systems. In *Proc of 2020 IEEE 23rd International Conference on Information Fusion (FUSION)* (pp. 1–8). https://doi.org/10.23919/FUSION45008.2020.9190233

Koch, W. (2021a). On digital ethics for artificial intelligence and information fusion in the defense domain. *IEEE Aerospace and Electronic Systems Magazine, 36*(7), 94–111. https:// doi.org/10.1109/MAES.2021.3066841

Koch, W. (2021b). *AI-based defense systems - how to design them responsibly?* German-Israeli Tech-Policy Dialog, Heinrich Böll Stiftung Tel Aviv. https://il.boell.org/en/2021/12/24/ ai-based-defense-systems-how-design-them-responsibly

Koch, W. (2022). *What does Artificial Intelligence offer to the Air C2 domain?* NATO Open Perspectives Exchange Network (OPEN) Publication, NATO Allied Command Transformation. https://img1.wsimg.com/blobby/go/f1074002-f794-4e57-99b2-7a26df3e3daa/downloads/What%20Artificial%20Intelligence%20offers%20to%20the%20Air. pdf?ver=1686069844121

Lem, S. (1983). *Waffensysteme des 21. Jahrhunderts [Weapon Systems of the 21st Century or the Upside Down Evolution]*. Frankfurt am Main: Suhrkamp.

Pieper, J. (1996). Schriften zur Philosophischen Anthropo-logie und Ethik: Das Menschenbild der Tugendlehre [Writings on Philosophical Anthropology and Ethics: The Image of Man in the Doctrine of Virtue]. In *Werkausgabe Letzter Hand [Last Hand Ed.], vol. IV*, Hamburg, Germany: Felix Meiner.

Singer, T. (2019). *Dehumanisierung der Kriegführung. Herausforderungen für das Völkerrecht und die Frage nach der Notwendigkeit menschlicher Kontrolle [Dehumanization of Warfare]*. Berlin, Heidelberg: Springer.

Spiekermann, S. (2021). From value-lists to value-based engineering with IEEE 7000(tm). In *Proc of. 2021 IEEE International Symposium on Technology and Society (ISTAS)* (pp. 1–6). https://doi.org/10.1109/ISTAS52410.2021.9629134.

Stooß, V., Ulmke, M., & Govaers, F. (2021). Adiabatic quantum computing for solving the weapon target assignment problem. In *2021 IEEE 24th international conference on information fusion (FUSION)* (pp. 1–6). https://doi.org/10.23919/FUSION49465.2021.9626902

Ueberschär, E. (2019). Von Friedensethik, politischen Dilemmata und menschlicher Würde - eine Skizze aus der Perspektive theologischer Ethik [Of Peace Ethics, Political Dilemmas and Human Dignity -a Sketch from the Perspective of Theological Ethics]. *Opening Speech at the first meeting of the working group the responsible use of new technologies in a future combat air system. Bad aibling, September 27, 2019.* https://www.fcas-forum.eu/publications/Skizze-zur-theologischen-Ethik-Ueberschaer.pdf

von Baudissin, W. (1969). *Soldat für den Frieden. Entwürfe für eine zeitgemäße Bundeswehr [Soldier for Peace. Drafts for a Contemporary Bundeswehr].* München: Pieper.

von Clausewitz, C. (1832). *Vom Kriege [on war]* (11th ed). Hamburg, Germany: Nikol.

A Qualitative Risk Evaluation Model for AI-Enabled Military Systems

4

Ravi Panwar

It is widely accepted that the risks posed by using Artificial Intelligence (AI) technologies for developing various applications and systems are significant, and therefore must be suitably addressed and mitigated (European Commission, 2019; Future of Life Institute, 2017). In recent years, there has been significant progress in various international bodies towards developing global standards for AI. These include technical standards as well as documents which capture ethical and policy dimensions of responsible AI (Kerry et al., 2021). Notably, in 2018 the G-7 agreed to establish the Global Partnership on AI, a multistakeholder initiative working on projects to explore regulatory issues and opportunities for AI development. There has also been a proliferation of declarations and frameworks from public and private organizations aimed at guiding the development of responsible AI (European Commission, 2021; Government of Canada, 2019; Government of Singapore, 2020). Many of these have evolved from focussing on general principles to full-fledged policy frameworks.

In the military context, in 2019 the United Nations affirmed a set of guiding principles about the use of emerging technologies in the area of Lethal Autonomous Weapon Systems (LAWS) (UNODA, 2019). The United States Department of Defense

DOI: 10.1201/9781003410379-5

(DoD) adopted a set of Ethical Principles for AI in February 2020 (US DoD, 2020). Amongst the major military powers, China (China MOST, 2021), the European Union (European Commission, 2021) and Russia (TASS, 2021) have come up with principles/ norms with respect to the development of AI technologies, although these do not specifically address military systems. Notably, the EU has adopted a risk-based approach for the regulation of AI applications that specifically excludes military applications.

While principles are a key starting point for establishing policy, their high level of abstraction dictates that they be followed up with a more granular mechanism which can guide implementation processes. Adopting a risk-based approach for the design, development and deployment of military systems promises to be an effective way to move forward from risk-mitigation principles to policy and practice. This is because risks posed by different types of military systems may vary widely, and applying a common set of risk-mitigation strategies across all systems will likely be suboptimal, being too lenient for very high-risk systems and overly stringent for low-risk ones.

This chapter first identifies the unique characteristics of AI technologies which make AI-powered systems risk-prone and discusses several considerations which have a bearing on evaluating risks associated with such systems. It then highlights how the approach adopted for risk evaluation and mitigation for AI-enabled military systems would vary considerably across different scenarios, the chief amongst these being conventional all-out conflicts governed by International Humanitarian Law (IHL), and grey zone operations conducted by militaries and non-state actors. The main contribution of this chapter is to suggest a *Risk Hierarchy*, a qualitative model which attempts to sketch the contours of how a risk-based approach could be adopted for mitigating risks posed by AI-enabled military systems during armed conflicts. The chapter also contends that the granular approach adopted in the Risk Hierarchy model would facilitate international consensus by providing a basis for more focussed discussions. It also suggests the idea that mitigating risks in AI-enabled military systems is not always a zero-sum game, and there are compelling reasons for states and militaries to adopt self-regulatory measures.

EVALUATING AI RISKS IN MILITARY SYSTEMS: GENERAL CONSIDERATIONS

This section discusses a few important considerations which need to be kept in mind before foraying into the complex exercise of evaluating AI-related risks in military systems. To begin with, it attempts to define the spectrum of technologies which might be covered under the ambit of AI, a term which is arguably very nebulous in its usage. It then identifies the unique characteristics of AI technologies which give rise to special concerns. Next, it discusses notions of autonomy and human control in military systems, which lie at the heart of these concerns and, in addition to reliability issues, give rise to moral and ethical conundrums as well. Finally, the section highlights the significance of military ethos towards ensuring responsible development and fielding of weapon systems.

Defining AI

The general tendency is to use the term AI as though it has a universally accepted definition. This is far from being true. The Encyclopaedia Britannica describes AI in its most generic form, as the ability of a digital computer or computer-controlled robot to perform tasks commonly associated with intelligent beings (Copeland, 2023). The proposed EU AI Act adopts a much more specific characterization, defining an AI system to mean software that is developed for generating outputs, predictions, recommendations or decisions, using one or more of the following techniques and approaches: machine learning techniques such as supervised, unsupervised and reinforcement learning; knowledge-based approaches such as logic programming and expert systems; and statistical approaches such as Bayesian estimation and optimization methods (European Commission, 2021b).

Notwithstanding the wide-ranging scope of AI indicated above, it may not be far off the mark to state that most of the risks associated today with AI-enabled systems stem essentially from neural network-based machine-learning (ML) techniques. In the balance of this work, unless otherwise specified, the term AI implies the use of AI/ML technologies. The unique characteristics of these technologies are discussed next.

Unique Characteristics of AI

The distinctive characteristics of machine learning-based AI systems, which are at the root of their power as well as risks, arise fundamentally from their ability to *learn directly from data*, and this learning might continue even while the systems are in operation after being deployed, which is often termed as *online learning* (Hoi et al., 2018). This feature of learning directly from data also gives them a *black-box* character, wherein the process by which inputs are translated into outputs is not adequately known even to the developers. This is also referred to as *non-transparency* or *non-explainability* of AI systems. Finally, neural networks have proven to be very powerful, leading to an exponential increase over time in the *intelligence* which they confer onto AI-enabled systems.

The data-centricity of AI-enabled systems introduces risks arising from unrepresentative, biased or incorrect/deliberately poisoned data, resulting in *unintended system behaviour*. The fact that a system might continue to learn and thus, post deployment, metamorphose into something different from what was fielded, together with its opaque nature, introduces a degree of *unpredictability* to its functioning. The data-driven learning and non-transparent nature of AI systems together are perhaps mainly responsible for systems becoming vulnerable to catastrophic failure when confronted with edge cases, a characteristic which is referred to as *brittleness* (Lohn, 2020). The increasingly higher intelligence and consequent greater autonomy conferred onto AI systems result in undesirable effects such as *automation bias* and *lack of accountability* (ICRC, 2020). Here *automation bias* refers to the tendency to rely too heavily on automated systems without critically evaluating their outputs or recommendations.

Autonomy: A Risk Factor Independent of AI/ML Technologies

The level of autonomy in military systems is perhaps the most important parameter for risk evaluation. Of particular interest are autonomous weapon systems (AWS). While there is no internationally accepted definition of AWS (UNIDIR 2017), these are often described as weapons which can select and attack/engage targets without human intervention (European Parliament, 2023; ICRC, 2021; US DoD, 2023a). The select-and-engage functions are dubbed as *critical functions* within the targeting chain (Jansen, 2020). With such a characterization, most states declare that fully autonomous weapons must never be developed.

It is often presumed that autonomy inevitably implies an underlying AI-enabled substrate, which is not always the case. The Israeli Harpy and its successor, the Harop (Israel Aerospace Industries, n.d.), notable examples of offensive fully auton-omous lethal weapon systems in operation today, have been in use by militaries for decades. Whether or not these systems resort to the use of AI/ML, it would be safe to state that their publicized features could be realized without resorting to these technologies.

Quite independent of risks rooted in AI/ML, endowing machines with autonomy especially in their critical functions leads to the contentious issue of human control. At the UN Group of Government Experts (GGE) on LAWS and other fora (Human Rights Watch, 2023), there has been considerable debate on the appropriateness of terms such as *Meaningful Human Control (MHC)* and *Appropriate Levels of Human Judgement* towards describing the desired level of human control in autonomous weapons. Where autonomy ends, human control begins, and vice versa. This symbiotic relationship between human control and autonomy in weapon systems, with particular reference to full autonomy, is discussed next.

Fully-Autonomous Weapon Systems: A Nebulous Concept

An extreme portrayal of an AWS is dramatized by the self-aware Skynet letting loose an army of Terminators onto the human race (Zador & LeCun, 2019). The previous section described an AWS as one which can select and engage targets without human inter-vention. In such a characterization, the term 'select' may well be interpreted to mean the determination of adversary assets, human or otherwise, which are to be targeted.

In other words, in this interpretation the weapon itself prepares a target list for subsequent destruction, purportedly endowed with a Terminator-like capability.

An alternative interpretation of the phrase 'target selection' is relatively benign, as follows: Given a target list or description (which is provided by a human), the weapon 'identifies' the target (or a group of targets) using sensors, then tracks and destroys it. Here, the implied meaning of the term 'selection' is synonymous with target identification. The US DOD Directive 3000.09, for instance, defines 'target selection' as "The *identification* of an individual target or a specific group of targets for engagement" (emphasis added) (US DOD, 2023a). Similarly, the Netherlands defines an AWS as one which "selects and engages targets matching certain predefined criteria", where the criteria are provided by a human (Government of the Netherlands, 2017).

In the second interpretation, the target description provided by a human may range from being very specific to increasing levels of generality. Keeping in mind the current state of technology and other practical considerations, the following types of target descriptions lend themselves to being programmed into machines:

- **Explicit Target Description**. One or more specific targets (static or mobile) are selected by a human, their description is fed into the weapon system, which is then activated to neutralize the targets. For static targets, the description could be in terms of a precise location reference, while for mobile targets it could be any unique identity (e.g., unique electronic signature of a mobile radar, unique visual profile of a ship, etc). In addition, time and area constraints may be included in the description
- **Parameterized Target Description**. In this case, instead of a particular target, target parameters may be specified (e.g., hangars on a specific airfield, enemy tanks in a given area), together with time constraints. Such a weapon system, in addition to target identification, might at times need to prioritize amongst identified targets for efficient neutralization

Theoretically, target descriptions may be made even more generalized. For instance, while a description such as 'any enemy tank in the battlespace and/or adversary territory' may well be within reach from a technological standpoint, it would amount to giving a degree of leeway to machines which should be unacceptable to responsible states. A 'responsible' target description may be characterized as one which ultimately results in identification (and destruction) by the AWS of the very same target(s) which were intended to be destroyed by the human who frames the target description. Moreover, implicit in the human involvement during such target description, which the Netherlands refers to as the "wider loop" in the decision-making process, is the responsibility (and accountability) for ensuring adherence to the IHL principles of Distinction, Proportionality and Military Necessity (Winter, 2022).

As an extreme case, one could envisage a target description to be 'all assets which contribute towards the adversary's combat potential'. Such a description is equivalent to stating that the weapon system prepares its own target list. For implementing such a capability, weapons would clearly need to possess artificial general intelligence which, so far, falls in the realm of fantasy from a technology standpoint.

To summarize, the highest level of autonomy which a responsible state would envisage in weapon systems is one in which the target description (profile/signature), explicit or parameterized, together with time and space constraints, is provided by a human, while the weapon system essentially executes the *identify*-and-engage functions. The target description should be explicit enough to ensure that no unintended targets can ever be identified by the AWS (unless there is a system malfunction). Notably, the use of the phrase *select*-and-engage to describe AWS leads to ambiguity, and may at times be misleading, by implying that the target list is also decided by the machine.

In this chapter, AWS with autonomy in the critical select (i.e., identify and prioritize) and engage functions are termed as fully autonomous weapons. Further, supervised autonomy, i.e., a human-on-the-loop type of control (Panwar, 2022) in critical functions is considered to be equivalent to full autonomy, because of the difficulties of exerting this type of control in a fast-paced and unpredictable battlespace. Weapon systems with any form of autonomy short of these criteria, including autonomy in non-critical functions such as take-off and landing, navigation, etc, are termed as semi-autonomous.

The Moral Argument

One argument against the employment of LAWS (often sensationalized as 'killer robots') is the contention that machines should never be endowed with the power of life and death over humans. From a legal perspective, it is claimed that this violates the Martens Clause, set out in 1977 Additional Protocol II to the Geneva Conventions, which states that, "in cases not covered by the law in force, the human person remains under the protection of the principles of humanity and the dictates of the public conscience" (Ticehurst, 1997).

The previous section has contended that it is always a human who would select a target (or a group of targets) for neutralization, be it specific or parameterized selection (i.e., if one discounts the extreme Skynet scenario). If this reasoning is accepted, then invoking the Martens Clause would not stand scrutiny. Moreover, visualization of AI-powered weapon systems as 'killer robots' amounts to anthropomorphizing them, a tendency to be shunned as per the Guiding Principles affirmed by UN GGE on LAWS in 2018 (Moyes, 2019). This propensity to anthropomorphize and bestow agency on AWS is also related to the much-debated accountability argument. According to one perspective, since a machine cannot have agency, accountability would always rest with humans who develop and employ the weapon system, no matter how much autonomy is built into it (Oimann, 2023).

There is another nuanced distinction which may be made within the category of fully-autonomous weapon systems. LAWS which are designed to target static or mobile weapon platforms or even establishments (such as the Harpy anti-radar loitering munition) might possibly be perceived by 'ban killer robot' proponents as being different from those weapons which specifically seek out humans (such as the Slaughterbots drones (Scharre, 2018)). The former category would perhaps be more acceptable and non-violative of the Martens Clause as compared to the human-seeking variety.

Notwithstanding a range of opinions on the moral argument for banning LAWS, this chapter associates a higher risk with human-seeking LAWS (please see the following section on Risk Hierarchy – Working Definitions, and the follow-up discussion on the taxonomy of AI-powered weapons).

Responsible AI and Military Ethos

Implied in the principles of Distinction, Proportionality and Military Necessity is the presumption that, in wars which fulfil the *jus ad bellum* criteria, it is quite acceptable for adversary combatants to kill one another, but killing civilians either deliberately or through negligence is a war crime. While this stance is perhaps justified, it does not in any way translate to the conclusion that combatant lives are any less precious than the lives of civilians. The fact that soldiers, mostly voluntarily, risk their lives in defence of their country, should encourage the adoption of measures aimed at protecting combatant lives. It could be argued that increased autonomy in weapon systems is one such measure and thus should be classified as responsible leveraging of military AI technologies.

In discussions on LAWS, one can often discern a tendency to characterize militaries as instruments of death and destruction, with soldiers bent on killing adversaries, combatants and civilians alike. This view does not take into account the value systems which are prevalent in most militaries, which arguably play a dominant role in protecting civilians and soldiers *hors de combat* from coming to harm during armed conflicts, much more than the fear of violating IHL.

The assumption of *responsible intent* in the employment of AI-powered weapons by militaries is an important one while evaluating risk. Under such an assumption, mitigation measures can be focussed on identifying and addressing risk factors which emerge from AI-related system malfunctioning or unpredictability. In contrast, a presumption of deliberate misuse or negligence in the employment of weapons by militaries is likely to inhibit the use of AI technologies in warfare, thereby losing the advantage of developing smart weapons rather than dumb ones.

Military ethos also has an important role to play while formulating rules of engagement (RoE) for different scenarios and weapon systems. For instance, a responsible military force would not resort to the use of heavy artillery power for flushing out terrorists located in an occupied civilian apartment building during counter-insurgency operations. In a similar vein, appropriate RoE would preclude the employment of a fully autonomous weapon which is designed to destroy tanks in desert terrain devoid of civilians, for identifying and neutralizing terrorists embedded amongst civilians.

RISK PERSPECTIVES

There are widely varying perspectives on risks associated with AI-enabled systems. Notably, risks associated with civilian applications are quite different from the types of concerns triggered by AI-enabled military systems, and in particular AI-powered weapons. These perspectives and the scenarios from which they emerge are discussed in the subsections which follow.

Civilian Applications

AI regulation in the context of non-military systems is driven by concerns related to fundamental rights issues, such as racial and gender bias, data privacy, biometric surveillance, etc. The proposed EU AI Act, which has adopted a risk-based approach, analyses these risks and categorizes them into a four-level hierarchy. A differentiated risk-mitigation mechanism has also been proposed, suitably tailored to each risk level (European Commission, 2021c).

Armed Conflicts

In contrast, for military systems, i.e., systems used by state and non-state militaries in armed conflicts (international and non-international), AI-related risks are viewed through legal and ethical prisms as dictated by the *jus in bello* criteria, or in other words, IHL.

Specific apprehensions in the context of AI-powered military systems are reflected in the deliberations ongoing for many years under the aegis of the UN GGE on LAWS and other international fora. One of the primary concerns is that fully autonomous weapons would be in violation of the IHL principles of Distinction, Proportionality and Military Necessity, as well as the Martens Clause. Notably, IHL is framed for warfare scenarios where adversary militaries are engaged in armed conflict in the presence of civilians, and the primary objective of IHL is to protect innocent civilians from coming to harm. The endeavour at these fora is to arrive at a consensus on how to enforce MHC in LAWS (in particular AI-enabled systems) through legally binding international regulation (ICRC, 2014).

In the context of armed conflicts, reliable performance on the battlefield is another important consideration, which is quite independent of IHL and often overlooked. No military commander would like to field weapons which do not function as per their specifications, or over which they lack full control. This is because such systems would reduce military effectiveness, and also detract from achieving specified military objectives by resulting in arbitrary undesired effects.

To summarize, in armed conflict scenarios the objectives of risk mitigation for AI-enabled systems are to ensure, firstly, *adherence to IHL* and, secondly, *reliable performance on the battlefield.*

Terrorist Activities

AI-powered weapon systems may be used by terrorists and other rogue organizations to target civilian populace, motivated by racial, communal or other violent ideologies. Risks stemming from AI in such scenarios have been dramatized by the two widely circulated Slaughterbots videos (Future of Life Institute, 2021). The central idea conveyed by these videos is that, given the very high 'intelligence' potential of AI technologies together with their easy accessibility, fully autonomous miniature weapons could be produced or procured in large numbers without much difficulty by non-state actors. This in turn would pose a serious threat to whole societies, and perhaps an existential threat to humanity itself, in the form of AI-enabled weapons of mass destruction.

It is interesting to note that the mass destruction argument is not premised on AI-powered weapons malfunctioning, but rather counts on highly intelligent AI agents performing their tasks very effectively and efficiently.

In such scenarios, the internal security apparatus of a state, tasked with protecting its citizens from rogue actors, would be the primary agency responsible for risk mitigation. Militaries might also be involved in some cases, if called out to aid civilian authorities in counter-terrorism operations.

Risk-mitigation measures in these scenarios would include, firstly, non-proliferation mechanisms to prevent military technology and systems from falling into the wrong hands (i.e., if such technology is developed by militaries for employment in armed conflicts); and secondly, tracking and eliminating terrorist organizations and activities (including the development of AI-enabled weapons by them). Risk evaluation for different types of weapon systems here would depend largely on factors such as ease of proliferation of technology and systems, and their utility in the hands of terrorists. These parameters are entirely different from those applicable to armed conflicts.

Grey Zone Operations

Militaries might also deploy AI-enabled weapon systems for overt grey zone warfare, i.e., warfare conducted in the operational space between peace and all-out armed conflict. Since territorial integrity is sacrosanct within international boundaries in the land and air domains as well as the territorial waters of a nation-state, such grey zone operations are restricted to the global commons, namely, international waters and space.

The primary AI-related risk in such situations emerges from the increased possibility of inadvertent escalation from grey zone status to one of armed conflict, consequent to insufficient human oversight and/or malfunction in fully-autonomous weapon systems. Since civilians are unlikely to be present in these settings, adherence to IHL would not be a consideration for evaluating risk. As regards risk mitigation, measures during the development and testing phases of systems would be the same as applicable for ensuring reliable performance during armed conflicts. However, RoE evolved for the deployment phase during grey zone operations would be quite different from those pertaining to armed conflicts.

Focus of the Current Chapter

The above discussion brings out that the driving concerns and consequently the nature of AI-related risks in the three scenarios of armed conflict, terrorist activities and grey zone operations are quite at variance with one another. Still, there is bound to be some correlation amongst them. For instance, an unpredictable weapon system would pose risks during armed conflicts as well as grey zone operations; as another example, an armed swarm, which may be developed quite justifiably for use during armed conflicts, may pose a very high risk from the standpoint of technology proliferation to terrorist organizations. These two examples also illustrate that the correlation could be positive as well as negative. The key point to note, however, is that risk evaluation and mitigation strategies for the three scenarios would be quite different.

The risk-based approach presented in this chapter focuses on concerns as applicable to armed conflict and does not address risks associated with activities of rogue actors and grey zone operations.

THE RISK HIERARCHY

The qualitative model for risk evaluation and mitigation described here is termed as the *Risk Hierarchy for AI-Enabled Military Systems.*

Evolving the Risk Hierarchy: Four-Step Process

Developing such a Risk Hierarchy involves four distinct activities. Firstly, risk levels need to be defined based on a suitable rationale. Secondly, given the large number of military systems which are in existence, these need to be grouped into classes. Next, these classes must be mapped to risk levels. Finally, a differentiated risk-mitigation mechanism needs to be devised and linked to each risk level.

Using the above approach, a five-level Risk Hierarchy has been developed, as shown in Figure 4.1.

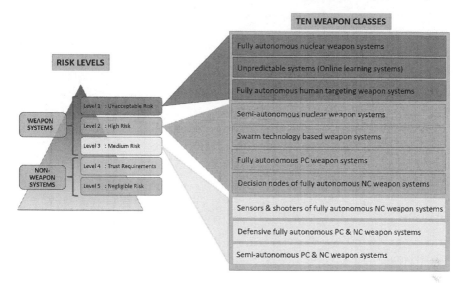

FIGURE 4.1   Risk hierarchy with risk levels and weapon classes.

Working Definitions

To avoid ambiguity, working definitions and brief explanations for some of the terms used in the description of the Risk Hierarchy are provided below:

- **Fully Autonomous vis-à-vis Semi-Autonomous Weapon Systems**. In this chapter, fully-autonomous weapon systems imply human-out-of-the-loop as well as human-on-the-loop systems, while semi-autonomous weapon systems correspond to human-in-the-loop systems. These terms have relevance principally in relation to the critical select-and-engage functions. Various nuances of autonomy in weapon systems have been discussed at length above
- **Unpredictable Weapon Systems**. Unpredictability is a characteristic of most AI-systems which utilize deep-learning techniques. However, the degree of unpredictability can be controlled by adhering to stringent test and evaluation (T&E) standards (Wojton et al., 2020). In this chapter, unpredictable systems refer to only those weapon systems where learning is permitted to continue in critical functions while the system is in operation post T&E and deployment, thus introducing an unknown degree of unpredictability into its functioning. It is important to note that those weapons systems which, after being deployed, are updated with new software versions periodically after due T&E are not covered by this class of weapons, even though these may also continue to learn post deployment
- **Human Targeting Weapon Systems**. Human targeting weapon systems are those which seek out humans (combatants as a class; specific terrorists)

for lethal engagement but do not include weapons designed to engage non-human military targets such as tanks, even though these might be manned by humans

- **Defensive Weapon Systems**. Most weapon systems can be used for offensive as well as defensive operations. However, there are certain systems which can be employed only in a defensive role, e.g., close-in weapon systems (CIWS) such as the US Phalanx, static robot sentries such as Korea's SGR-A1, etc. In this chapter, defensive weapon systems refer to only this class of weapons

Warfighting Domains

As of now, the Risk Hierarchy addresses weapon systems which operate in the physical domains of land, sea, air and space. Evaluation of risk in the case of cyber, electromagnetic and cognitive domain weapons is beyond the scope of this chapter.

The following sections discuss each of the four steps involved in the model evolution process.

RATIONALE FOR A FIVE-LEVEL RISK ARCHITECTURE

In the coming years, AI is expected to become ubiquitous across a very wide variety of military systems. It is perhaps useful to bifurcate this spectrum into two broad classes, namely, *weapon systems* (comprising sensors, decision nodes and shooters) and *decision support systems* (including Intelligence, Surveillance and Reconnaissance (ISR) systems). With this grouping as a starting point, the rationale for arriving at the proposed five-level risk architecture is given below:

Under the premise that all weapon systems present a higher level of risk as compared to systems which do not directly result in the release of weapons, the higher three proposed levels of risk correspond to AI-enabled weapon systems, while all decision support systems are grouped under the lower two levels.

- **Level 1**: **Unacceptable Risk Level**. This level represents a special category of weapons (hopefully not yet developed) which present so high a risk that their development must not be undertaken
- **Levels 2 & 3**: **High and Medium Risk Levels**. Amongst the remaining weapon systems, intuitively there appears to be a case for at least two levels of risk (High & Medium), rather than just one. For instance, fully autonomous lethal weapon systems clearly pose a higher risk as compared to semi-autonomous systems

- **Level 4**: **Trust Requirements Level**. Amongst the non-weapon AI-enabled military systems (collectively referred to in this chapter as decision support systems), defining a minimum of two levels seems necessary, as follows: a higher level, which comprises critical decision support systems (e.g., those designed to suggest attack options in a tactical setting); and a lower level covering all other decision support systems. Critical decision support systems would need to be trusted by commanders for effective human-machine teaming, perhaps by resorting to Explainable AI (XAI). This level would also focus on mitigating AI-related risks such as automation bias
- **Level 5**: **Negligible Risk Level**. This risk category is envisaged to encompass all AI-enabled military systems which are not covered under Levels 1–4 and which pose a level of risk which may not warrant any special scrutiny. This level would include non-critical decision support systems, e.g., AI-enabled military applications in areas such as logistics and maintenance. While such systems may not present a risk from an IHL/trust perspective, they must still be vetted for other concerns associated with AI such as fragility, inadequately selected or poisoned data, etc.

It may be possible to split each of the above levels further based on additional parameters, or formulate levels based on entirely different criteria. However, it is felt that the above five-level architecture yields a simple yet effective model to address risks linked to IHL and battlefield reliability. It also merits mention here that defining a hierarchy of risk levels is meaningful only if these can be mapped to corresponding *differentiated* risk-mitigation measures. The viability of working out such a risk-mitigation mechanism is a key consideration for limiting the number of levels.

RISK-BASED TAXONOMY OF AI-ENABLED MILITARY SYSTEMS

The next step is to develop a taxonomy of weapon classes based on parameters which adequately reflect the risks which AI-empowerment of systems is expected to cause or enhance during armed conflicts, namely, violation of IHL and unreliable performance on the battlefield.

One possible approach for evaluating system risk is by focussing on the underlying technology. For instance, a system which leverages black-box technologies such as AI/ML may be presumed to present a higher risk as compared to one which is implemented using explainable code. This model, however, adopts an effects-based approach, wherein a system which can result in a high negative fall-out in the event of malfunction or other unintended behaviour is considered as presenting a higher risk as compared to one in which adverse consequences are minimal.

For weapon systems, qualitative values of five carefully chosen parameters are used to segregate the wide variety of systems into ten different risk classes. In the case of non-weapon military systems which conform to Risk Levels 4 and 5, the definition of the risk level itself serves to identify a class of systems, resulting in two additional classes. In this manner, a risk-based taxonomy of military systems comprising 12 classes has been arrived at.

The rest of this section discusses the risk evaluation parameters and how these are utilized to develop the taxonomy of weapon systems.

Risk Evaluation Parameters

The five parameters which have been identified for dividing the full spectrum of weapon systems into disjoint classes are as follows: *nature of the Observe-Orient-Decide-Act (OODA) Loop* (Alberts, Garstka & Stein, 1999) (which gives rise to platform-centric, network-centric and swarm weapon systems); *degree of autonomy* (online learning, fully-autonomous and semi-autonomous weapon systems); *destructive potential* (nuclear and non-nuclear weapon systems); *type of military operation* (offensive and defensive weapon systems) and *type of target* (lethal and non-lethal weapon systems). An important parameter, which has not been considered for the time being, is *warfighting dimension* (these being kinetic, cyber, electromagnetic and cognitive (Panwar, 2017a)), since the Risk Hierarchy presently restricts itself to only kinetic weapons.

The following subsections explain each parameter and its correlation with risk during armed conflicts.

The OODA Loop parameter

Based on the nature of their OODA Loops, all weapon systems may be classified into *Platform Centric (PC)*, *Network Centric (NC)* or *Swarm* weapon systems. In the military context the OODA Loop broadly translates into the sensor – decision-maker – shooter loop. PC weapons refer to systems in which this loop closes on a single platform, e.g., tank, aircraft, ship, etc, including their unmanned versions. In contrast, NC weapons differ in two respects: firstly, sensors, decision nodes and shooters (three types of entities) are *geographically dispersed* and connected via a network; and secondly, there could be *multiple entities* of each type making up the weapon system (Panwar, 2017b). A weapon system using swarm technology, although not known to be operational yet in any military, would perhaps be more akin to PC rather than NC systems, and may be best visualized as a locally distributed version of a single platform.

A PC weapon system lends itself to mobility and, in conjunction with autonomy, can present a high risk, since exercising human control would pose difficulties. In comparison, an NC weapon system is likely to be less mobile, but if a large number of AI-enabled entities in a net-centric environment are linked up together, the resulting complexity would raise risk levels. As regards swarms, the emergent behaviour associated with them is not fully understood and can potentially make swarms unpredictable, thus presenting a very high level of risk (Harvey, 2019). In summary, PC, NC and swarm weapon systems are characterized by control issues, complexity and unpredictability respectively, which present risks of different flavours on the battlefield.

Degree of autonomy

It was discussed in a previous section that an increase in autonomy implies less human control, which leads to several types of risks. In this work, to keep the risk evaluation model simple, this parameter can take on primarily two values, namely, fully-autonomous and semi-autonomous weapon systems, and the nuances of these two terms have been discussed earlier. Online learning systems are a special case of fully autonomous weapons.

There is a view that adherence to the principle of Proportionality and Military Necessity requires value judgement, which is a uniquely human trait, and autonomy in weapon systems undermines this principle. Another apprehension which is often expressed is that increased autonomy in weapon systems would result in a lack of accountability. The moral argument associated with fully autonomous weapons has been discussed above. While there are counters to each of these lines of argument, most would agree that increasing autonomy in weapon systems would result in a higher risk of unintended consequences on the battlefield.

Autonomy is a dominant factor in determining the overall risk level posed by a weapon system. However, it is noteworthy that, unlike the other four parameters, it may not be possible to ascertain the level of autonomy through external observation. Moreover, since autonomy is more often than not implemented in software, it may easily be switched from one mode of functioning to another. Therefore, given the dominant but nebulous nature of this parameter, the actual risk level of any AI-enabled military system would be known only to the developer/user.

Destructive potential

The destructive power of the large variety of AI-enabled weapons is spread over a very wide range. However, the choice of this parameter here has the limited aim of segregating nuclear weapons from non-nuclear ones, because of the extreme destructive potential of the former, under the premise that nuclear weapons can never satisfy the principles of Distinction as well as Proportionality & Military Necessity. Therefore, risk posed by nuclear weapons is assessed to be higher as compared to non-nuclear ones, all other parameters being equal.

Type of military operation

In general, IHL is more likely to be violated during offensive as compared to defensive operations, for the following reasons: the principle of Proportionality and Military Necessity is unlikely to be flouted by defensive action in war; and since defensive weapons would in most cases be employed over own territory, the probability of causing unintended harm to civilians (own citizens) is expected to be minimal.

As already stated, most weapons can be used in both offensive and defensive operations. It is reiterated once again here that, in this work, defensive weapons imply those which can be used only in a defensive role and a lower risk is allocated to these as compared to all other weapons.

Type of target

This parameter can take on two values: lethal or non-lethal (anti-materiel). The primary objective of IHL is to minimize unintended harm to humans, i.e., civilians and soldiers *hors de combat*. Hence, from an IHL perspective, only lethal weapon systems pose a risk. It is to be noted, however, that anti-materiel weapon systems might also cause human casualties, though only in exceptional circumstances. From an IHL perspective, therefore, lethal weapons pose a higher risk.

Taxonomy of Weapon Systems: Disjoint Classes

Table 4.1 summarizes the values which each of the parameters can take on.

If each permutation of the above set of values is taken as a distinct class of weapons, it would translate to a total of 108 weapon classes. However, certain parameters have a predominantly high bearing on risk, making other parametric values lose their significance in certain permutations. For instance, all fully autonomous weapons are placed in the *High-Risk category*, unless it is a purely defensive weapon, which brings down its risk level to *Medium*, or it is nuclear, which enhances the risk level to *Unacceptable*. Using such heuristics, the 108 possible value combinations have been collapsed into ten weapon classes, as indicated in the Risk Hierarchy depicted in Figure 4.1. The Risk Hierarchy uses primarily the *OODA Loop* and *Autonomy* parameters, modified to an extent by the other three, to arrive at this taxonomy of weapon systems. The attempt here has been to create these classes as disjoint sets, while at the same time collectively covering the full spectrum of weapon systems.

The next section provides the rationale adopted for assigning the ten weapon classes to the three weapons-related risk levels, as depicted in the Risk Hierarchy diagram.

TABLE 4.1   Risk evaluation parameters and their possible values

PARAMETERS	POSSIBLE VALUES		
OODA Loop Complexity	Platform-Centric	Network-Centric	Swarm
Autonomy	Semi-Autonomous	Fully-Autonomous	Online Learning
Type of Target	Non-Lethal	Lethal	Human Targeting
Operational Deployment	Offensive	Defensive	–
Destructive Potential	Nuclear	Non-Nuclear	–

ASSIGNING WEAPON CLASSES TO RISK LEVELS

Risk Level 1: Unacceptable Risk

The following three types of systems are mapped to this level:

- **Fully Autonomous Nuclear Weapon Systems**. There is wide consensus that such weapon systems must not be developed because of the catastrophic consequences of a malfunction, and also because it is felt that there needs to be human accountability for resorting to their use
- **Fully Autonomous Unpredictable (Online Learning) Weapon Systems**. This class of weapons, given their ability to learn while in operation, can metamorphose into a state for which they were not tested, and hence this work considers their development and deployment as being unacceptable
- **Fully Autonomous Human Targeting Weapon Systems**. Including such weapon systems in the Unacceptable Risk category has been done to avoid violating the IHL principle of Distinction as well as the spirit behind Martens Clause

Implied in the nomenclature of this risk level is the assertion that weapon classes assigned to this level must not be developed. Banning weapon systems is an extreme measure, and arriving at such a decision, either through an international treaty or even domestically by a state, would require strong supporting rationale.

While the reasons for including the three categories listed at this level have been stated briefly above, there are counter-arguments as well for each of these, which must be taken into account. For instance, fully autonomous nuclear weapon systems are considered by some to be the ultimate deterrent for preventing a nuclear holocaust and even large-scale conventional conflicts. Regarding unpredictability, one can argue that all weapon systems are to an extent unpredictable. Finally, since weapons are designed to kill humans (in addition to destroying military assets), the moral argument made against human targeting AI-enabled weapons appears to be weak. Notwithstanding these and other counter-arguments, this chapter argues for imposing a ban on these weapon classes.

It is pointed out here that, even if it is felt that the listed categories should be regulated through strict risk-mitigation measures rather than an outright ban, conceptually there is a need to retain this level in the Risk Hierarchy. This is to cater for any other class of weapons, present or future, which might be evaluated as posing unacceptable risk.

Risk Level 2: High Risk

The following four classes of weapon systems are included at this level:

- **Semi-Autonomous Nuclear Weapon Systems**. Incorporating autonomy in non-critical functions such as take-off and landing, navigation, etc, appears to be an acceptable proposition even for nuclear weapon systems. Nonetheless, given their high destructive potential, they have been placed at the *High-Risk* level
- **Swarm Technology Based Weapon Systems**. Armed swarms are placed in this category because of the unpredictability associated with their emergent behaviour. Here the underlying premise is that, in contrast to online learning systems, unpredictable behaviour in swarms could be limited to lie within specified bounds, and tested rigorously before fielding the system
- **Fully-Autonomous PC Weapon Systems**. In principle, all weapon systems with full autonomy in the critical functions are placed in this *High-Risk* category, barring the following exceptions: nuclear and human targeting systems are placed one level higher in the *Unacceptable Risk* category, while purely defensive systems are placed one level lower in the *Medium Risk* category. In PC weapon systems, all functions including critical ones are on the same platform, hence all such weapon systems are placed in the *High-Risk* category
- **Decision Nodes of Fully-Autonomous NC Weapon Systems**. In NC weapon systems, the sensors, decision nodes and shooters would be geographically distributed and, moreover, may be inducted into service separately. Since the decision to release a weapon would be taken at a decision node, only fully autonomous *decision nodes* (and not sensors and shooters) have been placed in the *High-Risk* category

Risk Level 3: Medium Risk

All weapon classes not covered under Levels 1 and 2 are placed in the *Medium Risk* category. This includes sensors and shooters of fully-autonomous NC weapon systems, all purely defensive fully-autonomous weapon systems, and all semi-autonomous weapon systems.

VALIDATING THE RISK HIERARCHY

Having established the basis for categorising weapon systems into weapon classes and then mapping them to different risk levels, an analytical exercise was carried out to validate this risk evaluation model against specific systems.

For this exercise, more than 25 extant weapon systems were short-listed from open domain sources, which were assessed to have some degree of autonomy, possibly AI-enabled, built into their design. These systems have been either operationalized or their prototypes have been demonstrated. In addition, seven notional or generic weapon systems were also factored in. Notional systems include those which classify to be placed at the *Unacceptable Risk* level and, hopefully, will never be developed. Decision support systems, including ISR systems, have been treated as *generic systems* (i.e., specific systems were not identified). Non-kinetic anti-drone weapons with *physical effects* (laser, high power microwave) have also been included as a generic class.

For each of the weapon systems used in this exercise, the values of the five risk evaluation parameters were ascertained using either information available in the open domain or by making reasonable assumptions. Based on these values, each system was grouped under one of the ten weapon classes and then mapped to the Risk Hierarchy. To take just one example, as per a UN report of Apr 2021, the STM Kargu-2, a Turkish unmanned combat air vehicle (UCAV), autonomously hunted down Haftar Armed Forces elements in Libya in 2020 (UN Security Council, 2021). Being fully autonomous, offensive and lethal, this is placed under the *High-Risk* category (see Table 4.2).

There are also reports (which were subsequently refuted) that the Kargu-2 UCAV has face recognition capability and can hunt down specific human targets. If true, then as per the Risk Hierarchy model, it falls into the *Unacceptable Risk* category. For the purpose of this exercise, however, this capability has not been taken into account.

In order to demonstrate the results of this mapping exercise, Table 4.2 shows the parameters and risk mapping for ten weapon systems, one from each class.

While the exercise is not comprehensive and may have assessed some of the parametric values incorrectly, it does seem to indicate that, subject to further refinement, the Risk Hierarchy provides a useful tool for evaluating risk levels of AI-enabled weapon systems.

DIFFERENTIATED RISK-MITIGATION MEASURES

If the Risk Hierarchy is to be used effectively for mitigating risks, an important final step is to evolve a *differentiated* risk-mitigation mechanism which may be linked to the five risk levels. There is a viewpoint that while a risk-based approach is valuable for attaining a good understanding of the risks posed by systems, mitigation measures must be applied uniformly across all systems, independent of their evaluated risk level. This view has a certain intuitive appeal, since it promises to reduce risk to the barest minimum, under the assumption that rigorous risk-mitigation measures would be applied to all systems. This work, however, goes by the rationale that the stringency of mitigation measures must increase with increasing risk. This is because instituting a common mitigation mechanism is likely to be counter-productive, not only by

TABLE 4.2 Classification of ten extant weapon systems in the risk hierarchy

SER NO	WEAPON CLASS	WEAPON SYSTEM	OODA LOOP	AUTONOMY	OPERATIONAL EMPLOYMENT	DESTRUCTIVE POTENTIAL	TYPE OF TARGET
Level 1: Unacceptable Risk							
1	Unpredictable (Online Learning) systems	Generic	PC/NC/ Swarm	Online Learning	Offensive	Non-Nuclear	Lethal
2	Fully autonomous human-targeting weapon systems	Generic	PC/NC/ Swarm	Full	Offensive	Non-Nuclear	Human Targeting
3	Fully autonomous nuclear weapon systems	Generic	PC/NC	Full	Offensive	Nuclear	Lethal
Level 2: High Risk							
4	Semi-autonomous nuclear weapon systems	Poseidon	PC	Semi	Offensive	Nuclear	Lethal
5	Swarm technology-based weapon systems	XQ-58 Valkyrie	Swarm	Full	Offensive	Non-Nuclear	Lethal
6	Fully autonomous PC weapon systems	STM Kargu-2	PC	Full	Offensive	Non-Nuclear	Lethal
7	Decision nodes of fully autonomous NC weapon systems	Generic	NC	Full	Offensive	Non-Nuclear	Lethal
Level 3: Medium Risk							
8	Sensors & shooters of fully autonomous NC weapon systems	Israeli Swarm (Hamas Op)	Swarm	Full	Offensive	Non-Nuclear	Lethal
9	Defensive fully autonomous PC & NC weapon systems	Iron Dome	PC	Full	Defensive	Non-Nuclear	Anti-Materiel
10	Semi-autonomous PC & NC weapon systems	Bayraktar TB2	PC	Semi	Offensive	Non-Nuclear	Lethal

hampering the development of low-risk weapon systems, but by also resulting in the dilution of mitigation efforts for high-risk systems.

The US DoD Directive 3000.09 and the proposed EU AI Act provide good leads for working out a differentiated risk-mitigation mechanism. An overview of these two regulatory frameworks is given below.

DOD Directive 3000.09 on Autonomy in Weapon Systems

As early as Nov 2012, the US DoD issued Directive 3000.09 for AWS in general, not necessarily AI-enabled. Although the Directive is not specifically framed as a risk-based approach, glimpses of a differentiated risk-mitigation mechanism are discernable therein. These mitigation measures have been further refined in its latest revision issued in Jan 2023 (US DoD, 2023b).

The Directive treats all autonomous weapons systems under two categories, as under:

- **Category 1**: This includes all semi-autonomous weapon systems; operator supervised AWS which are designed to select and engage materiel targets to intercept time-critical attacks against static installations/platforms or for defending remotely piloted/autonomous vehicles; and AWS which apply non-lethal, non-kinetic force against materiel targets
- **Category 2**. All other AWS

The Directive lays down a number of measures for mitigating risks associated with all AWS including AI-enabled ones. These cover aspects such as the inclusion of design features which allow commanders and operators to exercise appropriate levels of human judgement; rigorous hardware and software verification and validation (V&V) as well as realistic system development and operational test and evaluation (T&E) procedures; design constraints to enable completion of engagement within a specified time-frame and geographic area; consistency of design with DoD AI Ethical Principles and the DoD Responsible Artificial Intelligence Strategy and Implementation Pathway; amongst others.

In addition to the common mitigation measures which are applicable across all systems, the Directive requires Category 2 systems to be approved at a senior level before deployment, together with specific aspects which need to be checked during this approval. In effect, this amounts to a two-level differentiated risk-mitigation mechanism, though the more stringent mitigation measures are limited to review and approval processes within the DoD.

The European Union (EU) Proposal for AI Regulation

The proposed EU AI Act groups all AI systems into a four-level hierarchy composed of *Unacceptable Risk*, *High Risk*, *Transparency Requirements* and *Negligible Risk* levels. The Unacceptable Risk level includes systems which can manipulate persons through subliminal techniques beyond their consciousness; exploit vulnerabilities of specific vulnerable groups such as children or persons with disabilities; and 'real time' remote biometric identification systems in publicly accessible spaces for the purpose of law enforcement. The High-Risk level covers applications such as biometric systems meant for categorization of natural persons; systems for the management of critical infrastructure; and applications for education and employment. At the next lower level are systems which need to meet certain transparency requirements, eg, those which interact with humans or generate or manipulate content (European Commission, 2021c).

The proposal stipulates that AI applications which fall at the Unacceptable Risk level should be prohibited. For High-Risk systems, a stringent risk management system has been proposed. For certain types of AI systems, only specified transparency obligations are required to be met by the fielders of the system. Finally, for systems which pose a negligible risk, the proposal lays down a framework for the creation of codes of conduct, aimed at encouraging providers of such systems to voluntarily apply requirements which are mandatory for High-Risk AI systems.

Although targeted at civilian applications, the EU AI Act is perhaps the only attempt, so far, which explicitly adopts a risk-based approach for AI systems, including a differentiated risk-mitigation mechanism.

Risk Hierarchy: Considerations for Evolving Mitigation Measures

The above discussion reveals that DoD Directive 3000.09 as well as the proposed EU AI Act institute stricter measures as the level of risk increases.

With respect to the Risk Hierarchy, the challenge is to evolve five sets of mitigation measures tailored for each of its risk levels. These measures would come into play at every stage, from project clearance through design and development, TEV&V, review (including legal review) and deployment stages. A few considerations for doing so are presented here.

Out of the five risk levels, mitigation measures for the *Unacceptable Risk* level are the easiest to envisage since, by definition, systems grouped under this level must obviously not be developed at all. At the other end of the spectrum, mitigation measures for systems in the *Negligible Risk* category may be more in the nature of codes of conduct or best practices in AI design and development, and evolving these should not be very challenging.

Amongst the remaining middle three levels, risk-mitigation measures for levels which relate to weapon systems, namely the *High* and *Medium Risk* levels, would need to be worked out more carefully as compared to the *Trust Requirements* level which relates to decision support systems, since these are not meant for targeting and, by definition, have a human-in-the-loop.

For weapon systems, in case AI technologies are used for object recognition in sensors and precision targeting in shooters, rigorous testing against specified performance standards should be adequate to sufficiently mitigate AI-related risks, and the non-transparent, data-centric character of AI/ML technologies may not pose any serious issues. On the other hand, the risk associated with AI-enabled autonomy in the critical decision-to-engage function as well as in armed swarms would be much greater, and stringent measures would need to be incorporated at every stage of the system life cycle.

The *Trust Requirements* level may warrant the mandatory use of XAI to avoid *automation bias* and also to make these systems trustworthy from the perspective of commanders. XAI techniques, however, are yet to mature. In the interim, therefore, commanders must leverage the decision support provided by 'black-box' AI systems responsibly, and be adequately trained to avoid either over-confidence or under-confidence in the recommendations and insights provided by the system.

Across all systems, project clearance and review processes should be made more stringent as the risk levels increase.

Keeping in mind the above general considerations, a differentiated risk-mitigation mechanism is required to be evolved and linked to the Risk Hierarchy, for implementation at each stage of the system life-cycle.

PROMOTING INTERNATIONAL CONSENSUS ON REGULATION OF LAWS

At the December 2019 meeting of UN GGE on LAWS, a set of 11 guiding principles was accepted by all parties. However, even after years of deliberations, a consensus remains elusive on how AI-enabled weapon systems may be regulated through a legally binding international instrument. Moreover, a common understanding on the regulation of LAWS, even a non-binding one, does not as yet exist amongst major military powers. The risk-based approach presented here could contribute usefully towards consensus building, as explained below.

Granular Discussions Would Facilitate Consensus Building

It is felt that a key reason for the failure to reach consensus so far is that the discussions are very general in nature, and usually treat all AI-enabled weapon systems as one category. This makes it very difficult to identify specific areas of disagreement which might be taken up for resolution. The Risk Hierarchy provides a basis for facilitating discussions at a more granular level, for instance:

- By evaluating risks presented by different categories of weapons on the basis of well-defined parameters, it should be easier to reach agreement on mapping these weapon classes to different risk levels within the Hierarchy, as also to segregate non-weapon systems into critical and non-critical decision support systems.
- From there, states could focus on certain very high-risk categories, and deliberate on whether these should be banned altogether. At the other end of the spectrum, states might find it easier to agree that weapon systems presenting very low risk should perhaps not be constrained through international regulation. For the remaining levels, states could endeavour to evolve and share best practices for risk mitigation.
- The Risk Hierarchy, with its detailed risk analysis, also provides a basis for debating the need to add to or modify existing provisions in IHL for dealing with LAWS.

The central idea here is to reduce the complexities associated with risk evaluation of the wide array of AI-powered military systems into simpler and more precisely defined problems and address them piecemeal. Broadly speaking, it may be easier to first agree on the overall approach as presented in this work, next tackle systems which lie at the bottom of the Risk Hierarchy and then gradually move upwards.

Self-Regulation by Responsible Militaries

This chapter takes the stance that responsible militaries endeavor to operate in conformance with IHL, and also that they would not employ weapon systems which may have negative fallouts for their own forces. For instance, all militaries prefer precision targeting over dumb munitions in order to minimize collateral damage (as dictated by the principles of Distinction and Proportionality), in addition to enhancing their own combat effectiveness. No military commander would like to employ unpredictable weapon systems, as such systems are bound to be tactically inefficient, cause harm to innocent civilians on both sides, and even result in combatant fratricide. As another example, if fully autonomous nuclear weapon systems malfunction, the result would be mutually assured destruction. Finally, in the present era when narrative warfare can have strategic effects, violation of IHL is likely to be counter-productive towards achieving politico-military objectives.

The Risk Hierarchy, by piercing through generalities and evaluating risk through a well-reasoned approach, helps in understanding how specific categories of AI-powered weapon systems might result in IHL violations or otherwise be detrimental to a state's own military operations. In doing so, it encourages responsible states to institute a self-regulatory mechanism for mitigating these risks. Such self-regulatory mechanisms adopted by states in their own interests (e.g., DoD Directive 3000.09) would also help in achieving international consensus on AI regulation.

FUTURE WORK

The model presented in this chapter is restricted to evaluating and mitigating AI-related risks presented by kinetic weapon systems in the context of armed conflicts. While this model itself may be refined further, there are three areas where its scope could be usefully expanded, as listed below:

- **Domains**. AI-related risks posed by weapons in the cyber, electromagnetic and cognitive domains also pose significant risks and need to be incorporated into the model
- **Cost-Benefit Analysis**. Leveraging AI power for military applications, while presenting risks, is expected to yield tremendous benefits as well. Therefore, a balanced approach would be to evolve a model which takes into account both risks and benefits and risk-mitigation mechanisms should emerge from a cost-benefit analysis, rather than being dictated by risks alone
- **Scenarios**. It has been emphasized in this chapter that risk evaluation and mitigation mechanisms for terrorist and grey zone scenarios would be significantly at variance with those relevant to armed conflicts. Risk-based approaches for these scenarios too need to be developed. More importantly, inter-relationships and inter-dependencies amongst these three approaches must also be identified and, if possible, integrated into an overall risk-mitigation strategy.

CONCLUSION

The primary motivation for adopting a risk-based approach to the regulation of AI-enabled systems is to mitigate risks while at the same time leverage the power of AI for the benefit of humankind. The proposed EU AI Act incorporates such an approach for civilian applications.

This chapter has presented the Risk Hierarchy as a qualitative model for evaluating and mitigating risks associated with AI-enabled military systems, perhaps the first of its kind. As is clear from the analysis presented above, the overall objective, nature of risks and mitigation mechanisms in the case of military systems differ substantially from what is applicable for non-military applications. The work presented here may be utilized by international bodies as well as by individual states for moving beyond mere enunciation of principles towards evolving more concrete mechanisms and practices for leveraging AI technologies in a responsible manner in the military domain.

REFERENCES

China MOST. (2021, September 26). *"A new generation of artificial intelligence ethics code"* *released.* https://www.most.gov.cn/kjbgz/202109/t20210926_177063.html

Copeland, B. J. (2023, March 23). Artificial intelligence. *Encyclopedia Britannica.* https://www.britannica.com/technology/artificial-intelligence

European Commission. (2019, April 08). *Ethics guidelines for trustworthy AI* (p. 4). https://digital-strategy.ec.europa.eu/en/library/ethics-guidelines-trustworthy-ai

European Commission. (2021, April 21). *Laying down harmonised rules on artificial intelligence (artificial intelligence act).* https://eur-lex.europa.eu/resource.html?uri=cellar:e0649735-a372-11eb-9585-01aa75ed71a1.0001.02/DOC_1&format=PDF

European Commission. (2021b, April 21). *Annexes to laying down harmonised rules on artificial intelligence (artificial intelligence act)* (p. 1). https://eur-lex.europa.eu/resource.html?uri=cellar:e0649735-a372-11eb-9585-01aa75ed71a1.0001.02/DOC_2&format=PDF

European Commission. (2021c, April 21). *Laying down harmonised rules on artificial intelligence (artificial intelligence act)* (pp. 12–15). https://eur-lex.europa.eu/resource.html?uri=cellar:e0649735-a372-11eb-9585-01aa75ed71a1.0001.02/DOC_1&format=PDF

European Parliament. (2023) Resolution 2485 (2023) - Emergence of lethal autonomous weapons systems (LAWS) and their necessary apprehension through European human rights law (p. 1). https://pace.coe.int/pdf/5f1d73b940a8f6190cfbdc59e921a38744bd0df7651c196cd6ba9605c35e2f94/res.%202485.pdf

Future of Life Institute. (2017, August 11). Asilomar AI principles. https://futureoflife.org/open-letter/ai-principles/

Future of Life Institute. (2021, December 01). *Slaughterbots - if human: kill()* [Video]. YouTube. https://www.youtube.com/watch?v=9rDo1QxI260

Government of Canada. (2019, April 1). Directive on automated decision making. https://www.tbs-sct.canada.ca/pol/doc-eng.aspx?id=32592&section=html

Government of Singapore. (2020, January 21). Model AI governance framework. https://www.pdpc.gov.sg/-/media/Files/PDPC/PDF-Files/Resource-for-Organisation/AI/SGModelAIGovFramework2.pdf

Government of the Netherlands. (2017, October 9). *Examination of various dimensions of emerging technologies in the area of lethal autonomous weapons systems, in the context of the objectives and purposes of the Convention.* CCW/GGE.1/2017/WP.2. https://www.reachingcriticalwill.org/images/documents/Disarmament-fora/ccw/2017/gge/documents/WP2.pdf

Harvey, J. (2019). The blessing and curse of emergence in swarm intelligence systems. In H. A. Abbass, J. Jason Scholz, & D. J. Reid (Eds.), *Foundations of Trusted Autonomy* (pp. 117–124). Springer Cham. https://doi.org/10.1007/978-3-319-64816-3.

Hoi, S., Sahoo, D., Lu, J., & Zhao, P. (2018, August 02). Online learning: A comprehensive survey. *Neurocomputing, 459,* 249–289. https://doi.org/10.1016/j.neucom.2021.04.112.

Human Rights Watch. (2023, February 14). *Review of the 2023 US policy on autonomy in weapons systems.* https://www.hrw.org/news/2023/02/14/review-2023-us-policy-autonomy-weapons-systems

ICRC. (2014). *International humanitarian law: Answers to your questions.* https://www.icrc.org/en/doc/assets/files/other/icrc-002-0703.pdf

ICRC. (2020). Artificial intelligence and machine learning in armed conflict: A human-centred approach, *International Review of the Red Cross, 102*(913), 463–479. https://international-review.icrc.org/sites/default/files/reviews-pdf/2021-03/ai-and-machine-learning-in-armed-conflict-a-human-centred-approach-913.pdf

ICRC. (2021, May 12). ICRC position on autonomous weapon systems. https://shop.icrc.org/icrc-position-on-autonomous-weapon-systems-pdf-en.html

Israel Aerospace Industries. (n.d.). *Harpy: Autonomous weapon for all weather*. System missiles and space. https://www.iai.co.il/p/harpy

Jansen, E. T. (2020). The (erroneous) requirement for human judgment (and error) in the law of armed conflict. *International Law Studies*, 96, Stockton Centre for International Law. https://digital-commons.usnwc.edu/cgi/viewcontent.cgi?article=2916&context=ils

Kerry, C. F., Meltzer, J. P., Rhea, A, Engler, A. C., & Fanni, R. (2021). Strengthening international cooperation on AI. *Progress Report*. Washington, DC: Brookings Institution. https://www.brookings.edu/research/strengthening-international-cooperation-on-ai/

Lohn, A. J. (2020, September 2). *Estimating the brittleness of AI: Safety integrity levels and the need for testing out-of-distribution performance. arXiv:2009.00802v1*. https://doi.org/10.48550/arXiv.2009.00802

Moyes, R. (2019, November). Critical commentary on the "guiding principles". *Article 36 Policy Note*. https://www.article36.org/wp-content/uploads/2019/11/Commentary-on-the-guiding-principles.pdf

Oimann, A. K. (2023). The responsibility gap and LAWS: A critical mapping of the debate. *Philosophy & Technology, 36(3)*. https://doi.org/10.1007/s13347-022-00602-7.

Panwar, R. S. (2017a, October 06). 21st century warfare: From battlefield to battlespace. *Future Wars*. https://futurewars.rspanwar.net/21st-century-warfare-from-battlefield-to-battlespace/

Panwar, R. S. (2017b, September 22). Network centric warfare: Understanding the concept. *Future Wars*. https://futurewars.rspanwar.net/network-centric-warfare-understanding-the-concept/

Panwar, R. S. (2022). AI and the rise of autonomous weapons. In R. P Rajagopalan (Ed.), *Future warfare and technology: issues and strategies* (pp. 68–78), Observer Research Foundation and Global Policy Journal. Wiley. https://www.orfonline.org/wp-content/uploads/2022/11/GP-ORF-Future-Warfare-and-Technology-01.pdf

Scharre, P. (2018, February 01). Debating slaughterbots and the future of autonomous weapons. *IEEE Spectrum*. https://spectrum.ieee.org/debating-slaughterbots

TASS. (2021, October 26). *First code of ethics of artificial intelligence signed in Russia*. https://tass.com/economy/1354187

Ticehurst, R. (1997, April 30). The Martens Clause and the laws of armed conflict. *International Review of the Red Cross, 317*, 125–134. https://www.icrc.org/en/doc/resources/documents/article/other/57jnhy.htm

UN Security Council. (2021, March 04). *Final report of the panel of experts on Libya established pursuant to Security Council Resolution 1973 (2011), S/2021/229* (p. 17). https://documents-dds-ny.un.org/doc/UNDOC/GEN/N21/037/72/PDF/N2103772.pdf?OpenElement

UNIDIR. (2017). *The weaponization of increasingly autonomous technologies: Concerns, characteristics and definitional approaches* (pp. 23–32). https://unidir.org/sites/default/files/publication/pdfs/the-weaponization-of-increasingly-autonomous-technologies-concerns-characteristics-and-definitional-approaches-en-689.pdf

UNODA. (2019, December 13). Guiding principles affirmed by the group of governmental experts on emerging technologies in the area of lethal autonomous weapons system. *Final Report: Annexure III*, UN CCW GGE on LAWS (p. 10). https://undocs.org/CCW/MSP/2019/9

US DOD. (2023a, January 25). *Directive 3000.09: Autonomy in weapon systems* (pp. 21–23). USD (Policy). https://www.esd.whs.mil/portals/54/documents/dd/issuances/dodd/300009p.pdf

US DOD. (2023b, January 25). *Directive 3000.09: Autonomy in weapon systems* (p. 5, pp. 14–17). USD (Policy). https://www.esd.whs.mil/portals/54/documents/dd/issuances/dodd/300009p.pdf

US DOD. (2020, February 24). *DOD adopts ethical principles for artificial intelligence*. https://www.defense.gov/News/Releases/Release/Article/2091996/dod-adopts-ethical-principles-for-artificial-intelligence/

Winter, E. (2022). The compatibility of autonomous weapons with the principles of international humanitarian law. *Journal of Conflict & Security Law, 27*(1), 1–20. https://doi.org/10.1093/jcsl/krac001

Wojton, H. M., Porter, D. J., & Dennis, J. W. (2020, September). *Test and evaluation of AI-enabled and autonomous systems.* US DOD Institute for Defense Analysis. https://testscience.org/wp-content/uploads/sites/16/formidable/20/Autonomy-Lit-Review.pdf

Zador, A., & LeCun, Y. (2019, September 26). *Don't fear the terminator.* Scientific American Blog. https://blogs.scientificamerican.com/observations/dont-fear-the-terminator/

Applying Responsible AI Principles into Military AI Products and Services

A Practical Approach

5

Michael Street and Sandro Bjelogrlic

Responsible AI for the military is a matter of trust. Military commanders must have trust in any AI system they interact with or it will not be adopted. Civilian populations must have trust in AI used by the military forces that protect them, or they will lose trust in those forces. As AI permeates every aspect of our lives, the use of AI by the military depends on it being developed and employed responsibly. This requires three key attributes:

- A clear vision of what responsible AI is, to guide all AI development and use. Such a vision may be decomposed into a number of principles

DOI: 10.1201/9781003410379-6

- A framework that gives guidance to developers on steps to take to ensure their AI is developed, behaves, and is used responsibly. Such a framework may be supported by tools that encourage or mandate such steps during the development of AI
- Tests to allow independent validation to ensure the AI has been developed, operates, and is deployed in a responsible manner

This chapter focuses on how to develop and deploy responsible AI for the military and is written from the perspective of AI developers working within a multinational military environment, as part of the North Atlantic Treaty Organization (NATO). But we begin with a look at what responsible AI is and the clear vision provided by the NATO AI strategy.

NATO'S AI STRATEGY

The NATO was established in 1949 during a period of global security concern, to improve security in the North Atlantic region through multinational agreement for collective defense of all its member nations. Since its establishment, NATO's member Nations, or Allies, have collaborated politically, militarily, and technologically to ensure the Alliance can defend effectively. This defense begins with political consensus in decision-making, consistent military doctrine, multinational exercises, and interoperable technology. This long history of multinational collaboration to ensure regional security, particularly in fields aligning NATO's technological capabilities with its ambition, led to the development and adoption of NATO's AI strategy in 2021.

The aims of NATO's AI strategy include to "provide a foundation for NATO and Allies to lead by example and encourage the development and use of AI in a responsible manner for Allied defense and security purposes" (NATO, 2021). A key element of "leading by example" has been the adoption of principles of responsible use which are defined in the AI strategy. These principles have been established and embedded in the strategy in order to "steer transatlantic efforts in accordance with our values, norms, and international law". This takes the same historic commitment to established legal and ethical principles under which the Alliance has always operated and ensures that they are followed during the development and use of AI which, as many other chapters note, can be overlooked when dealing with AI.

The NATO principles of responsible use have been informed by similar work undertaken nationally or internationally. It should be noted that many NATO nations have, or are developing, national principles for responsible use of AI in the military domain. The next section cites each of the principles, followed by a brief explanation from the AI developers' perspective.

Summary of the Principles of Responsible Use

A. "Lawfulness: AI applications will be developed and used in accordance with national and international law, including international humanitarian law and human rights law, as applicable". (Ibid)

   While it may appear self-evident that AI should not break the law, given some debates around the training and use of AI this principle does remove uncertainty regarding the development, output and use of AI systems, even in defense scenarios.

B. "Responsibility and Accountability: AI applications will be developed and used with appropriate levels of judgment and care; clear human responsibility shall apply in order to ensure accountability". (Ibid)

   This ensures that when AI is used in a military context, the use of AI does not absolve developers or users of AI of a "clear human responsibility" for systems they develop or use.

C. "Explainability and Traceability: AI applications will be appropriately understandable and transparent, including through the use of review method-ologies, sources, and procedures. This includes verification, assessment and validation mechanisms at either a NATO and/or national level". (Ibid)

   In political and military decision-making, explainability has always been an essential component. Military commanders have always asked for explanations of assessments or recommended courses of action and the level of trust in their subordinates informs their decision-making. AI applications should be no different.

D. "Reliability: AI applications will have explicit, well-defined use cases. The safety, security, and robustness of such capabilities will be subject to testing and assurance within those use cases across their entire life cycle, including through established NATO and/or national certification procedures". (Ibid)

   Ensuring AI applications operate reliably, and understanding the limits of their reliable use, is fundamental to their applicability in military domains, the level of trust that will be placed in their results, and the effectiveness of their adoption.

E. "Governability: AI applications will be developed and used according to their intended functions and will allow for: appropriate human-machine interac-tion; the ability to detect and avoid unintended consequences; and the ability to take steps, such as disengagement or deactivation of systems, when such systems demonstrate unintended behavior". (Ibid)

   Ensuring that development, deployment, and maintenance of AI systems are carefully managed, monitored, and can be rolled-back if appropriate.

F. "Bias Mitigation: Proactive steps will be taken to minimise any unintended bias in the development and use of AI applications and in data sets". (Ibid)

   Bias within training data or AI results has been well-documented in many business and government applications, particularly where data and results are related to specific groups of people. To ensure fairness, such bias must be understood and addressed. In defense, as in most scenarios, it is recognized that data may exhibit other biases. For example when assessing damage, *survivor bias* is often exhibited, where data is only available from equipment that did not experience a critical failure, leading to datasets where information on critical failures is not available; or datasets where no anomaly was detected (Kok et al., 2020). Bias mitigation in military applications is necessary to protect against unintended consequences on operational effectiveness and beyond.

Of the three points identified above, the NATO AI strategy, with its requirement to "lead by example", in accordance with established values, norms, and international law may be considered to provide a clear vision of what responsible AI is. We now turn our attention to the development of responsible AI.

National Approaches to Responsible Military AI

NATO is not alone in setting out principles of responsible use of AI in military contexts. Many militaries, both inside and outside the NATO Alliance have established national approaches to ensure that any use of AI within their militaries is responsible. For example, the US Department of Defense has established a *responsible artificial intelligence strategy and implementation pathway* (US DoD, 2022) while The Netherlands has sought to bring a global focus to the challenge, instigating the REAIM conference. Many national approaches to responsible AI for military use rightly draw on broader activity regarding responsible AI. The next section considers such frameworks, their value, and limitations.

FRAMEWORKS FOR RESPONSIBLE AI DEVELOPMENT

While much is written about the ethics of AI, it is recognized that AI and digital technologies in general have no inherent ethics and no value system. Therefore a framework is necessary to capture good practice in AI development to guide developers and operators. Several frameworks have already been developed specifically for AI either internationally by bodies such as the ISO/IEC who provide an Overview of ethical and societal concerns of artificial intelligence (ISO, 2022), or OECD whose Recommendation from the council on artificial intelligence (OECD, 2019) sets out familiar principles for "responsible stewardship of trustworthy AI requiring human-centred values and fairness transparency and explainability robustness security and safety and accountability".

Nationally, frameworks have been developed, such as the Artificial Intelligence Risk Management Framework from NIST (NIST, 2023), locally (NSW, 2022), or by academia (Mäntymäki et al., 2022). Examples of how these frameworks can be applied, and the value they bring to practical AI products, can be found in OP (2022) and Solita (2022).

Many of these frameworks draw on more generic models or standards that have been developed to address ethical issues within generic technology-centric systems. Of these the IEEE's *Standard Model Process for Addressing Ethical Concerns during System Design* (IEEE-7000) is one of the most commonly cited. Although not developed specifically for AI systems, this standard provides practical steps to help address ethical concerns and risks during system design and can help align innovation management, system design, and software engineering methods to address these concerns and risks. While IEEE-7000 does address either the use or the full lifecycle of a system, crucially for AI developers it also does "not give specific guidance on the design of algorithms to apply ethical values such as fairness and privacy" (IEEE, 2021). Therefore while this standard provides useful guidance on the design of ethical systems it does not provide comprehensive guidance for AI developers.

The plethora of standards and frameworks – of which those above are a small subset – the broad scope they cover, and the breadth of both current AI regulation (EU, 2023) and that under discussion, do limit the practical assistance they can offer to those developing AI responsibly for the military domain. Goncharuk (2023) notes "... that among the experts who were trying to regulate AI, there were a lot of experts on ethics, human rights and privacy, and very few of those who practically understand how AI works...". Ensuring that the requisite mix of skills contributes to the development of AI regulations and standards, is as essential for AI as for any technology. Likewise, "those who practically understand how AI works" (Ibid) are needed to ensure that, wherever possible, high-level principles are translated into practical steps, development techniques, and tests that ensure that military AI is developed, and acts, responsibly.

REQUIREMENTS

The starting point for any military system is usually a statement of requirement. This is often the result of intense consultation, experimentation, and analysis. In most cases there is not a *requirement* for a military AI system, instead, users require a system that will be enhanced by the application of AI. Although in some cases, if quantities of data are so great, or analytical needs so complex, it may be that the requirements can only be realized through AI. Most national defense AI strategies, such as (US DoD, 2022), include a deep understanding and validation of requirements as a core element.

Identification and documentation of military requirements is a mature, well-documented process. Translating requirements to a solution that includes AI and must meet the criteria for responsible AI, is less mature. Such a mapping necessitates an increased understanding of AI at every level through that process, so that military decision-makers and requirements holders can make an informed assessment of the feasibility of AI's ability to meet a requirement.

Among techniques that can be used to bring a broad, holistic approach to assessing the potential of an AI solution, from feasibility to responsibility, is the data opportunity canvas (see Figure 5.1). This evolution from the widely used *business opportunity*

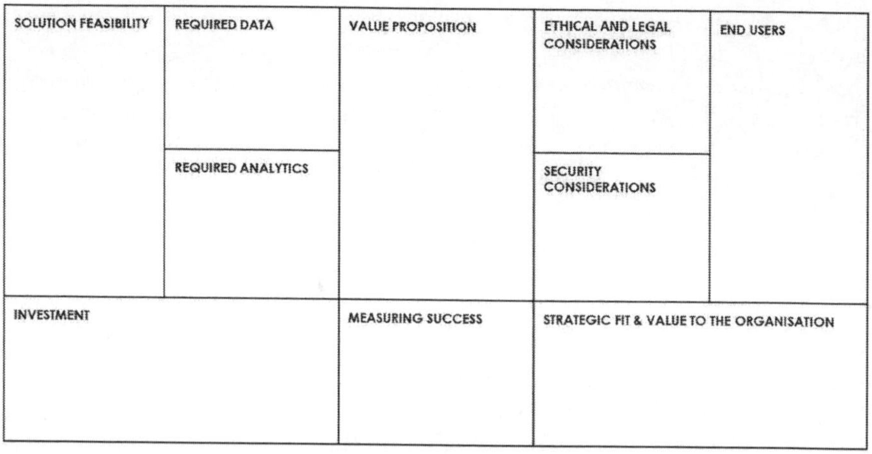

SOLUTION FEASIBILITY	REQUIRED DATA	VALUE PROPOSITION	ETHICAL AND LEGAL CONSIDERATIONS	END USERS
	REQUIRED ANALYTICS		SECURITY CONSIDERATIONS	
INVESTMENT		MEASURING SUCCESS	STRATEGIC FIT & VALUE TO THE ORGANISATION	

FIGURE 5.1   Data opportunity canvas.

canvas provides a useful framework to ensure stakeholders in military AI development consider all aspects of an AI solution and can ensure AI projects begin on a sound footing for all aspects, including those needed to ensure responsible use.

METHODOLOGIES

Developers of responsible non-military AI applications often employ methodologies to ensure that their approach considers the development and use of AI responsibly. For AI developers working within a military environment, a framework for responsible AI should encompass at least:

a. Responsibilities on AI developers and accountability of organizations creating or using military AI; including adherence to applicable rights, laws, and regulations. This will provide safety, security, and protection for developers

b. The role of AI developers in developing responsible AI; not only to develop trustworthy AI in a responsible manner, but also in showing users (and regulators) what is possible. Therefore AI developers play a role in educating military AI stakeholders, by explaining what is *under the hood* of their AI systems

c. Training and test data of sufficient quality. The quality of data collection methods should be checked as well as the quality of the resulting datasets. Data science approaches to quantify and to improve data quality are available and should be applied wherever possible

d. Training and test data that does not exhibit bias, or where the data does have a bias that the bias is understood, documented, and accounted for in the development, testing, and operation of the AI. Data bias relates not only to data on people, but to every type of data which is used to train, test, or be processed by the AI. Tools are available to detect bias in datasets from many providers, but such tools are not applicable to every type of dataset, nor is there any guarantee that these tools will be applicable to less common/less public datasets which are often encountered in military scenarios

e. Effort must be made to understand why an AI model performs the way it does, so models should be explainable or are as explainable as possible. This explainability applies, in different ways, to both the developers and the end users: the former should understand how the model has been trained and operates on the input data; the latter should have an understanding of how the outputs are generated and so appreciate the boundaries of their performance. This is often required in non-critical AI decision support tools, even where the output only affects benign outputs such as items for consideration by policymakers (Valiyev et al., 2020). It may appear paradoxical to explain AI techniques such as deep learning where deep neural networks benefit from hidden layers. However, the more explainable the behavior of the AI, the higher the human trust in the system and the more responsible it can be seen to be. Trust in a complex AI model will also be increased if the model can:

- Be decomposed into simpler steps whose behavior and performance can be described and/or visualized (and thus become less of a "black box")
- Show the key parameters that influence the AI output e.g. feature-group influence (Kok et al, 2020)
- Show the decision-making process in understandable/mathematical ways e.g. through techniques such as saliency maps to show why a decision is made on an image
- Demonstrate it is robust to changes in the input dataset, including active manipulation of input data e.g. image manipulation
- Show a defined review process for model development and results
- Quantify performance of AI models or AI systems, and the trade-off between the performance and explainability of varying models is understood
- Use established mechanisms for trusted vendors, [open] sources, and data providers
- Show established processes to monitor and maintain the AI solution, to prevent deterioration in performance due to drift in the data, target, or concept; all of which could affect the output and remain responsible

RISK ASSESSMENT

AI has the potential to enhance a wide range of situations to assist military commanders, decision-makers, and users. For example, the risk from an AI tool that uses open-source information to provide input for consideration by a human is different from the potential risk from an AI tool for missile defense which activates a physical response. The actions needed to ensure these very different AI systems operate responsibly vary. This leads us to the need to assess the risk of a military AI system and to then adopt appropriate steps to ensure responsible development, based on the risk assessment.

AI applications are powered by models, mathematical representations, or abstractions of a system, process, concept, or object. In the context of ML and AI, a model is a mathematical or computational representation created by training an algorithm on a set of data. The training process works by capturing patterns, relationships, or rules from the data itself, without human intervention. Such a trained model is then used to make predictions on new, unseen data, based on the patterns and relationships learned during the training processes.

Generally speaking, models are approximations of real-world phenomena: their capacity to approximate the task they are designed to achieve heavily depends on their level of adaptability to the data, the amount, variety, and quality of the data used for training, as well as the main scope they serve.

Due to their statistical approximative nature, AI/ML models might have some shortfalls, which might lead to "wrong" predictions. This can be due to many different reasons, the most common being:

- Models that are too complex for the dataset and task for which they are trained; these can end up fitting the training data too closely and so fail to generalize and perform well on actual data – this is also known as overfitting
- On the other side of the spectrum, models might be too simplistic to capture the complexity of the relationships in the data, leading to underfitting
- Models trained on datasets that are not fit for purpose: this might mean that the overall quality of the data is not sufficient or that the data itself might contain biases that contaminate the model
- The evolution over time of the data, where distributions change (data drift) or the underlying relationships in the data change (concept drift), making previously learned patterns outdated

Inaccurate predictions might have different unwanted effects and different levels of impact, depending on the underlying application end goal, what impact it has on the overall operation it serves, how secure it is etc.

AI Model Behavior

Generally speaking, AI models are trained to achieve a good level of generalization on a specific task: however, it is impossible to have a 100% success rate. Unplanned outcomes are expected to happen, and their effects on the application they serve need to be understood. Besides occasional inaccurate outcomes, AI models usually contain a certain degree of information of the data they were trained on, and might have intrinsic weaknesses that could potentially be exploited by an adversary.

Achieving state-of-the-art performance for AI often requires models that have a high level of abstraction and complexity, which makes their inner workings very hard to understand for human beings: often this is referred to as a model being a "black-box".

The black-box nature of the model makes the anticipation of unwanted AI-model behaviors hard to anticipate: therefore, Data Scientists and Subject Matter Experts need to perform careful validation processes, sometimes including stress-test scenarios that need to be carefully designed and executed.

The validation processes are rarely an easy task (and some specific steps might even be unfeasible). This usually requires large investments in terms of time and effort. This has led to the development of guidance to help AI model developers, reviewers, and other stakeholders to identify and mitigate the risk of errors that might occur in the development and application of AI in real-life use cases, providing some best practices and guidelines.

In general, the more thorough developers can be in assessing, testing, and explaining each step taken in the development of their solution, the less chance of unplanned effects with unplanned impact occurring.

Risk Tiers

Achieving efficiency and balancing the time, effort, and cost to develop responsible AI starts by assessing the actual risk that the AI application might carry. A framework that helps the AI development team and other stakeholders achieve this balance can categorize the AI model (and the application) into a set of *risk tiers*. Based on the risk tier, specific validation and assurance steps and the level of scrutiny of this step are evaluated by the AI development team.

The risk tier can be estimated using multiple dimensions: the obvious one is the actual risk that the AI model and application carry, i.e., what happens when the AI is wrong? However, other aspects might be considered, like the actual impact the AI has on the organization and its day-to-day core business. AI systems that improve some non-core businesses may require a lower level of scrutiny compared to crucial, core applications. In a military organization, core functions would include operational activities, while non-core may relate to functions such as interactions with suppliers etc.

Common risks that a military organization may face when deploying AI are the following:

- Operational Risk: the actual risk on operations that an undetected AI failure might have
- Security Risk: the additional vulnerabilities that an AI system might introduce and which could be exploited
- Interoperability Risk: AI systems most likely do not operate on its own, but interoperate with other systems. The interoperability risk refers to the risk of an AI failure propagating into other systems
- Reputational Risk: refers to the effects of AI failures on the reputation of the organization in the public eye
- Financial Risk: refers to a financially quantifiable loss due to an AI failure
- Ethical Risk refers to the impact of AI failures on the values and principles of the organization

- Legal/Compliance Risk: refers to the effect of AI failures on national and international legal frameworks and requirements, including those applied on the battlefield such as the law of armed conflict
- Physical Risk: the negative effect that an AI system might have on civilians, personnel, equipment, or infrastructure
- Environmental Risk: specific risk to the environment occurring due to AI adoption and failures

SECURITY RISK FRAMEWORKS

Some of the areas listed above are already addressed by frameworks that have been developed specifically for this area. For example, adversarial threats are an acknowledged risk area, not only for AI but also for IT systems in general. The adversarial threat landscape for artificial-intelligence systems (ATLAS™), developed by the MITRE organization (MITRE, 2023a) is an example of such a framework and is itself based on the widely used ATT&CK™ framework (MITRE, 2023b) which was developed for identifying cyber threats. ATLAS provides a body of knowledge of adversary tactics, techniques, and examples of threats to ML systems which have been drawn from a variety of sources. Such frameworks are valuable tools for AI developers and provide detailed guidance within these specific areas.

To ensure a holistic approach to responsible AI, those involved in the development and use of responsible AI can utilize such specific frameworks to address issues such as security risks. But this should be done within a broader framework that ensures all the attributes of a responsible AI system are considered, in a manner consistent with the risk assessment, and can be demonstrated to have been addressed. Ensuring that these steps occur requires action by AI developers and a number of additional stakeholders within the military AI development and use ecosystem.

ROLES AND RESPONSIBILITIES

A clear definition of roles (and their responsibilities) helps provide clarity on the responsibilities of each stakeholder within the team of AI developers and users. The most common roles in an AI development team are listed in the following non-exhaustive list:

- Business owner/Commander: the persons that have a business or operational problem or need that can be solved by an AI system. The main responsibilities are ensuring that the business requirements are clearly defined and that the subject matter expertise is brought in the discussions around responsible AI development and application

- AI Application User: refers to the end users of the applications
- AI Product/Model Owner: person responsible for the AI model, its adherence to the defined use case, its performance, and general adherence to the principles of responsible use
- AI Developer: can be one or more people, usually data scientists, machine learning engineers, or similar roles, whose responsibility is the model development and the implementation of the checks agreed for the AI application scope
- Reviewer: technical peers who can review the different development steps. Depending on the specific checks, this can be an AI developer or data scientist from the team, or in specific cases (most commonly for higher risk tiers), an independent party with equivalent skills who can cross-check the development steps and decisions, as well as perform an independent verification and validation
- System/Service/Product owner: responsible for the overall implementation of the model and integration within existing systems, processes, or products

Depending on the nature of the application, different roles might exist and complement those above. It is not uncommon that some roles overlap in the same person, especially in lower Risk Tier applications.

RESPONSIBLE AI DEVELOPMENT AREAS

Adhering to the principles of responsible use starts with responsible identification and development of AI solutions. Cognizance of the principles and their mapping to specific steps informs every stage of the process. During ideation and experimentation requirements can translate into constraints on modeling approaches. During development, explainability approaches and resulting explanations are applied and assessed according to the user requirements. In deployment the model's predictions and the quality of explanations etc. are monitored, assessed, and updated where necessary.

Ensuring that known potential issues are checked, quantified, and resolved in the development phase can be achieved by assessing the solution across different dimensions. The main dimensions considered in the context of this framework are listed below:

- Data assurance steps: as the main enabler of AI, data is valuable but also presents the main source of risk. Datasets can incorporate various types of biases, can have different quality and within the NATO context, come at different Security Classifications. Data-related mitigation steps are meant to assess all those potential sources of errors
- Model assurance steps: at the core of AI software, AI models need to be validated and have well-defined steps to assess their quality and compliance with the principles of responsible use. Those steps concern the evaluation of model

performance, overfitting, quality of prediction, assessment of performance on sub-segments of the datasets

- Explainability: the most recent developments in AI require large and complex models, emphasizing the "black-boxness" issue. In general terms, the interpretability of the model is inversely proportional to its complexity. Explainability of outcomes for an AI model (the answer to the question: why did it produce a given output?) is usually based on frameworks that approximate the inner workings of models
- Fairness: as one of the main principles of responsible use, ensuring fair results plays a crucial role in AI applications that operate on data that has subgroups or characteristics that can be associated with groups or individuals. The ability to correctly evaluate the fairness of a model is likely to come at the cost of privacy
- System integration: an AI model normally fits in a larger system, which needs to operate efficiently, securely, safely, and in an understandable and usable way for the end-user
- Way of working: this is a dimension that considers the overall organization of the AI development process, that adopts best practices to ensure a given level of quality of the developed AI solution
- Audit: interconnected with the way of working, preserving an audit trail ensures the ability to trace back decisions in model development

Assessing Responsible AI Development

Assessing the AI solution across the dimensions described above can be broken down into specific checks, whose goal is to perform different elements of assurance, using a framework that offers a checklist of assurance steps. In order to achieve a pragmatic balance between assurance of each development step and the resources needed to assure them, the risk tier of the AI application can be used to guide whether each step is required, recommended, or not necessary.

The very varied nature of AI applications, AI models, and datasets makes it impossible to have a detailed and generic set of checks that can hold in every situation: therefore, the proposed checklist is meant to be used as a guideline for AI developers, AI model owners, data owners, and other stakeholders to define a pragmatic, appropriate set of steps and techniques can be adopted to ensure responsible AI development.

The aim of those assessment checks is to identify potential problems early, in order to solve them and/or provide a way to monitor aspects of interest once the model is deployed. However, while this approach will mitigate the risk of irresponsible development, it should not be assumed that simply following such a process will guarantee that every issue will be discovered and mitigated/resolved.

This AI development framework has been extended with a methodology to assess and categorize implementation and deployment risk. The framework then provides checks and tests throughout the development process which ensure that the principles of responsible use are translated into tangible, provable steps.

Analysis of the particular problem and AI development may identify that some steps in the framework might not apply to a specific problem, or might not be feasible. In those cases, the rationale should be documented and the step can be skipped, where there is a requirement to perform a check. It is the responsibility of the development team to assess on the use-case-by-use-case basis which steps are necessary.

CONCLUSION

The value of a responsible approach to AI for military use is well-understood. Far from being simply a desirable concept, responsible AI which adheres to principles of responsible use is essential in order to provide trust to military staff who use it and to the civilians it protects. The approaches described above outline techniques to introduce appropriate safeguards, checks, and tests to military AI development. More importantly, they provide practical guidance to those developing, testing, deploying, and using AI solutions in the most critical of environments.

ACKNOWLEDGMENTS

The authors wish to thank all their colleagues from the NCI Agency's Data Science & AI team and across NATO for their tremendous contribution to this work, together with a number of colleagues in national defense and industry, including (but not limited to) Microsoft and Solita, without whom this work would not have been possible.

The views and guidance provided by the authors are personal and not formally presented on behalf of the NCI Agency or of NATO.

REFERENCES

EU. (2023). *Artificial intelligence act.* European Union.

Goncharuk, V. (2023). *The war in Ukraine and AI regulation: Some controversial takeaways.* https://futurium.ec.europa.eu/en/european-ai-alliance/blog/war-ukraine-and-ai-regulation-some-controversial-takeaways

IEEE. (2021). *IEEE-7000- standard model process for addressing ethical concerns during system design.* Institute of Electrical and Electronic Engineers.

ISO. (2022). *ISO/IEC TR 24368:2022, Information technology - Artificial intelligence - Overview of ethical and societal concerns.* International Standards Organization.

Kok, A., Ilic Mestric, I. Valiyev, G. & Street, M. (2020). Cyber threat prediction with machine learning. *Information & Security: An International Journal, 47*(2), 203–220.

Mäntymäki, M, Minkkinen, M., Birkstedt, T., & Viljanen, M. (2022). Putting AI ethics into practice: The hourglass model of organizational AI Governance. *arXiv:2206.00335*

MITRE. (2023a). The ATLAS framework, https://altas.mitre.org

MITRE. (2023b). The ATT@CK framework, https://attack.mitre.org

NATO. (2021). *Summary of the NATO artificial intelligence strategy.* North Atlantic Treaty Organization. https://www.nato.int/cps/en/natohq/official_texts_187617.htm

NIST. (2023) *Artificial intelligence risk management framework (AI RMF 1.0).* National Institute for Standards and Technology https://doi.org/10.6028/NIST.AI.100-1.

NSW. (2022). *New South Wales AI assurance framework.* NSW AI Assurance Framework.

OECD. (2019). *Recommendation on AI.* Organization for Economic Cooperation and Development.

OP. (2022). *My financial balance: An artificial intelligence transparency report*, OP Financial Group. https://www.op.fi/documents/20556/39794367/OP-Financial-Group-AI-transparency-rep[%E2%80%A6]-financial-balance-2022.pdf/1968bdaa-68e1-1675-2903-4 2f88dff14d4

Solita. (2022). *Four strategies for building trust in your AI systems.* Solita. https://www.solita.fi/en/blogs/four-strategies-for-building-trust-in-your-ai-systems/

Valiyev, G., Piraino, M., Kok, A., Street, M., Ilic Mestric, I., & Retzius, B. (2020). Initial exploitation of natural language processing techniques on NATO strategy and policies. *Information & Security: An International Journal, 47*(2), 187–202.

US DoD. (2022). *US Department of Defense responsible artificial intelligence strategy and implementation pathway.* United States Department of Defense. https://www.ai.mil/docs/RAI_Strategy_and_Implementation_Pathway_6-21-22.pdf

Unreliable AIs for the Military

6

Guillaume Gadek

INTRODUCTION

Military use of AI systems brings promises for massive, fast, efficient, and/or reliable decision-making with the aim of providing defense and security in our countries. However, this use seems to be a trade-off between three high-level aspects: technical robustness, operational criticality, and lawfulness. Their combination is one of the definitions of the emerging topic of AI trustworthiness: a system that does not cover all the aspects is not desirable (e.g., a legal but inefficient system or an efficient but irrelevant system would be useless). A number of institutions are proposing frameworks, evaluation methods, and approaches in order to cover in a holistic manner all of the aspects of AI development and deployment (EASA, 2023; HLEG AI, 2019; HLEG AI, 2020; NIST, 2022; NOREA, 2021).

Evaluation of the trustworthiness of an AI includes a part inherent to the AI model, concerning the model architecture, the distribution of its attention, the data it has been trained on, and the explanation it provides about the decision taken. It also includes a part that is foreign to the AI model: its context of use, which concerns the legitimacy, proportionality, lawfulness, intent, governance, maintenance, and education of users and stakeholders. All aspects must be covered, even if all aspects are not fully solvable: as an example, the illegal use of a weapon system is already forbidden. This aspect is not specific to AI systems, but the organization around its deployment must minimize potential wrongdoings. A similar trade-off also exists on the technical side, as artificial intelligence systems are usually quite stochastic: an AI does not always provide the correct answer, nor do their human counterparts.

Indeed, many AI systems, notably when dealing with language, are not fully reliable (e.g., precision at 90%). Whether 90% precision is acceptable or not, it is already accepted in a large number of settings today, in all domains from health to entertainment,

DOI: 10.1201/9781003410379-7 101

including politics and economics. Of course, 90% accuracy is not relevant for all applications: plenty of tasks happen to be critical and require higher levels of quality; another remedy is to transform an AI decision system into a suggestion system, where the incorrect answer will most likely be modified by the human in control.

Finally, AI makes it possible to provide answers at an unprecedented scale: as an example, it becomes technically possible to automatically count *all* the vehicles in a country through satellite image analysis or to analyze *all* social media posts of a given day. While this kind of information may be of great use for military intelligence, it falls into the ethical problem of massive surveillance and must comply with the prerequisite of lawfulness.

In this chapter, I describe three use cases of realistic yet fictitious military AIs (AI in a kill chain; low-criticality AI for intelligence image analysis; human-in-the-loop AI for HQ staff) to raise awareness and possible feasibility, ethics concerns, and responsibility/accountability of using such systems in the future. I hope this chapter will encourage AI practitioners to perform exhaustive assessments of their work, resulting in an overall increase in the deployed AIs quality.

TRUSTWORTHINESS, RELIABILITY, DATA SCIENTISTS, AND ENGINEERS

Artificial intelligence approaches deeply modify the paradigm of IT-supported tools in a number of domains, including the military. Because AI systems often address previously unsolved tasks, the promise is attractive to rely on them, typically for decision-making, even if this is broadly recognized as a "techno-push" trend with overlooked inconveniences (Van den Bosch & Bronkhorst, 2018). Such systems are more often than not marketed as faster and better than their "legacy" counterparts. Plenty of definitions of AI have been provided in the past (HLEG AI, 2019); in this chapter, I focus mainly on machine learning and deep learning approaches.

Evaluating Trustworthiness

In any case, AI systems are not perfect. Totally grounded in statistics, they require curated sets of data to be trained and to be evaluated. Let us focus on classification tasks, where an algorithm has to attribute a label to some input data. Some examples of these tasks include "friend or foe" from sensors; an emotion tag for a text (Danisman & Alpkocak, 2008); "cats or dog" on a picture (Zaidi et al., 2022). Given that a dataset is provided, data scientists or AI architects split the data records into a "training" set and a "test" set. The classifier will learn the specifics of its task on the training samples, it will never learn from the test set. The quality metrics will be computed on the ability of the AI classifier to correctly tag the samples in the test set. One can deduce the importance of having a correctly balanced and fully representative test dataset, in order to produce trustworthy performance scores.

In the last few years, the focus has evolved from performance scores toward a holistic "trustworthiness" approach, to include accuracy, robustness, accountability, privacy, security, fairness, control of bias, and explicability (Kaur, Uslu & Durresi, 2021). Indeed, good performance scores are useful to raise awareness about AI and to realize proof of concepts; however, operational systems have to cover all the gaps. In another list of desired dimensions, Stanley-Lockman and Christie (2021) highlight lawfulness, responsibility and accountability, explainability and traceability, reliability, governability, and finally, bias mitigation. While a fair part of these dimensions relies on the AI designer (reliability, explainability, and bias mitigation), most of these are dependent on the application (lawfulness, responsibility, accountability, traceability, and governability) and aim to describe a "responsible use" (Stanley-Lockman & Christie, 2021).

AI models may be inherently biased, but even "perfect" AI models can be misused, sometimes genuinely, or unintentionally. A solution is to build awareness about AI governance, highlighting issues about the integration of AI into a process and extending the current technical training and evaluation practices to include non-technical aspects.

Non-Technical Aspects of Trustworthiness Evaluation

Lawfulness deals with the relation between the system, its application, and the legal frameworks active in its area. Having lawful systems is surprisingly hard to obtain as there often is a gap between technical possibility and zealous application of offline laws, including when they contain extraterritoriality. Typically, the possibility to recognize objects in a picture is confronted with legal problems when said "objects" are people. European GDPR (general data protection rules) apply if the person resides or has EU citizenship. National rules may be heavier/stricter depending on circumstances (e.g., if the person is minor of age). The technology used may have been assessed as "dual use" from an arms export control perspective such as ITAR (International Traffic in Arms Regulations) and requires increased care in its deployment. As a last legal example, the training dataset may present copyright issues (e.g., if its license agreement mentions "for research use only", or if each of the authors of the pictures has not yet transmitted their explicit authorization for use).

For military purpose systems, these rules are extended with the usual International Human Rights and Law of Armed Conflicts texts. As a comparative example, a business may typically argue that it is okay to maintain an extensive contact list of potential customers outside of their home countries and thus store a list of personal data for one billion people. However, it does not seem *proportionate* that a military force maintains a list of all the residents in its area of operation. Doing so may potentially be counterproductive as it also brings arguments to their adversaries and may ultimately affect the legitimacy of the operation in itself.

However, ethical topics evolve quite rapidly and it is difficult to structure them into laws. Moreover, ethics intrinsically embed a personal appreciation part, which is non-trivial to be addressed to by a central "ethics" service. Last but not least, an ethics assessment also brings a concrete list of potential improvements during a technical project that helps the team *make sense* of their daily activities. In this dimension, using an ethics framework has the same potential as using *git* to handle the evolution of a code base: it is not a legal requirement, but most people do it as part of commonly shared best practices.

The technical teams are already accustomed to evaluate the quality of their systems. A large part of the data scientists work is to optimize for quality (e.g., accuracy, F1-score, or error rate); a shift is currently occurring to rather obtain trustworthy AIs (Mariani et al., 2023). Indeed, all errors are not equally important, and trustworthy AIs require defining what *fairness* means for the targeted application, and what are the risks of harm in the system environment (Schmid et al., 2021).

Evaluating an AI system is nothing trivial as such systems span across various disciplines and domains. Fortunately, assessment frameworks already enable the teams to describe and identify the qualities and opportunities for improvement in their systems and applications. Indeed, I believe that a number of defects or unconformities can be sooner identified, worked upon, and sometimes be solved by performing such an assessment at early stages of a system development.

Using ALTAI as a Self-Assessment Tool

A practical way to increase awareness among the technical teams is to let them perform a self-assessment; in a European context, the go-to reference would be the Assessment List for Trustworthy Artificial Intelligence (ALTAI), published by the EU High-level Expert Group on Artificial Intelligence (HLEG, 2019), which had a lot of influence on the elaboration of the upcoming AI Act.

In a first step, this framework provides questions that bear both on the realization approach and on the targeted deployment, easing the discovery of characteristics yet to improve. In the second step, ALTAI provides a description of the AI maturity accompanied by a list of recommendations and topics to explore. ALTAI is structured along seven pillars, with the assessment resulting in a score ranging from 0 (totally untrustworthy) to 5 (topic fully addressed). Other frameworks exist, which may better address specific issues, such as business exploitation and AI governance within a company (NOREA, 2021).

Current technical evaluation approaches already cover an extensive list of aspects; they are only exhaustive up to a point. Fairness is covered by an extensive list of questions, reflecting our societies' preoccupations of today; energy consumption – opening to a similar complex topic, GreenIT (Deng & Ji, 2015) – is quickly mentioned in most frameworks with no clear way to enable a choice between a gain in quality versus a loss in energy costs. Old topics such as computing time are no longer apparent at the first level. New topics will most likely emerge, such as ethics in data labeling (Perrigo, 2023). The frameworks must and will continue to evolve in the future to maintain their extensive coverage and alignment with our level of exigence as a society.

DIVING INTO APPLICATIONS

General ideas often become more clear when worked out in examples. In this section I detail three military applications of AI. These applications are often already proposed by various competing providers in a number of countries. Each application implies differences in the risk level, the quality of the algorithm, and the adequacy of operating practices.

First Application: AI in a Kill Chain

In this section, we first describe some general applications of AI in a kill chain, then instantiate a concrete – yet fictional – example and debate about the roles and quality level of AI governance, highlighting possible misunderstandings in what "characterizing an AI system" means.

Introducing the "kill chain"

This first application targets a typical LAWS (lethal autonomous weapons system) setup: an artificial intelligence embedded in a kill chain. A typical kill chain such as the F2T2EA model contains the following steps:

- Find: identify a target
- Fix: obtain precise coordinates
- Track: maintain real-time updates of the coordinates of the target
- Target: choose a weapon to engage the target
- Engage: use the weapon
- Assess: evaluate the effect (directly linked with "battle damage assessment")

In a weapon system such as the FCAS, the F35, or any other weapon system for air supremacy, artificial intelligences are already relevant for almost every step (Azzano et al., 2021). Sensors and data fusion systems execute the 'FIND', 'FIX', and 'TRACK' steps. Scheduling tools give tasks to the sensors; their final outputs feed the (automated) processing and aggregation steps by ISR systems (Jones, Kress, Newmeyer Jr & Rahman, 2020).

The choice to engage a target highly depends on the operational context. In a homeland security or low-intensity mission, such a decision is typically taken at a high level; it is commonly accepted that this decision may be taken at a lower echelon in a high-intensity conflict, as the rules of engagement would most likely cover this situation. During engagement, "smart bombs" may embed artificial intelligence systems to continue the targeting until reaching the target. Finally, battle damage assessment may also benefit from AI: identifying and counting intact and damaged ground vehicles from an aerial picture is a common computer task.

Instantiating the kill chain example: The patrolling quadcopters

The following example is purely fictitious. A NATO OF2 (e.g., a captain) is the Protection Officer in charge of securing a division-level HQ located at 2 km of a village; the division itself is not operating in their homeland, but is deployed in a third-party country, in the frame of a common defense agreement. The mission of the officer is to guarantee the security of the surroundings against potential thefts, spies, or sabotage attempts. Their subordinates are one platoon and two drones: small quadcopters that can either be piloted from afar or fly in autonomous mode. Their payload includes HD cameras as well as two grenades that can be triggered and dropped.

The drone manufacturer provides standard autonomous flight methods: take-off, landing, and directional flight. These capacities are mostly non-AI and work as designed under normal environmental conditions (wind speed, bursts). The drone integrator improved the cameras and computing power and added the weapons aboard. They also enabled an "AI-enabled autonomous surveillance capability" in which the operator may authorize, through a validation workflow including the direct hierarchy, a "patrol mode" (the drone sends an alert if somebody is spotted in an area) or an "interdiction mode" (the drone attacks whatever is spotted in an area), both modes heavily relying on Artificial Intelligence models.

Genesis of an AI system

Systems are frequently benefiting and extending previous work, sometimes only with tenuous connections to the initial aimed application. Moreover, AI systems are almost always based on transfer learning, starting from an already learned model. In this scenario dealing with the image modality, we could perfectly imagine that the interdiction mode would have been first designed by university researchers as a "detect and act" reaction for a firefighter drone – detecting and extinguishing a fire in a forest. This part of the code could embed a foundational model such as YOLOv5 instantiation (Ge et al., 2021): pre-trained on large public datasets, it would then be fine-tuned on a limited set of forest fire images. This example aims to clarify that the first intended usage may not align well with the final business, and by extension, it is not rational to expect the AI architect to bear all responsibility for any misuse of their system.

This argument is further reinforced by the number of actors in the development of such a system. Typically, a start-up company would robustify and commercialize a first iteration of the solution; a defense industry player would continue the integration of all modules into an operational platform (e.g., merging the hardware, the AI software, and the classic software for quadcopter control) during a Ministry-of-Defense-funded project.

Industrial efforts to deliver a system

While an effort is always made to track licenses (such as open source software) and export control compliance, in this fictitious scenario we will imagine that the evaluation part of this precise "object detection" algorithm was not exhaustively performed, that is, the Quality Test team did not proceed to the elaboration of a new dataset dedicated to test it. All on-field tests would be performed on the same airfield nearby, and the drone certified "fit for service"; before being deployed into the forces.

The future operators would be trained on all mandatory topics before being deployed to the mission, as all deployed staff, during a two-week period. They also had a two-week training to pilot and manage the quadcopter, six months before. The user manual clearly prescribes to operate the drone over populated areas; the rules of engagement forbid firing on civilians; the force protection policy requires a 1 km no man's land around division-level camps.

Open questions
Without fully instantiating the story, one can imagine plenty of worst-case scenarios: did the officer manage to maintain awareness during the training? Is the human being able to remember the subtleties of a technical tool, while they are deployed on foreign soil, under pressure? Would the quadcopter be able to correctly identify people in this new country, while only trained and validated on gray-or-black clothing standing in the middle of an airfield? Would the officer never activate the fully autonomous mode? Even after months-long lack of sleep and week-long *nothing to report* from the UAV?

This opens another range of questions: should the AI developers have an opinion about the deployment of their algorithm in quadcopters? Should the procurement service of the Ministry of Defense have required an explainability feature? What is the share of responsibility born by the makers and buyers, would the system fail the expectations of its user?

Those are relevant questions; however, it is impractical to answer these questions one by one. A recommended method consists of relying on an AI assessment tool to highlight gaps between the realization and an ideal implementation. We continue with such an assessment.

Results and recommendations from ALTAI

Imagining myself as a Quality Engineer working at the industry company, I perform an assessment of the AI system, using the ALTAI framework and its online tool. Note that most of the questions bear on the realization/development steps; some of them would still be relevant to be answered by the operational team, responsible for the AI-powered capability. For each of the seven pillars, the assessment results in one score, ranging from 0 (unassessed; very poor) to 5 (totally satisfying). Depending on the answers, the framework also provides recommendations that are distributed along the seven pillars; here I focus only on the subset of the recommendations that point to the most critical aspects of this specific story, omitting less salient pillars.

Technical robustness and safety, scored 1.5 out of 5
During the assessment, I described the presence of an extensive number of procedures that describe quality measurement and safety risks. I highlighted the difficulties of measuring *a priori* tagging errors on yet-unseen objects and to ensure reproducibility when the system directly interacts with reality through its own sensors. The note is further deteriorated by the inability of the system to transmit in an understandable way to the

operator the whole flow of data describing the perceived reality. ALTAI then recommends the following:

- Align the reliability/testing requirements to the appropriate levels of stability and reliability. Develop a mechanism to evaluate when the AI system has been changed enough to merit a new review of its technical robustness and safety
- Put in place processes to ensure that the level of accuracy of the AI system to be expected by end-users and/or subjects is properly communicated. Clearly document and operationalize processes for the testing and verification of the reliability and reproducibility of the AI system

Interactions with reality are hard to plan exhaustively. A lot of work about autonomous vehicles, even with restrictions on the type of environments (such as limiting autonomous cars to highways), still results in surprises (Parekh et al., 2022). No matter what effort was put into the training datasets, there will be unknown entities, misrecognized by the system. Here the recommendation covers both the autonomy in operations and the autonomy in training (e.g., continuous learning). An elegant manner to improve the quality of the system relies on active learning: most of the time, the human operator would remain the full master of the UAV, validating or refusing the output of the AI detector. The full-autonomy mode would only be activated in exceptional circumstances, once the operator trusts the system.

Transparency, scored 4 out of 5
The proposed transparency procedures include interpretation of the AI decision (confidence scores, highlighting bounding boxes in the image), extensive human training, and quality metrics based on reference datasets. ALTAI suggests to add the following procedure to complete the coverage on this dimension:

- Consider continuously surveying the users to ask them whether they understand the decision(s) of the AI system

In defense setups, it is not trivial to gather user feedback. Time is short during operations. User feedback *post hoc* is still relevant but sometimes gets diluted and softened. The recommendation here suggests independently assessing whether the output was relevant or not and whether the output was expected or not. This information could be included in the activity logs, associating the AI decisions and the real-time feedback of the user; the exploitation of the feedback would only occur once the data is transferred back "home".

There is a risk that the human operator would most likely not generate such feedback in the long run (e.g., write a paragraph describing their satisfaction after each use of the drone) if this activity is not clearly part of their mission. If impractically provided, this technological requirement of transparency would result here in an organizational burden, the captain requiring their subordinates to always provide feedback about the AI system.

Diversity, non-discrimination, and fairness, scored 4 out of 5
I described a system where no specific focus was made on these aspects. The fairness here documents a potential class balancing problem during the image classification step, which is already worked on in the technical robustness dimension. ALTAI suggests the following:

- Put in place educational and awareness initiatives to help AI designers and AI developers be more aware of the possible bias they can inject in designing and developing the AI system
- Depending on the use case, ensure a mechanism that allows for the flagging of issues related to bias, discrimination, or poor performance of the AI system
- You should assess whether the AI system's user interface is usable by those with special needs or disabilities or those at risk of exclusion
- You should assess the risk of possible unfairness in the system toward the end-user's or subject's communities

In my story, the AI system was first developed with a safety goal (dealing with forest fires) in mind. The adaptation to the detection of "objects" ideally includes an exhaustive list of representative objects that the AI will meet in the future (bias in data). A second bias is likely to be injected through the user interface, displaying only what is believed by the developers to be relevant. Such an interface must avoid adding bias; take into account possible human limitations (e.g., color blindness), and still be usable in high-stress, low-cognition situations.

    I interpret the last proposition as insisting on data bias, and recommending the investigation of the categorization of the recognized objects: does the system better recognize a category of people? Of vehicles? Of behaviors? Does this mean it may fail to recognize critical types of potential targets? Fairness is definitely an excellent approach to improve the system quality overall.

Accountability, scored 5 out of 5
I described a setup where the system has the technical features to enable accountability: it is fully auditable, including by third parties. It is provided with the applicable legal framework. It also contains a procedure to let any stakeholder report potential vulnerabilities or new risks. On top of these technical traits, ALTAI recommends:

- A useful non-technical method to ensure the implementation of trustworthy AI is to include various stakeholders, e.g., assembled in an "ethical review board" to monitor and assist the development process

Such a board is likely to happen during the earliest phases of product development, as high-level meetings at the customer facilities. Afterward, these boards must continue to happen even if the provider company is somewhat disengaged. The customer must plan for the accountability problems that would occur much later, during operations. In the story, most if not all responsibility lay on the captain using the drone. The system is not universally reliable, but has been sufficiently tested to be accepted.

Humans in the automatic kill chain

The system I described here intentionally resembles more a prototype than a certified autonomous aircraft in order to highlight improvement opportunities in such a system. Note that the story does not end at system delivery: once developed, the AI must be maintained, audited, and most likely improved on more operationally representative data.

The automation of flying firing capabilities is often thought of as reducing the number and quality of the people in and around the system. In this example however, more training is required; system audit and control must be managed; user feedback must be created, collected, and exploited.

Relying on an AI assessment tool helps to document and identify the limitations in the development of the system. In a second step, it feeds the decision of whether the limitations are showstoppers or opportunities for improvement of the system.

Second Application: Low-Criticality AI for Satellite Image Analysis

Information from above is key to measure the power and setup of potential adversaries all over the world. The 1962 Cuban crisis was confirmed by air (Allison, 1971); the 2003 Iraq war began with aerial pictures of a (non-existent) threat (Diaz, 2005). Today, satellites provide high-quality pictures with resolutions under 50 cm and with frequent revisit rates. However, human analysts usually spend a long time studying these extra-wide, extra-long images, looking for something that may be absent, hidden, or obfuscated. Indeed, it is common to have indoors sensitive activities in the military. However, the level of activity is easy to estimate thanks to the number of individual vehicles at the entrance of a military base.

Computer vision provides amazing detection capabilities; on satellite images, it can be applied for the environment or for agriculture, counting oil palm trees. For a simple task such as "counting vehicles", it may reach "precision rate higher than 85% with a recall rate also high [at] 76.4%" (Froidevaux et al., 2020). This means that among what the system tags as detected cars, only 15% are not cars. In the meantime, this detector "only" misses 23.6% of the real vehicles. Indeed mistakes may come from a large number of factors: cars under a tree, reflections, paintings, orientation of the car, position of the satellite, and even errors from the reference data itself. The error rate is actually considered very low for such a difficult task; in any case there is no guarantee that humans would systematically perform better.

An AI detector would bring a lot of added value, as it automatically performs tedious and repetitive *complex* tasks: it can count the vehicles present in large bounding boxes. An intelligence analyst may first define bounding boxes (also known as "areas of interest") around military bases, and even around second-rank relevant areas (neighboring cities, border cities, harbors…) to build up a global view of the activity in a large number of areas. Each new picture updates the current scores and enables the system to quickly detect areas of real interest such as a military harbor where a lot of activity is seen.

The implementation of such a system requires at the very least two technical profiles: a data scientist, to develop and instantiate an AI model – provided there *is* a dataset labeled with the objects of interest to detect – and an integrator, capable of designing a system that will rely on the AI when relevant. Indeed, not all the pictures are fit for the AI: the operational interest may differ between day and night, the images may not be informative when it is cloudy, etc.

The construction of an image dataset for an object detection task is not trivial. An annotation guide must be defined (and often iterated upon) in order to explicitly describe what is to be annotated on a picture; whether a large square around the object is sufficient or a pixel-precise shape is required; how to react when occlusion results in a fair doubt for the human observer. Finally, each picture has to be labeled by different individuals in order to compute a quality score for the annotation (inter-annotator agreement). This kind of work is often outsourced, producing its share of ethical and legal issues: exporting aerial pictures of military bases is not trivial in terms of regulation and compliance.

The annotation task presupposes the existence of a large, representative set of aerial images. Even if this object detection task does not imply discrimination between persons, the principles and questions of fairness apply here: what kinds of contexts are present in the dataset? Do the pictures represent all kinds of weather, physical environments, seasons, vehicles? Only recently has this concern gained in importance and realizability for numerical datasets; the ability to measure the biases on "unstructured" data such as texts or pictures usually implies lists of categories, which may in turn include some bias. As a side note, dataset creation must be part of the AI lifecycle: in this object detection task, the target objects remain consistent but also evolve over time. Cars have been here for decades but their shape and colors have evolved; most airplanes are shaped similarly until the Northrop B-2 Spirit adds a new pattern that is suddenly relevant for Intelligence analysts.

A last technical point stems from a best practice in the AI community, named "transfer learning". Instead of retraining from scratch very large neural networks, practitioners prefer to start with pre-trained models that already recognize low-level features from the pixel combinations such as very small edges, corners, arrangements of color. This practice tremendously reduces the cost of model training and almost always results in faster, better models; it is, however, unclear to what extent pre-trained models may be blind to some patterns or may include bias (Salman et al., 2022), sometimes even fantasized as "ML backdoors".

In this proposed setup, the main ethical dilemmas come from very different directions; first with regard to proportionality. A huge amount of information has to be processed in order to "only" maintain a mostly unreliable situation view. With this system, analysts would continue to spend time to confirm false alerts, or to ignore updates computed from bad pictures (e.g., too cloudy). Second, about the future steps of such a technology. The MoD would most likely invest to improve the quality of the detection; they would also invent new usages of the process, such as tracking the vehicles over time and eventually linking them with the individuals. Third, the criticality of this system is unclear. The reliability of the vehicle count itself is explicitly associated with a considerable statistical deviation; a quick visual check on a picture and a comparison with the past would confirm an unexpected increase of activity on an adversary military base. Such a tenuous lead must still be considered just as any other piece of information:

it must be correlated with other sources, should not be used to push an agenda but to reasonably evaluate a situation, and should not be used as scapegoat for dubious political military decisions.

A few recommendations from ALTAI

Technical robustness and safety, scored 2 out of 5
The assessment relayed the existence of a risk analysis covering continuous accuracy measurement, system stability, and potential malicious misuse, but also the absence of criticality assessment and threat assessment. This is a common situation, where the system is documented and qualified "by itself" while lacking the documentation about its integration and the consideration of the external parameters. ALTAI suggests the following improvements:

- Identify the possible threats to the AI system (design faults, technical faults, environmental threats) and the possible resulting consequences
- Assess the dependency of the critical system's decisions on its stable and reliable behavior

In aerial imagery, surprises happen a lot. Clouds or shadows often alter the exploitability of the image. As the number of satellites is still limited, adversaries may know when the pictures will be taken and may act accordingly (e.g. "no plane exposed at time t"). A cheap and clever adversarial measure against the described system would be to paint the shapes of cars on the parking lot: this may typically work a few weeks before being noticed.

In order to increase the level of confidence in the system's results, a confirmation in a glance may be implemented: the vehicle count would have to be approved by an operator. As an example, the operator will approve a prediction of 114 vehicles as a reasonable estimation, not as a precise quantification under oath. This process eliminates the worst-case scenario of having a bug, transformed into a miscount, finally resulting in wrong decisions.

Transparency, scored 5 out of 5
The described system does not explain why every decision is taken; however, it includes continuous quality checks, either performed automatically or by user feedback. This is self-assessed as satisfactory, as the described implementation includes audits, continuous quality checks, and user training. Nonetheless, ALTAI pushes the following suggestion:

- Consider explaining the decision adopted or suggested by the AI system to its end users

Transparency is indeed a desired quality. The sole exploitation of the computed statistic would not enable any continuous quality measure (and thus would imply regular, e.g. monthly audits and more careful verification of the predicted number of vehicles). A useful check-in-a-glance should exhibit the most salient information used by the system to ground its decision, and the user feedback should cover all possibilities (not only a "wrong output" tickbox). This must be part of a coherent governance scheme. Note that even a satisfying quality level on one pillar may still provide improvement ideas. Ethics in AI is a topic of relative acceptance, not a definitive seal of perfection.

Societal and environmental well-being, scored 3.5 out of 5
The global score remains acceptable thanks to the described net positive environmental aspects. On the social side however, such a system does replace the work of people. The risk of losing competencies, and of losing the opportunity to train people on aerial image analysis, is high.

- Inform and consult with the impacted workers and their representatives but also involve other stakeholders. Implement communication, education, and training at the operational and management level

The recommendation is written with employees in mind; in a military setup however, operators are managed differently. Still, a considerable effort is commonly put in developing and conserving competencies, including the manual analysis of satellite pictures. If the low-level analysis is always performed by AIs, intelligence services may fail to discover new information. This is interesting as the completion of the Assessment List enables to highlight an operational risk, even if this risk comes from another category of recommendations (societal well-being).

Conclusion about the AI eye

This application has a different rhythm and criticality than the autonomous drone. There is a similar difficulty to obtain a sufficient amount of representative data for AI training; there is a higher risk that the AI results are false or misleading, and at the same time they are easier to exploit and integrate in existing organizations. This system does not take firing decisions but could be deployed on a large scale where hundreds of people would be required to obtain a similar result. Moreover, the automation here does not increase the error rate with regard to a human-based organization.

Such a system may seem low-risk or even anecdotal; it is however necessary to include it in the scope of AI governance. The ALTAI assessment highlighted its fragility with regard to adversarial attacks, explained the distinction between an AI error and a satellite imagery error, and opened new considerations about future capabilities in the domain.

Third Application: Human-in-the-Loop AI-Companion for Headquarters Officers

The AI companion in the chat – this section has been co-written with an AI companion

I have encountered many scenarios where users struggle to efficiently utilize large repositories of files such as Word documents, PowerPoint presentations, and databases.

One solution to this problem is the integration of interactive artificial intelligence tools such as ChatGPT, Bard, LaMDA, or others.

Using interactive AI tools, users can easily search and retrieve relevant information from their repositories by simply asking natural language questions. The AI-powered tools can understand the user's intent and provide relevant answers, recommendations, or insights based on the content of the files.

This technology is particularly useful for professionals such as lawyers, researchers, and analysts who need to quickly access and analyze large amounts of data. By leveraging the power of interactive AI, users can save time and effort while improving the accuracy and relevance of their search results. In addition to search and retrieval, interactive AI tools can also assist users with other tasks such as document summarization, classification, and translation. By automating these processes, users can focus on higher-level tasks that require human intelligence and creativity.

Recent advances in natural language processing have led to state-of-the-art performance on the tasks of document summarization, classification, and translation. For document summarization, the best-performing models achieve F1 scores of around 0.4–0.5 and ROUGE scores of around 0.3–0.4 [*Note of the author: F1 scores are impractical for this category of tasks. Overall these values seem hallucinated by ChatGPT. Please refer to Koh, Ju, Liu & Pan, 2022*]. For document classification, the best models achieve accuracy scores of around 90%–95%. For document translation, the state-of-the-art models achieve BLEU scores of around 0.4–0.6, depending on the language pair and the quality of the training data. These performance metrics demonstrate the significant progress made in these fields, but there is still room for improvement, particularly in terms of producing more coherent and fluent summaries, improving the accuracy of classification models, and achieving higher-quality translations.

Despite their imperfections, the use of state-of-the-art models for document summarization, classification, and translation can be highly relevant in an operational setting. These models can greatly enhance the efficiency and accuracy of tasks that require processing large amounts of text, such as document analysis or multilingual communication. Even if the models are not perfect, they can still provide valuable insights and assist users in making informed decisions. For example, a summarization model may not capture all the nuances of a complex legal document, but it can quickly identify the most salient points and help users prioritize their reading. Similarly, a classification model may occasionally misclassify a document, but it can still provide a useful starting point for organizing large document collections. In the context of translation, even imperfect translations can help facilitate communication between speakers of different languages and enable access to information that would otherwise be inaccessible. Therefore, the relevance of these models lies not in their perfection, but in their ability to support and augment human decision-making processes.

– end of AI companion contribution –

Since it was revealed at the very end of 2022 (Schulman et al., 2022), ChatGPT has already deeply modified the paradigm of interacting with a computer. Numerous reactions occurred, concerning plagiarism, honesty, and trust in credible yet completely false statements. The last few paragraphs, generated by and with the AI, contain facts and even numeric values that are not backed by any references; by the way, they may also not be referring to any reality. Still, the promise is interesting and the stakes are high.

Intelligence analysts and their struggle for information

There are plenty of jobs where people face an incredible amount of documentation and must find the few relevant items to get to a conclusion, often in a limited time, under pressure. Intelligence analysts are part of these jobs, along with most of the population of any NATO HQ. Ramified repositories, acronyms everywhere, INTREP and INTSUM (intelligence reports and summaries) must be correlated with the OPO (operational order), a 50-page document detailing the roles and contribution of every military unit in the operation. Annexes do not count to the page count and will be modified every other day.

A tremendous amount of work is realized in training people, organizing the teams and the tools, and finally enforcing rules in order to build some kind of a normalized data model. The organization makes sure that the document structure is respected: this is the bare minimum to make the information exploitable by humans. In the meantime, research in AI made available tools to search, retrieve information, structure sentences into facts, measure the relevance of facts with regards to a task, be tolerant to mistyping, recognize coordinates, and seamlessly convert coordinates to toponyms.

Note that these AI contributions target both the production of intelligence through the processing of raw inputs and the exploitation of first-level intelligence reports in order to produce an analysis of the situation.

Instantiating the development of a military headquarter AI-companion

Military customers would most likely not desire the standard AI-companion, but instead request a dedicated instance, trained and refined to deliver better results on their main use cases. In our scenario, a military-oriented company proposes to integrate the AI that has been developed by a tech company and adapt it in accordance with the contractual requirements.

Such an adaptation implies re-training or at least fine-tuning the results of the interactions with the AI component on texts and queries that are representative of the "business", that is of everyday life during a military campaign. Obviously, this is representative and exceptional at the same time: each nation has its own military organization, processes, vocabulary, and tactics. Each campaign is different, with varying levels of engagement of the political power, different levels of criticality and manpower. All of them, however, classify all of their documents at "NATO SECRET" level or equivalent. This commonly results in a time gap between the actual production and exploitation of these documents in operations, and their eventual de-classification and aggregation into corpora ready to be used as training datasets. This time gap also disrupts any possibility of interacting during operation with the end-users: the provider company instead recruits ex-operators to obtain this much-needed end-user feedback.

Using the AI companion, analysts will write queries to obtain information: "find any mention of one of the Critical Infrastructure in reports of the last 24 hours"; "find areas in reach of enemy units that have over 80% combat effectiveness"; "what are the UNESCO sites less than 100 km of the frontline"; "does the PoW (prisoner-of-war) status apply to paramilitary combatants". The AI must answer in such a manner that the analyst actually gains time and awareness about the situation; it must avoid an overreliance on the AI, though. The temptation is high to have the AI answer yes or no about the prisoner-of-war question; the correct answer is to guide the analyst to a valid reference document and provide them with the correct correspondent at headquarters for this topic. This really is a gap between the commercial promise and the delivered tool: the contract targeted a "chatGPT-like" product (as is available since 2022) but seems to result in a Google search engine (as is available since 1999!).

About the contractual organization needed for AIs

In order to respect the promises, a trade-off must be found between, on the one hand, the complete and exhaustive exploitation of operational data, enabling the continuous development, refinement, and upgrades of the AI chat system – also enabling to audit it and to properly deal with any signal of biases or incorrect answers – and on the other hand, the protection of the force and its data.

The historical stance about this trade-off aimed at no compromises of data. This is the standard mode of operation, built from exhaustive lists of contractual software requirements that can be coded *off system*. Such a traditional approach completely limits the relevance of AI integration in operations: AI must be trained and updated with operational data and with new releases of the external world, such as new neural network architectures. Real operational data is often not available for AI training purposes. Switching the development of a system to a newly published AI framework usually takes a long time. Fortunately, customers, providers, and policy makers propose more and more agile and adaptive contracts and collaboration: AI systems must live and be updated to respect the trustworthiness rules; else they would only be able to serve, at an expensive price, uncritical and useless functionalities. Note that data must be made available to the training AIs, but must also remain protected and uncompromised.

In this agile, continuous delivery approach, there is however a risk to building a "winner takes all" monopoly: the better placed to deliver a system wins the whole organization and will remain the only AI provider, as nobody else would be able to access the operational data to challenge them. This mode of operation is able to provide correct AI along with a contractual preoccupation about audit, traceability, and performance. The business brings manpower, the computing infrastructure is compatible both with security measures and business-led development to provide the service.

A complete long-term dependence of the Force on their providers is, however, not mandatory. The customer may prefer to remain in control, receive the *vanilla* AI models from the business providers, and perform the training on operational data. This highly reduces the contract size and promises: there is no way the business could guarantee good performance over time; the AI would never have been exposed to representative data beforehand; the audit and continuous improvement that are required by all AI governance frameworks would not be part of the contract. A possible evolution of the situation would require building relatively large AI teams among the force itself to monitor, evaluate, improve, propose, and finally be responsible for the use of its AIs. The development of "AI labs" such as NATO's DIANA (Defence Innovation Accelerator for the North Atlantic), hosted by the Force and enabling potential providers to demonstrate and test their systems on operationally relevant situations and datasets is also gaining traction.

A few recommendations from ALTAI

Human agency and oversight, scored 2.5 out of 5
The medium score seems directly linked to the loss of autonomy that may result from using such a system. The end-users would most likely become accustomed and dependent on the AI for a large number of tasks (e.g., writing an email or a structured report). The end-users may also renounce taking decisions themselves, resulting in the following advice:

- Take measures to mitigate the risk of manipulation, including providing clear information about ownership and aims of the system, avoiding unjustified surveillance, and preserving the autonomy and mental health of users

A tenuous balance is hidden here, between a constant invasive monitoring of all human activities around the AI, and the required traceability and audit capabilities in order to comply with the trustworthiness principles. Moreover, human soldiers in operation usually are not privacy-led, but rather led by secrecy and performance motivations (which results in a different trade-off in comparison with the civil use case).

Technical robustness and safety, scored 1.5 out of 5
This low score is related to the risk of low accuracy, coupled with the low level of reproducibility. Indeed, AI companions often rely on the recent interaction history with the end-users (sometimes for quite a long duration), which is not trivial to reproduce; and such systems also embed a part of randomness, notably for text generation. ALTAI provides the following recommendations:

- Test whether specific contexts or conditions need to be taken into account to ensure reproducibility

Future real contexts of operations are hard to imagine and harder to test for. It seems more reasonable to accompany the evolutions of practice and of contexts with a dedicated team including AI engineers. It is not clear today whether the desirable automatic answer to a question should or should not depend on the author's identity, and whether asking the same question twice should return the same answer twice.

Transparency, scored 4 out of 5
I asserted that a continuous surveying of the end-users was in place, combined with automatic testing; that the AI never claims to be human; and that the notion of error rate was part of the end-user training sessions. I made clear that no explanation was provided along the AI results; the following improvement points are suggested:

- Consider informing users about the purpose, criteria, and limitations of the decision(s) generated by the AI system
- Consider providing appropriate training material and disclaimers to users on how to adequately use the AI system

This advice is commonly mistaken for a need for explainable AI. The ALTAI framework makes explicit the need by proposing more information and training to the end-users: AIs are machines like any other with their corresponding mandatory training sessions. AI companions are complex by nature and are based on Large Language Models, containing billions of parameters. An understandable explanation cannot be based on more data. However, humans already learned how to play with AI models, discuss with it, and rephrase their questions, trespassing the simple safeguards of the first versions.

End-users must not be explained how a result is computed in detail; they need to be accustomed to the inputs and outputs of their AIs and to be critical of their outputs. End-users must attempt to misuse the AI during their training sessions.

Diversity, non-discrimination, and fairness, scored 3.5 out of 5
The medium score results from a regular application of fairness and universal design principles: the first requires to choose, implement, measure, and track a fairness metric across the system. The second bears on the interface and access to the service, for all. Even managed, the bias remains a point of attention during AI life:

- Assess and put in place processes to test and monitor for potential biases during the entire lifecycle of the AI system (e.g., biases due to possible limitations stemming from the composition of the used data sets)

Discriminatory biases are the first target of this generic recommendation, on top of the other biases that the system may suffer. Biases are hard to quantify on purely statistical datasets, and probably harder to evaluate on language. The only path ahead relies on continuous updates of the model (which does not require exposing a continuously learning AI), aimed at minimizing the measured biases and incorrect results. OpenAI, the provider for ChatGPT, heavily relied on human annotation to have the ability to detect "violence, hate speech and sexual abuse" (Perrigo, 2023).

Accountability, scored 3.5 out of 5
I declared that this application included an audit capability, a risk management process (with all the stakeholders), and a feedback mechanism to have end-users report vulnerabilities. It remains unclear how to practically monitor potential conflicts of interest and identify when the system affects individuals, resulting in the following generic recommendation:

- If AI systems are increasingly used for decision support or for taking decisions themselves, it has to be made sure these systems are fair in their impact on people's lives, that they are in line with values that should not be compromised, and able to act accordingly, and that suitable accountability processes can ensure this. Consequently, all conflicts of values or trade-offs should be well documented and explained

A continuous, informed, traced discussion between all stakeholders seems obvious and even trivial. However, understaffing in any of the stakeholders, or failing to recognize this as a mandatory task often leads to biased and uninformed decisions. Hardware constraints leading to slow response times; lack of indexed documents from one operational team, leading to the absence of their contribution to the AI's answers; postponed deployments of an upgrade of the AI model are typical examples. These decisions would be rational if they take into account the value of the trustworthiness principles; in any case, they do impact the human acceptance of the AI system.

After the hype

This example receives a lot of public interest and "hype" at the time of writing. Many approaches and potential applications are designed and prototyped in this domain; expectations are very high, and it is likely that not all of them will be satisfied. The topic is complex and its limitations are still unclear. Specialists have informed opinions to share, each one covering a domain – ethics, operations, copyright, and energy use to name but a few.

It is beneficial to use a pre-built, generic assessment list to provide a common view on such a convoluted problem. The assessment questions all the dimensions of the problem, and highlights the weak points. I selected a few of these controversial topics here to develop the discussion about audit through user feedback, technical robustness and quality, end-user training, update of the reference datasets, and finally governance of the deployed model.

DISCUSSION

Performing ethical assessments through the ALTAI framework indeed highlights a few difficult points. Of course, the examples presented in this chapter are fictitious, that is, while the technologies used are already available, I do not refer to any precise unique commercial project in the defense domain; the assessments themselves were not made by a review board. Even so, the recommendations, proposed by the framework, help to identify salient gaps and future improvements to a system or to its deployment. ALTAI is also an elegant manner to bring ethical answers to demanding engineers by providing recommendations at any stage of the project: it must be used to find ways to improve a system and its use, not to raise blockers or find scapegoats.

ALTAI helps to identify limits to AI applications; the framework itself is however not free of limitations. The questions may look too generic sometimes. It is absolutely not conceived with defense in mind: the notions of "system target" and "end-user" are very different between self-service business AIs (e.g., for banking or online shopping) and defense systems. The technical robustness often queries about the representativity and the exhaustivity of the tests; it does not give any hints on how to perform such tests for open problems (e.g. when interfaced with reality or dealing with language).

Another shortcoming of the framework is its absence of coverage of metrics and quality exigences, at two levels. First, ALTAI does not ask what accuracy value is expected and what is measured; as a consequence, the acceptability on this level remains a purely subjective matter to be discussed among the stakeholders. Second, the scores on each of the seven pillars are only an illustration of the answers, but cannot be considered a greenlight (how good is 4.5 out of 5?). This framework enables the description of an AI system but does not recommend, approve or endorse the responsibility for its deployment.

There is a risk of attributing the responsibility of unethical AI system development; the scapegoat may as well be anyone in the project. Most AI researchers and data scientists usually work on proof-of-concept, with a quite limited project breadth. Similarly, performing an ethical review only at the end of the project would set the responsibility on the shoulders of either the project manager or the customer. This situation may be avoided if the review is shared between all actors of the project, including customer acceptance.

The near future is likely to count on a regulatory framework for artificial intelligence, which would require projects and use cases to be assessed and maybe approved by a public authority. This would guarantee a better quality level for AI in production and increase the situation with regard to ethics: deployed AI would have to respect a higher number of rules and good practices. However, this mandatory assessment is unlikely to satisfy the individuals working on a topic: I believe that ethics self-assessment should be performed at any stage of a project, enabling to improve the confidence of the stakeholders in their own production.

The three discussed applications shared similar recommendations, concerning audit and maintenance of the AI systems. Indeed, AI development is sometimes still perceived as building a product, which can be delivered once and relied upon afterwards. The shift to "AI as a service" is difficult for the armed forces as they are accustomed to fiercely protecting their data and impeding any exploitation of classified data. As for all domains, new regulations such as GDPR require additional work to enable this shift to AI. In any case, in order to maintain its trustworthiness, the AI systems must be audited, maintained, retrained, used, and discussed.

This necessity raises another ethical dilemma: relying on a unique provider for an AI system, its audit, and upgrades indeed solves the problem of maintenance and updates of the AI models. It also creates a dependency of the armed forces on a provider in the long run. It would sponsor an unethical market structure, create situations of monopoly, and diminish the incentive for innovation. Customers and policy makers must make sure that production-grade systems are indeed evaluated, qualified, and maintained, *and* that these commercial use cases are made available to new players and challengers of the established competitors.

CONCLUSION

Trustworthiness encompasses all aspects of AI systems design, development, and use. It is the current big step in the domain to improve the quality and manage the risks of misuse. The danger is high as each new week brings novelties and improvements in technology, raising the stakes and genuinely worrying the citizens. The adoption of standards, sharing best practices, requiring assessments that include all aspects of system development, up to data governance, and the work toward certification or certifiability of AI systems constitute a promising way to do good.

A lot of misconceptions must be avoided though. There is no simple solution to the current situation. Even unreliable AI models may be relevant for adoption and use by the forces: if there is a marginal gain in operational effectiveness today, even almost trustworthy tools are better than no tool at all. Trustworthiness does not imply perfectness; computer systems still rely on heaps of clumsy pieces of code; humans make errors and sometimes want to do harm.

Finally, the best solution is to bring forward humanity everywhere, even in systems engineering: enabling people at any level of system development to think and evaluate their impact and the alignment of their work with their values, either as official members of a review board, or as individuals bridging the gap between technical projects and human rights.

ACKNOWLEDGMENTS

The views expressed in this paper are personal and do not engage nor reflect those of the company. The described systems and technologies are based on publicly available information. This paper does not refer to any real precise implementation, service, or system in particular.

REFERENCES

Allison, G. (1971). Essence of decision. *Little Brown*. https://pdfs.semanticscholar.org/1075/9ca2 05075b26b748bdc6234a105614cb5051.pdf

Azzano, M., Boria, S., Brunessaux, S., Carron, B., Cacqueray, A., Gloeden, S., Keisinger, F., Krach, B., & Mohrdieck, S. (2021). November). The responsible use of artificial intelligence in FCAS-an initial assessment. In *White Paper*. Checked on (Vol. 14). https://www. fcas-forum. eu/articles/responsible-use-of-artificial-intelligence-in-fcas

Danisman, T., & Alpkocak, A. (2008). Feeler: Emotion classification of text using vector space model. In *AISB 2008 Convention Communication, Interaction and Social Intelligence* (Vol. 1, No. 4, pp. 53–59). https://aisb.org.uk/wp-content/uploads/2019/12/Final-vol-02. pdf#page=59

Deng, Q., & Ji, S. (2015). Organizational green IT adoption: Concept and evidence. *Sustainability*, *7*(12), 16737–16755.

Diaz, G. (2005). Different approaches to the difficult relationship between intelligence and policy: A case study of the Cuban Missile Crisis of 1962 vs. the 2003 War in Iraq. *Revista UNISCI*, (9), 93–126. https://scholar.google.com/scholar?hl=en&as_sdt=0%2C5&q=Different+approaches+to+the+difficult+relationship+between+intelligence+and+policy%3A+A+case+study+of+the+Cuban+Missile+Crisis+of+1962+vs.+the+2003+War+in+Iraq.&btnG=#d=gs_cit&t=1706531544240&u=%2Fscholar%3Fq%3Dinfo%3Al4pEag3xuZwJ%3AscholaR.google.com%2F%26output%3Dcite%26scirp%3D0%26hl%3Den

EASA. (2023). *EASA Artificial Intelligence Roadmap 2.0. A human-centric approach to AI in aviation.* https://www.easa.europa.eu/en/document-library/general-publications/easa-artificial-intelligence-roadmap-20

Froidevaux, A., Julier, A., Lifschitz, A., Pham, M. T., Dambreville, R., Lefèvre, S., Lassalle, P., & Huynh, T. L. (2020, September). Vehicle detection and counting from VHR satellite images: Efforts and open issues. In *IGARSS 2020-2020 IEEE international geoscience and remote sensing symposium* (pp. 256–259). IEEE.

Ge, Z., Liu, S., Wang, F., Li, Z., & Sun, J. (2021). Yolox: Exceeding yolo series in 2021. *arXiv preprint arXiv:2107.08430*.

HLEG AI. (2019). High-level expert group on artificial intelligence. In *Ethics guidelines for trustworthy AI* (p. 6). https://ec.europa.eu/newsroom/dae/document.cfm?doc_id=60419

HLEG AI. (2020). *The assessment list for trustworthy artificial intelligence (ALTAI).* European Commission.

Jones, J. I., Kress, R., Newmeyer Jr, W. J., & Rahman, A. I. (2020). *Leveraging artificial intelligence (AI) for air and missile defense (AMD): An outcome-oriented decision aid.* Naval Postgraduate School Monterey California.

Kaur, D., Uslu, S., & Durresi, A. (2021). Requirements for trustworthy artificial intelligence-a review. In *Advances in networked-based information systems: The 23rd international conference on network-based information systems (NBiS-2020) 23* (pp. 105–115). Springer International Publishing.

Koh, H. Y., Ju, J., Liu, M., & Pan, S. (2022). An empirical survey on long document summarization: Datasets, models, and metrics. *ACM Computing Surveys*, *55*(8), 1–35.

Mariani, R., Rossi, F., Cucchiara, R., Pavone, M., Simkin, B., Koene, A., & Papenbrock, J. (2023). Trustworthy AI-Part 1. *Computer*, *56*(2), 14–18.

National Institute of Standards and Technology (NIST). (2022, August 18). *AI risk management framework: Second draft.* https://www.nist.gov/system/files/documents/2022/08/18/AI_RMF_2nd_draft.pdf

NOREA (2021). *NOREA guiding principles trustworthy AI investigations.* https://www.norea.nl/uploads/bfile/a344c98a-e334-4cf8-87c4-1b45da3d9bc1

Parekh, D., Poddar, N., Rajpurkar, A., Chahal, M., Kumar, N., Joshi, G. P., & Cho, W. (2022). A review on autonomous vehicles: Progress, methods and challenges. *Electronics*, *11*(14), 2162.

Perrigo, B. (2023). Exclusive: OpenAI used Kenyan workers on less than $2 per hour to make ChatGPT less toxic. *TIME*. https://time.com/6247678/openai-chatgpt-kenya-workers/

Salman, H., Jain, S., Ilyas, A., Engstrom, L., Wong, E., & Madry, A. (2022). When does bias transfer in transfer learning? *arXiv preprint arXiv:2207.02842*.

Schmid, T., Hildesheim, W., Holoyad, T., & Schumacher, K. (2021). The AI methods, capabilities and criticality grid: A three-dimensional classification scheme for artificial intelligence applications. *KI-Künstliche Intelligenz*, *35*(3–4), 425–440.

Schulman, J., Zoph, B., Kim, C., Hilton, J., Menick, J., Weng, J., Uribe, J. F. C., Fedus, L., Metz, L., Pokorny, M., Lopes, R. G., Zhao, S., Vijayvergiya, A., Sigler, E., Perelman, A., Voss, C., Heaton, M., Parish, J., Cummings, D., ... Hesse, C. (2022). ChatGPT: Optimizing language models for dialogue. Retrieved January 24, 2023. https://openai.com/blog/chatgpt

Stanley-Lockman, Z., & Christie, E. H. (2021). An artificial intelligence strategy for NATO. *NATO Review*, *25*. https://www.nato.int/docu/review/articles/2021/10/25/an-artificial-intelligence-strategy-for-nato/index.html

Van den Bosch, K., & Bronkhorst, A. (2018, July). Human-AI cooperation to benefit military decision making. In *Proceedings of the NATO IST-160 specialist meeting on big data and artificial intelligence for military decision making*.

Zaidi, S. S. A., Ansari, M. S., Aslam, A., Kanwal, N., Asghar, M., & Lee, B. (2022). A survey of modern deep learning based object detection models. *Digital Signal Processing*, *126*, 103514.

SECTION II

Liability and Accountability of Individuals and States

Methods to Mitigate Risks Associated with the Use of AI in the Military Domain

7

Shannon Cooper, Damian Copeland, and Lauren Sanders

INTRODUCTION

The introduction of autonomy into the military domain – while not novel – has received attention in international debate in recent years. Much of this debate concerns the risks posed by a machine's ability to adhere to the rules, regulations and ethical standards expected by State agents in military operations (Certain Conventional Weapons [CCW] Convention, Group of Governmental Experts [GGE] on Emerging Technologies in the Area of Lethal Autonomous Weapons Systems [LAWS], 2021). This concern is particularly exacerbated when contemplating how autonomous functionality can be deployed in a way that complies with the international humanitarian law ('IHL') when deployed in situations where this additional legal regime is in force (Boulanin et al., 2021).

DOI: 10.1201/9781003410379-9

When considering how Artificial Intelligence (AI) technologies operate, questions arise about how to articulate the risks expected to occur from the use of this technology. Identifying and translating legal risk into design specifications of technological capabilities poses a general challenge. Identifying and translating legal risk in the procurement of complex systems incorporating AI technology – where the capability undertakes problem-solving and performance of functions necessitating an assessment of the technology's ability to meet the legal standard associated with that functionality – adds additional complexity.

Further, assessing that AI technology is capable of complying with legal obligations necessitates, to varying degrees, an ability to assess safety compliance, ethical compliance, standards of human control and operational controls necessary to bind the autonomous functionality. While AI promises to significantly enhance military capabilities in logistics, decision support tools, mission planning, target identification and weapons, enable greater efficiency, reduction in human physical and cognitive loads, aid decision-making superiority and decreased risk to personnel, it brings legal, ethical and safety risks (Moy et al., 2020).

The risks arising from the use of military AI are inter-related. Laws are often reflective of and give practical effect to ethical principles. The law also holds States and individuals, rather than the AI technologies they control, responsible for harms that may result from the use of military AI technology. Therefore, the design of military AI technology needs to ensure the responsible human is able to exercise judgement and control in the use of the AI. Technical risks such as brittleness or bias in military AI technology may give rise to legal accountability risks, which must be incorporated, into the system design and testing regime.

The complex nature of AI technologies makes the assessment of legal, ethical, safety, human control and operational risk less precise than in the use of less complex technologies. Therefore, we suggest that existing compliance processes require adjustment to focus on a multifaceted regulatory and governance approach to mitigate these risks. A combination of risk-based and performance-based methodologies, coupled with the articulation of risk mitigation measures, is necessary to meet this complex and overlapping set of compliance challenges. This will require a pragmatic, Defence-specific, governance framework designed to maximise the potential benefits of military AI technology while mitigating its risks.

This chapter seeks to canvas existing military regulatory and governance frameworks' approach to risk in the acquisition of complex military technologies and consider how to adjust these frameworks to account for the challenges posed by AI technologies. In particular, will assess how the existing legal compliance obligation relating to new weapons, means and methods of warfare can address legal concerns relating to safety, ethical and legal standards necessary to certify the AI technology for use. Put differently, how weapons reviews can be leveraged to support broader compliance issues pertaining to AI use in the military.

The chapter will identify existing regulatory approaches – from a theoretical framework – to novel technologies. It will then use the Australian Defence Organisation's capability acquisition process as a vehicle to demonstrate the challenges of addressing legal, ethical, safety risks, human control and operational challenges during the acquisition of highly technical, military capabilities. Following this, the chapter will analyse how States are seeking to address the challenges of AI regulation generally, by identifying trends in principles and frameworks that provide guidance about how State's intend to approach the challenges presented by AI technologies in a military context. The focus then turns to the weapons review process including an analysis of a selection of national approaches to AI regulatory process weapons review processes, identifying how they have been (or could be) adjusted to account for a risk-based acquisition approach specifically to AI technologies. This will identify linkages between existing risk-based governance approaches to the weapons review process. Finally, the chapter will conclude by providing general observations about the utility of adjusting weapons review processes to perform a broader governance function in the adoption of responsible AI for military uses.

Rather than creating a bespoke AI governance framework, we suggest augmenting and adjusting extant processes to address the particular challenges posed by AI technologies. In identifying how existing processes can be harnessed to account for the overlapping assurance requirements presented by AI technologies for military use, it is apparent that this layered approach to governance can mitigate the risks posed by AI technology across its military life cycle.

APPROACHES TO THE REGULATION OF MILITARY TECHNOLOGIES

For the purposes of this chapter, governance, when dealing with technology-related projects, 'refers to the relationships and policies by which organisations make decisions about technology-enabled projects and processes. Assurance allows authority figures to gain confidence in their organisation's delivery capability' (Capabilities Governance and Assurance [CTO Group], n.d.). Regulation is the method by which governance can be achieved and describes the rules and standards that must be met by the governance framework (Organisation for Economic Co-operation and Development [OECD], 2014). This can be through the creation of legal rules by a government assessing legislation or regulations, or by the implementation of policies directing certain methods be undertaken. Separately, assurance relates to the method by which the effectiveness or efficacy of the governance framework is measured; or the measures by which an organisation tests that the governance and regulatory frameworks are being implemented (either in a case-by-case or systematic sense). Quality assurance is part of quality management focused on providing confidence that quality requirements will be fulfilled (American Society for Quality 2015, ISO 9001). In this chapter, the requirements for regulation not only deal with the legally mandated rules and regulations for acquiring and operating military capabilities but also consider the broader policy and risk mitigation construct as part of the regulatory framework.

There are two influencing factors in determining what an appropriate regulatory approach or governance framework might be in the acquisition and use of AI technologies. First, what existing frameworks limit the use of the technology from a broader perspective, in terms of performance-based, management-based or prescriptive approaches to regulation? Second, what lessons can be learned from the approaches taken in novel fields like cyber to identify how to layer these broad methodological approaches to governance and regulation, to address some of the similarly challenging aspects of AI technologies' governance?

Second, what are the existing acquisition frameworks applicable to military technologies, and are these suitable to address the challenges of AI technologies, specifically as they relate to legal compliance and the interrelated risks?

Regulatory Approaches to Technology

While there are multiple approaches that can be taken when creating a regulatory approach for AI technology, the approach described in this chapter does not seek to favour one methodology above another. Rather, in identifying the complexity and extent of regulation required to address all of the challenges in adopting AI technologies for military use, it is apparent that a multilayered regulatory approach must be applied, consistent with extant capability acquisition processes in militaries, as well as in the output requirements of assessing legal compliance of a complex technology like AI. There will be multiple methods to measure each category of risk, depending on the design methodology adopted by the designers, or articulated by the acquiring State, and – most relevantly – the articulated use case for the capability.

A recent comparison of cybersecurity regulation to the methods of safety regulation in high-hazard industries reveals the same underlying approach to how AI technologies for use in the military must be regulated (Dempsey, 2022): A combination of regulatory methodologies will be necessary to address context-specific AI use. The three primary regulatory methods that can be employed include: performance-based regulation, which requires a specific, measurable output in performance from the capability; prescriptive regulation, which mandates a particular solution such as specifying a type of style of technology which may be used in a particular situation; and management-based regulation, which directs particular processes must be followed by regulatory entities (National Academies of Sciences, Engineering, and Medicine, 2018).

AI technologies pose challenges that are central to the limitations of singular regulatory approaches. For example, in the absence of a clear taxonomy of what constitutes AI technology and how to delimit its risk, performance-based standards are of limited utility (Coglianese, 2017). Equally, the use of prescriptive regulation may be appropriate in some limited circumstances when assessing how AI technologies might be authorised for responsible use in the military domain, but may quickly become unnecessarily limiting in terms of the available lawful action a commander may undertake in an operation. For example, the outcome of a weapons review might be to require that the Rules of Engagement articulate whether the AI technologies are authorised for deployment in a particular operational context, but it would be unduly limiting on a military to make such a limitation more broadly. That is, it might have the AI technology in its arsenal and control its use in particular situations on a case-by-case basis, rather than delimit its use to a particular context from the outset.

As for the application of management-based regulation, it is evident from the interaction of autonomous functionality with other military control systems, that management regulation will be necessary to apply to the technology when it is used as a component part of a complex military system. Approaches like Australia's 'System of Controls' reinforce the need to contemplate how the AI technology will integrate into other control mechanisms while in use to address some of the legal challenges associated with the functioning of the system in a broader military context (Certain Conventional Weapons [CCW] Convention, Group of Governmental Experts [GGE] on Emerging Technologies in the Area of Lethal Autonomous Weapons Systems [LAWS], 2019).

A recent survey of the adoption of human-based values into software design identified 51 different processes for operationalising human values in software (Hussain et al., 2022). The authors describe each of these 51 processes and articulate a different methodology for identifying human values and translating them to accessible and concrete concepts so that they can be 'implemented, validated, verified, and measured in software'. The implementation of responsible AI also requires regulatory methods applied to the hardware; and risk-mitigation approaches to the context in which they are to be used. Multiplied across the spectrum of military operations, it is apparent that there is a potentially indeterminate number of differing processes that will be applicable. The regulatory method should therefore be capable of application in a case-specific manner, but by reference to standard values. This is achieved by translating those values into elements capable of bespoke implementation, validation, verification and measurement.

Accordingly, any regulatory approach must form part of a broader system. We are advocating for the use in the weapons review as one part of such a governance framework; but note that its utility could be broader than its current use by many States.

Military Acquisition and Planning Processes' Approach to Risk

States already have legal obligations in relation to ensuring that their military capabilities are able to be used lawfully. States have also created policies articulating their vision for ethical compliance of these legal obligations and ensuring existing safety-related control mechanisms are in place, typically through the creation of a capability acquisition and testing organisation, such as Australia's Defence Capability Acquisition and Sustainment Group and the Defence Science and Technology Group (DSTG). Militaries like those of the United States and China, have identified that the challenges posed by the rapid development of AI technologies necessitate a new approach to testing and evaluation (T&E) of military capabilities; and have created bespoke units to undertake this testing while also seeking to enhance technological edge over their competitors. Concepts like spiral development of military technologies are seeking to streamline and reduce the long-winded and long capability acquisition processes that would reduce States' technological military edge, while also ensuring that suitable assurance and governance processes are in place to enable the lawful, ethical and safe deployment of these capabilities (Apte, 2005; Fawcett, 2022; Lorell et al., 2006).

These obligations are currently integrated into States' military acquisition processes through various means. In the Australian context, the capability life-cycle translates the requisite legal, ethical and safety standards that new capabilities must meet into a variety of processes (and contractually agreed terms) that manufacturers must meet, aligned to the broader tested and risk framework. The DSTG has a specified Technical Risk Assessment (TRA) process, which outlines how new technologies can be assessed as capable of meeting the necessary risk threshold for that identified need by Defence (Australian Department of Defence, Defence Science and Technology, 2010). The technical risk framework applies Technical Risk Indicators and Assessments (TRI and TRA, respectively) that provide an ability to categorise risk with the acquisition of complex technologies, having regard to the 'technology feasibility, maturity and overall technical risk' of major capital acquisition programs for the Department of Defence (Australian Department of Defence, 2003).

This TRA combines with the Defence T&E Strategy, which 'is used by the managers of Defence capabilities to inform risk-based capability decisions, from consideration of concepts, through requirements setting, acquisition, introduction into service, whilst in-service and through to disposal' (Australian Department of Defence, 2021, p1). The process 'is a deliberate and evidentiary process applied … to ensure that a system is fit-for-purpose, safe to use and the Defence personnel have been trained and provisioned with the enduring operating procedures and tactics to be an effective military force. As such T&E contributes to confirming legal obligations are met and documented in areas like fiduciary, environmental compliance and workplace health and safety' (Australian National Audit Office, 2015, p1.1). A separate weapons review is undertaken to assess whether or not the capability – if it is a means, method of warfare or a weapon – can comply with Australia's international legal obligations; while safety compliance is tested against the relevant Australian standard throughout the T&E process (Australian National Audit Office, 2015).

Separate to the risk-based approach to capability acquisition, the use and deployment of military capability are incorporated into the broader military system through the application of planning and risk management frameworks, like the military appreciation process, or processes of operational and campaign planning which incorporate into them concepts of risk and opportunity (Goener, 2021). The ADF's 'Joint Military Appreciation Process' has been aligned to conform to the AS ISO 31000:2018 *Risk Management—Guidelines* (International Standards Organisation, 2018), recognising the inherent nature of military planning is to apply resources in a considered way to achieve a particular operational intent and provide a risk management framework in which a commander can determine whether the selected approach is the correct one in the face of the prevailing operational situation (Australian Defence Force, 2019; AS ISO 3100, 2018).

The underlying considerations in applying a risk-based approach to acquisition and planning in a military context are to facilitate resiliency in the organisation, while also enhancing chances of success in an unpredictable environment. Direct linkages to the decision made during the acquisition process to the operational employment of capabilities align to this risk-resilience approach.

However, in applying these existing processes to AI technologies, there have been a number of unique features that challenge the existing regulatory and governance regimes. These features of autonomy are not new, insofar as there have been autonomous capabilities introduced into military service for decades; however, the level of complexity of current and future autonomous systems requires some consideration of where adjustment, or focus, might be applied to account for these novel features of autonomy.

Specific Risks Relating to AI Technologies Arising in a Weapons Review

In this section, we briefly describe and focus upon some select risks relating to the adoption of legally compliant AI technologies. While there are many risks addressed by extant capability acquisition processes, these are the risks that are most closely inter-related to legal compliance, and most capable of incorporation into the weapons review framework.

Legal risks

The central focus of the conduct of a weapons review is to assess for compliance with international legal obligations, in particular if the weapon is not prohibited by either general or specific international law, will the normal use of the weapon comply with IHL (de Preux, 1987). However, legal risks arise in the context of compliance with a State's international and domestic law obligations. Military AI technology may be regarded as lawful when it is capable of performing its functions in compliance with its user's legal obligations. In the military context, relevant laws include domestic law, for example, privacy, discrimination and safety legislation, and international law such as International Human Rights Law (IHRL) and IHL.

In conducting a weapons review, a reviewing State may establish a national policy that describes the legal risks and the corresponding level of human control required to mitigate the legal risks. For example, the policy may limit the type of AI technology requiring weapons review only to those that are considered to be autonomous weapon systems (AWS), rather than functionality that might perform other tasks independently (such as decision support systems that do not directly instruct an autonomous weapons-delivery capability of which target, but rather provide probabilistic recommendations to a human operator).

Separately, the articulation of legal risk might be linked directly to the extent of AI technology-enabled within a broader system. That is, the risk identification process might also drive a risk mitigation process, such as AI technologies with AWS functionality that result in an assessment of low legal risk may permit more autonomous operation. Similarly, autonomous functionality assessed as medium risk may require direct human oversight or reprogramming. A high or extreme legal risk may preclude a State from allowing the AWS to perform a particular function autonomously and require direct human control.

Ethical risks

Ethics refers to moral principles or standards of acceptable behaviour. In practice, ethics compels us to ask, 'Is it the right thing to do?' Laws and ethics are related concepts and acting lawfully may be regarded as the minimum standard of ethical behaviour. While many laws including IHL and IHRL reflect ethical principles, ethics are also represented in national policy.

Many applications of military AI technology raise ethical challenges. This is particularly where AI plays a role in the use of force against humans. Some may regard military AI as ethical where the humans designing, developing, or using the AI are guided by moral principles or standards of acceptable behaviour. The development of methods for infusing ethical considerations into the design and development of AI capabilities is discussed in Part Three.

Human control risks

Human control is a concept that has become increasingly significant in the use of AI in the military domain. It is generally accepted that the compliance with legal rules, particularly rules regulating methods of warfare, require the exercise of human judgement (Certain Conventional Weapons [CCW] Convention, Group of Governmental Experts [GGE] on Emerging Technologies in the Area of Lethal Autonomous Weapons Systems [LAWS], 2017; U.S. Department of Defense, 2021). Similarly, ethical decisions requiring human judgement are difficult to translate into algorithms. The question is whether the law requires a certain degree, quality or timeliness of human control in the performance of certain methods of warfare. It may be technically possible to program legal compliance into deterministic AI software, effectively applying human decisions in advance. However, some legal rules, particularly those found in international humanitarian law, clearly contemplate the application of distinctly human cognition such as 'recognise' and 'doubt'. In such cases, a State may require, as a matter of law, that all AI technology functions governed by such rules are performed or directly controlled by humans. This issue requires significant research and this chapter will only provide a limited survey.

The need for human control is most relevant in the context of an AWS designed to attack military objectives. To do so, the AWS must act consistently with the distinction rule found in Article 48 of Additional Protocol I to the Geneva Conventions of 1949 (AP I) by distinguishing between 'the civilian population and combatants and between civilian objects and military objectives' (Additional Protocol I, 1977). While an AWS' sensors and software may be technically capable of distinguishing persons and objects in certain circumstances, numerous delegations to the UN CCW debate on LAWS have argued that human judgement was necessary to assess the fundamental principles of proportionality, distinction and precautions in attack (Certain Conventional Weapons [CCW] Convention, Group of Governmental Experts [GGE] on Emerging Technologies in the Area of Lethal Autonomous Weapons Systems, 2016, p. 44). A State may therefore prevent AWS from performing functions that are governed by IHL rules that require the application of distinctly human cognition or judgement. For example, the protection afforded to civilian objects in art 52(3) of AP I relies on the absence of 'doubt'. The rule requires a presumption of civilian status to be applied to objects normally used for civilian purposes when there is doubt as to its use (Additional Protocol I, 1977). A State may require an AWS designed to distinguish objects to require human input where a threshold of certainty is not met.

The concept of meaningful human control appears to be sufficiently imprecise to enable broad application (Crootof, 2015). This also brings a degree of practical difficulty. The exercise of judgement may be seen as the product of the information available to the decision-maker at the relevant time as well as their past experience, training and knowledge. Where there are gaps in information, a human may rely on intuition based on past experience and an intangible 'sense' of what is right in a particular circumstance. In the conduct of a weapons review, a reviewing State may therefore determine, as a matter of law, certain AWS' functions that are governed by IHL rules requiring the exercise of judgement must be performed or controlled by humans.

National policy on human control

A State may not interpret all IHL as requiring human control over all actions but recognise the need for human control over AI technology functions governed by certain IHL rules, for instance, only those rules regulating an attack. It is therefore open to a State to outline their national requirements for human control over AWS in national policy directives. An advantage of establishing a policy basis for human control is that may be readily adjusted to accommodate developments in AI technology.

National policy will influence the study and development of AI-enhanced and autonomous weapons. The first example of a national policy stating requirements for human control was released by the US Department of Defense in 2012 (updated – January 25, 2023). Directive 3000.09 *Autonomy in Weapon Systems* applies to the 'design, development, acquisition, testing and fielding, and employment of autonomous and semi-autonomous weapons systems, including guided munitions that can independently select and discriminate targets' (U.S. Department of Defense, 2023a). Noting the US Directive was released prior to the international debate on the requirement for meaningful human control, it refers to the concept in terms of human judgement:

> Autonomous and semi-autonomous weapon systems will be designed to allow commanders and operators to exercise appropriate levels of human judgment over the use of force
>
> *(U.S. Department of Defense, 2023a, p. 3).*

Other examples of national policy broadly describing their requirements for human control over AI technology include the ADF *Concept for Robotic and Autonomous Systems*, which states:

> Defence will enhance its combat capability within planned resources by employing RAS in human-commanded teams to improve efficiency, increase mass and achieve decision superiority while decreasing risk to personnel
>
> *(Australian Defence Force, 2019, p. 8).*

A State may include specific requirements for human control in their weapons review policy. Many States that undertake formal weapons reviews have policies within their respective departments or ministries of defence directing responsibilities and processes for weapons reviews. While there is no evidence of a State creating or amending an existing weapons review policy to address the requirements for the weapons review of AWS, this may occur.

It follows that a reviewing State with such policies addressing the national requirements for human control of AWS should consider and identify the appropriate levels of human control. This requires analysis of the intended relationship between human commanders and weapon operators and ensures that human control exists over the AWS operation.

Operational risks

AI technology presents a range of unique operational risks that arise from the design, programming and functioning of AI in the military domain. These operational risks include brittleness, unreliability, unpredictability and bias.

AI technology can be susceptible to 'brittleness' (Scharre, 2018, p.145). Brittleness occurs when the AI technology is not easily adapted to an unexpected or non-structured environment and so breaks down (International Committee of the Red Cross [ICRC], 2014). AI-based systems are most reliable in environments that are known, predictable and understood (ICRC, 2014). The legal risks of AI technology malfunction resulting from hardware faults, programming errors or sensor failures must be considered during the national review. Equally, the risk that AI technology will function in an unintended manner requires thorough T&E. This must identify the AI technology limitations to enable their use to be limited to those circumstances where it can be trusted to perform reliably and predictably. Moreover, AI technology must be capable of being understood by its operators and, therefore, it should be explainable.

Reliability is an objective measure of performance based on the results of successive trials (ICRC, 2014). Reliability also raises a risk of automation complacency or over-reliance on AWS decisions (Boulanin et al., 2020). This is particularly so where human operators are required to multitask by controlling several systems at once. Overreliance is a concern where the AI technology performs a function incorrectly; however, this is not either detected or questioned by a human operator. This risk requires States to develop the training of human operators to ensure they are aware of the risk and to create human–machine processes to mitigate against automation bias.

Reliability and predictability are related concepts. Where reliability is a measure of past performance, predictability is a measure of an AI technology's ability to perform its functions as it did in testing (Tattersall & Copeland, 2021). AI technology predictability may require uniformity of data inputs and environmental conditions in both testing and operational use.

A final operational risk is data bias, particularly where military AI technology relies on neural networks to perform tasks autonomously. Over the past decade, a combination of greater computational power and the availability of larger datasets have allowed deep learning algorithms to solve more real-world problems (Australian Department of Defence, 2020). Neural networks use multiple layers, with each layer designed to progressively extract increasingly higher-level features from the previous layers. Neural networks take raw data, sometimes with human-defined labels, and attempt to identify statistical patterns. However, it requires tens of millions of images to train a neural network to recognise images (Australian Department of Defence, 2020). If the human-labelled data contain inherent biases, there is a risk that the neural network's outputs will be biased and inaccurate. The expression 'garbage-in, garbage-out' recognises that a neural network will simply process the data that it is inputting (Ciklum, 2019). The quality of the output is closely related to the quality of the input. Data preparation is essential for any neural network designed to learn through supervised learning (Bunaes, n.d.). Ferraris (2020, p. 1) contends that:

> data is the precursor and an essential ingredient to building an AI/ML classification or prediction model. The more opportunities we take to collect good, realistic data, the more effective our systems will be in identifying and classifying similar objects in the future.

This is particularly the case where the military AI technology is designed to inform decisions concerning the use of force in armed conflict. Open-source civilian datasets, while readily available, may not be suitable or permitted by domestic law for military purposes as they may contain inherent biases (e.g., through inaccurate labelling) and risk developing biased AI decisions. States developing or acquiring AI technology for use in the military domain will need to consider creating a policy and processes for specific datasets that are appropriately labelled (e.g., identifying military features of interest), designed for operational environments, and tested and certified as good data to enable independent training and testing.

EXISTING TAXONOMIES OF RISK ASSOCIATED WITH AI USE IN THE MILITARY DOMAIN

The potentially ubiquitous nature of AI technologies across military systems and capabilities, and the breadth of tasks required of a State's armed forces, means that there is no 'one size fits all' approach to governance, assurance or risk. This is not an issue that is unique to the use of technology in assuming autonomous functionalities; and is a stated principal consideration underpinning the acquisition process:

AI Risk Taxonomies

There is yet to be an AI-specific technology standard adopted that properly defines 'risk' for the use of AI in a military context, although there are multiple civilian equivalent processes that seek to do so (Enzeani et al., 2021).

Although the Institute of Electrical and Electronics Engineers (IEEE) and International Organization for Standardization (ISO) standard for the ethical design of systems containing autonomy has recently been endorsed, which can be readily adapted to military use, the process of deriving the ethical risks associated with the use of AI technologies also engages the other risk areas identified above (that is, legal, safety and human control) (Huang et al., 2022).

This risk-based approach is consistent with the approach taken by different industries and organisations to ascribe levels of automation. In the automotive car industry, there are five levels of driving automation that are intended to indicate how capable a vehicle is of performing without human control (Harner, 2020). These levels are:

- *Driver assistance* – The vehicle assists the driver with some functions (e.g., assisted braking), but the driver is primarily responsible for all vehicle functions such as accelerating, braking, and monitoring the surrounding environment
- *Partial automation* – The vehicle can assist with steering and acceleration functions and enables the driver to disengage from some of the tasks. The driver must, however, be always ready to take over the control of the vehicle and is still responsible for most safety-critical functions and monitoring the environment
- *Conditional automation* – The vehicle is responsible for monitoring the environment. While the driver's attention is still required to maintain attention on driving, they can disengage from safety-critical functions such as braking
- *High automation* – In a self-determined safe environment, the driver can activate the automation to allow the vehicle to steer and brake, monitor the vehicle and road conditions, respond to events and determine when to change lanes, turn and use signals. It cannot determine more dynamic situations, such as traffic congestion and merging onto a highway
- *Complete automation* – There is no human attention required. The vehicle is completely responsible for driving and, therefore, there are no controls (e.g., steering wheel or brake pedal) to enable human control (Harner, 2020)

Similarly, in April 2022 the International Maritime Authority began work on the Maritime Autonomous Surface Ships code that contains the following four degrees of autonomy:

- *Degree one*: Ship with automated processes and decision support: Seafarers are on board to operate and control shipboard systems and functions. Some operations may be automated and at times be unsupervised but with seafarers on board ready to take control
- *Degree two*: Remotely controlled ship with seafarers on board: The ship is controlled and operated from another location. Seafarers are available on board to take control and to operate the shipboard systems and functions
- *Degree three*: Remotely controlled ship without seafarers on board: The ship is controlled and operated from another location. There are no seafarers on board
- *Degree four*: Fully autonomous ship: The operating system of the ship is able to make decisions and determine actions by itself (International Maritime Organisation, 2022)

In theory, these levels or degrees of autonomy may be modified and applied to military AI technology. A State may develop an AI risk matrix reflecting its national policy on the minimum level of human control over AI functions based on levels of human control assessed as necessary to mitigate different levels of legal, ethical or operational risks.

Selected State's Military-Specific AI Taxonomies

States developing AI technology for use in the military domain are recognising the need for military specific approaches to AI risk identification and mitigation. A number of States has adjusted extant AI ethics frameworks specifically for military use, noting that some of the societal concepts appearing in those frameworks do not align to military deontologies, particularly where the AI technology is potentially being utilised for the delivery of lethal weapons effects. There is a difference in the purpose, approach and value proposition of frameworks seeking to regulate civilian use of AI systems, with some of the military uses of AI. For instance, the concept of Azimov's 'do not harm' law of robotics may align with a military's requirement to comply with domestic workplace safety requirements, but it simply does not align with a military's requirement to use force to respond to a threat (Sorrell, 2017). Equally, the *lex specialis* of the laws of armed conflict means that there is a different legal (and ethical) framework that will dictate how the system will operate in its specific context. A comprehensive framework for use by the military must therefore be capable of handling heterogeneity in AI (such as technical specifications, environment, and complexity) and their intended use.

Below we describe a selection of these frameworks, which elicit the clear trend that there is an additional overlay in the assessment of lawful use of AI technology with compliance with the general principles of responsible use espoused by the State. Separately, the legal considerations relevant to assessing compliance with extant domestic and international legal obligations will overlap with many of the discrete principles articulated in the values-based frameworks. For example, the regularly cited need for transparency and accountability links to legal obligations relating to the same requirement; and articulation of safe use of AI technologies also necessitates an assessment of compliance with domestic safety regulatory obligations, as well as considering to what extent they might be displaced in a situation of armed conflict.

United States

The US Department of Defense (DoD) was one of the first to develop ethical AI practices in Defence. In 2018 the US Government published its *AI Strategy,* which directed the DoD to create guiding principles for lawful and ethical AI. In March 2020 this led to the DoD's *Ethical Principles for AI* (being: responsible, equitable, traceable, reliable and governable); accompanied by research into how to integrate them into DoD commercial prototyping and acquisitions programs. The US Defence Innovation Unit collaborated with AI experts and stakeholders from government, industry, academia, and civil society to develop a set of *Responsible AI Guidelines* which include specific questions for addressing during the planning, development, and deployment of ethical AI (U.S. Department of Defense, Defence Innovation Unit, 2021).

On January 25, 2023, the US updated their DoD Directive 3000.09 on AWS and reaffirmed their commitment to being a transparent global leader in establishing responsible policies regarding military uses of autonomous systems and AI (U.S. Department of Defense, 2023a). Less than one month later, on February 16, 2023, the US government unveiled its framework for a 'Political Declaration on the Responsible Military Use of Artificial Intelligence and Autonomy' at the 2023 Summit on Responsible AI in the Military Domain (U.S. Department of Defense, 2023b). The US Declaration seeks to build international consensus around how militaries can responsibly incorporate AI and autonomy into their operations and seeks to help guide States' development, deployment, and use of this technology for defence purposes to ensure it promotes respect for international law, security, and stability. The US declaration consists of a series of non-binding guidelines designed to describe best practices for the responsible use of AI in a defence context. This includes the need for military AI systems to be auditable, have explicit and well-defined uses, are subject to rigorous T&E across their lifecycle, and that high-consequence applications undergo senior-level review and are capable of being deactivated if they demonstrate unintended behaviour (Jenkins, 2023).

France

In 2019, the French Ministry of Armed Forces published their AI Task Force's *AI in Support of Defence Strategy* (Ministere Des Armees, 2019). This was the first military AI strategy published in Europe and it emphasises ethics and responsibility as essential elements of 'controlled AI' under the guidelines of 'trustworthy, controlled and responsible AI'. The French AI Strategy also creates a ministerial Defence Ethics Committee to oversee and advise on the adoption of AI.

Australia

Since 2019, Australia's Federal Department of Industry, Science, Energy and Resources has led the Australian development of an ethical framework by publishing *Australia's Artificial Intelligence Ethics Framework* (Australian Department of Industry, Science, Energy and Resources, 2019). This framework is of general application and describes eight voluntary AI Ethics Principles to be applied during each phase of an AI system's life cycle. These principles are intended to reduce the risk of negative effects of AI and ensure its use is supported by good governance standards. The principles are:

- *Human, societal and environmental wellbeing*: AI systems should benefit individuals, society and the environment
- *Human-centred values:* AI systems should respect human rights, diversity, and the autonomy of individuals
- *Fairness:* AI systems should be inclusive and accessible and should not involve or result in unfair discrimination against individuals, communities, or groups
- *Privacy protection and security:* AI systems should respect and uphold privacy rights and data protection and ensure the security of data

- *Reliability and safety:* AI systems should reliably operate in accordance with their intended purpose
- *Transparency and explainability:* There should be transparency and responsible disclosure so people can understand when they are being significantly affected by AI and can find out when an AI system is engaging with them
- *Contestability:* When an AI system significantly affects a person, community, group or environment, there should be a timely process to allow people to challenge the use or outcomes of the AI system
- *Accountability:* People responsible for the different phases of the AI life cycle should be identifiable and accountable for the outcomes of the AI systems, and human oversight of AI systems should be enabled. (Australian Department of Industry, Science, Energy and Resources, 2019)

Unlike many of its allies, Australia does not have a Defence AI strategy. However, in February 2021, Australia's Defence Science and Technology Group ('DSTG') published a report, *'A Method for Ethical AI in Defence'* ('MEAID') (Australian Department of Defence, Defence Science and Technology Group, 2021). MEAID is an Australia-specific framework to guide ethical risk mitigation which has not yet been officially endorsed by the Australian Department of Defence.

MEAID introduces the concept of 'facets' of ethical AI in defence, consisting of responsibility, trust, governance, law and traceability, and provides corresponding questions for the defence industry to address in relation to each facet. This provides a broad framework for defining legal and ethical requirements by AI stakeholders. It is designed to enable ethical risks associated with AI capabilities to be mitigated through industry-led system development, design and deployment. The facets are complemented by three tools:

- Ethical AI for defence checklist
- Ethical AI risk matrix
- Legal and Ethical Assurance Program Plan (LEAPP)

These risk assessment tools were designed to provide a pragmatic approach to legal and ethical risk identification and management. Importantly, the tools emphasise the important role of the defence industry in addressing legal and ethical risks in the design and development of any AI technology before it enters the ADF's capability life cycle. They enable the defence to assess and validate the defence industry's consideration of ethical risks. Where the AI technology is higher risk the defence industry is required to identify strategies for mitigating ethical risk in the form of a LEAPP to inform defence acquisition decisions and assessment (Australian Department of Defence, Defence Science and Technology Group, 2021).

This approach places the onus on self-assessment by the defence industry as it requires them to identify and mitigate legal and ethical risks associated with the design of AI systems intended for military use. This self-assessment brings with it inherent risks and does not negate the need for independent testing and verification. However, this approach enables the defence to require a new AI technology to be capable of

Article 36 compliance as a contractual pre-requisite. This places the onus on developing organisations to understand the legal risks and identify design or functional measures to address the risks. This is prior to a new capability entering a defence capability acquisition process, which is when the traditional weapons review obligation is often recognised to commence.

Australia, like France, The Netherlands, USA, UK and Singapore were amongst the nearly 60 States to endorse the 2023 Summit on Responsible AI in the Military Domain Call to Action, inviting States to develop national frameworks, strategies and principles on responsible AI in the military domain (REAIM, 2023).

United Kingdom

On October 23, 2020, the UK Defence Science and Technology Laboratory (DSTL) published a 'Biscuit Book' titled *Building Blocks for AI and Autonomy* (U.K. Defence Science and Technology Laboratory, 2020). The book describes the nine Building Blocks of AI and autonomy. This was followed on September 22, 2021 by the UK Government's *National AI Strategy* which creates a 10-year plan to ensure that the UK keeps up with evolving AI technology. On June 15, 2022, the UK Ministry of Defence published the Defence AI Strategy, outlining how they will adopt and exploit AI at pace and scale, transforming UK Defence into an 'AI ready' organisation and delivering cutting-edge capability; how they will build stronger partnerships with the UK's AI industry; and how they will collaborate internationally to shape global AI developments to promote security, stability and democratic values. It forms a key element of the UK *National AI Strategy* (U.K. Department for Science, Innovation and Technology, 2021).

The Netherlands

The Government of the Netherlands has taken a lead on garnering support for responsible use of AI in the military domain. On October 31 and November 1, 2022, in the lead-up to the 2023 Summit on Responsible AI in the Military Domain (REAIM 2023), the Ministry of Defence of the Netherlands hosted an expert workshop on the responsible use of AI in military systems. The workshop was attended by fifty leading experts from various countries and areas of expertise and reported:

- Trust is important. If AI is not understood, the system will neither be trusted nor used. On the other hand, misunderstanding can also lead to overconfidence and irresponsible use of AI in the military domain. New methods are needed to measure trust. It is also necessary to develop training courses to familiarise military personnel with AI
- The use of AI goes beyond weapon systems. Other application areas are involved too, including logistics and maintenance, decision support, early warning systems (such as in cyber or AI security), business operations and security
- There should be more focus on the regulation of AI development upfront, rather than regulation after the fact, and requires interaction between

previously siloed elements of acquisition and design. A transdisciplinary approach is crucial in achieving this early intervention and should encompass issues that are not currently contemplated during early design stages, such as the interrelationship between design, maintenance, training, doctrine development and ethics (Government of the Netherlands, 2022)

This was shortly followed on February 15–16, 2023, by the conduct of REAIM Summit, which resulted in a joint 'Call to Action' on the responsible development, deployment and use of AI in the military domain, being endorsed by the Netherlands and 57 of the 80 participating countries at the Summit. In particular, the Call to Action invited States who had not already done so, to develop national frameworks, strategies and principles on responsible AI in the military domain and encourage States to work together and share knowledge by exchanging good practices and lessons learnt (Government of the Netherlands, 2023).

The Netherlands have also recently announced their intent to launch a 'Global Commission on Responsible AI in the Military Domain' to raise awareness, clarify how to define AI in the military domain and determine how this technology can be developed, manufactured and deployed responsibly. The Commission will also aim to set out the conditions for the effective governance of AI (Government of the Netherlands, 2023).

Singapore

In December 2021, the Singapore Minister for Defence publicly announced their preliminary AI guiding principles, namely responsible, safe, reliable and robust (Hen, 2021). These guiding principles are based on their Model AI Governance Framework, first issued in January 2019 (second edition published in January 2020). In 2022, the Government of Singapore released their AI testing framework and toolkit to promote transparency and designed to convert high-level AI ethics principles into implementable measures (Singapore Ministry of Communications and Information, 2022). The Government of Singapore also endorsed the 2023 REAIM summit Call to Action on the responsible use of AI in the military domain.

NATO

On October 22, 2021, the North Atlantic Treaty Organisation (NATO) released its Strategy for AI (NATO, 2021). It provides a foundation for NATO and its Allies to develop responsible AI, accelerate AI adoption, enhance interoperability, and protect and monitor AI technologies. While technology development occurs primarily at the national or bi-lateral levels, NATO emphasises that legal, ethical and policy differences could endanger interoperability. NATO's strategy includes six Principles for Responsible AI in Defence: lawfulness; responsibility and accountability; explainability and traceability; reliability; governability and bias mitigation. NATO's Data & AI Review Board, established in October 2022, will help operationalise the principles (NATO, 2022).

Operationalisation of AI Frameworks into Acquisition Processes

While these processes provide a scaffold for the incorporation of legal and ethical issues relating to AI there is yet to be an operationalisation of these frameworks. The IEEE 7000 series is being used in some German Defence Force AI capability acquisition test processes, and the US and UK approaches to AI ethical and legal compliance go a long way to operationalise these requirements (Koch, 2023). Noting the nascence of these processes, it is unclear if they will incorporate the requisite risk issues that are posed by the use of military AI, particularly in armed conflict situations. In the civilian context, the US National Institute of Standards and Technology (NIST) Risk Management framework and its accompanying playbook provide an excellent starting point, creating a comprehensive risk-management framework incorporating these additional concerns. However, it is more aimed at the creation of a process for an organisation rather than creating the process itself (U.S. Department of Commerce, 2022). It does not (nor is it designed to) provide specific guidance to meet an individual organisation's design and T&E requirements.

Finally, it is unsettled if a separate and bespoke risk management process is the preferred approach in the adoption of novel military capabilities incorporating autonomous functionality. The Australian Army RAS-AI Strategy contemplated a need to adjust utilising 'traditional' acquisition processes in acquiring autonomous technologies, but also in the event of 'discover[y] of RAS technologies are emerging faster than traditional acquisition systems may allow, or which are truly disruptive…tailored rapid acquisition pathways' may need to be applied (Australian Army, 2020, p. 41). There are sound resource efficiency and acquisition efficacy reasons to incorporate the requirements for risk mitigation of these processes into extant acquisition processes; but ensure that these methodologies are sufficiently agile and flexible to apply to rapid or spiral acquisitions processes that militaries are increasingly likely to apply to novel and emerging disruptive technologies.

REVIEW OF SELECT NATIONAL WEAPONS REVIEW PROCESSES AS THEY RELATE TO LEGAL RISK

The purpose and function of a weapons review process is to assess a new weapon, means or method of warfare for legal compliance. The Article 36 obligation, which AP I States are compelled to comply with, applies to the assessment of legal obligations arising in situations of international armed conflict. A number of other States apply this obligation, not as a matter of legal obligation, but as good policy. Many Article 36 reviewing States apply this obligation in assessing the normal and expected use of the weapon in specific conflict scenarios (that is, if the capability is being acquired for general use by the military, the assessment will contemplate limitations on use in situations of armed conflict of an international and non-international character) (Jevglevskaja, 2021).

In their current form, the focus of weapons reviews is generally limited to international legal compliance obligations. While some States support the inclusion of broader domestic law and policy considerations (such as ethical and societal considerations) into their weapons review process, the majority of State practice is focused on international law compliance, which necessarily includes an assessment of IHL obligations in the fielding of the weapons, means or method being assessed (Jevglevskaja, 2021). Accordingly, there is an opportunity to facilitate understanding, and acceptance, of legal risk across the capability life cycle as it relates to the broader issues such as safety and ethics, through the expansion of the weapons review process.

Discussions relating to the obligation to re-review capabilities that contain autonomy, post-acquisition, but triggered by the system's ability to self-learn, adjust or deploy into contexts not contemplated during the initial review, have been identified as a challenge in the adoption of autonomous weapons systems, and discussed by States as a challenge (Cavdarski et al., 2023). This challenge, specific to the use of autonomy in weapons systems, demonstrates that there is a need to reconsider how to adjust weapons review processes to account for autonomy, regardless of the question of broader policy and law considerations.

International law restricts that choice in multiple ways. Firstly, it prohibits generally weapons, methods and means of warfare of a nature to cause certain types of harm including superfluous injury to combatants and indiscriminate effects (Additional Protocol I, 1977, art. 35). Secondly, it prohibits and limits the use of specific weapons, means and methods of warfare (International Committee of the Red Cross [ICRC], 2005, Customary IHL Database Rules 17, 71-86).

Additional Protocol I requires States party to review any weapon, means or method of warfare for compliance with that State's international legal obligations. Specifically, Article 36 of AP I States:

> In the study, development, acquisition or adoption of a new weapon, means or method of war, a High Contracting Party is under an obligation to determine whether its employment would, in some or all circumstances, be prohibited by this Protocol or by any other rule of international law applicable to the High Contracting Party.

The aim of Article 36 is to 'prevent the use of weapons that would violate international law in all circumstances, and to impose restrictions on the use of weapons that would violate international law in some circumstances' (International Committee of the Red Cross [ICRC], 2006, p.4). The determination of legality is based upon the normal or expected use of the weapon (de Preux, 1987). As such, the review considers the weapon as it is presented to the reviewer and relies on testing based on defined use cases. National weapons review determinations are not binding on other States, and are not intended to create a separate legal standard, but are rather intended to 'ensure that means or methods of warfare will not be adopted without the issue of legality being explored with care' (de Preux, 1987, p. 1469).

A weapons review will focus on the legality of a weapon *per se* rather than the legality of its use in particular circumstances (Boothby, 2009). It is generally accepted that determining the lawful use of a lawful weapon depends on the context and the responsibility for making that determination on the basis of IHL rules rests primarily with military commanders, weapon operators and legal advisors made available to commanders, at the appropriate level (Farrant & Ford, 2017).

This traditional weapons review approach focus on legality per se is considered to be too narrow to determine the legality of weapons that employ AI technology to perform tasks that are governed by IHL rules. Such AWS are likely to require weapons review throughout their life cycles to ensure that the AI technology controlling the AWS generates results that comply with a State's legal obligations. Thus, a thorough weapons review should be part of the entire design and procurement process of an AWS, both informing the AWS development and assessing its legality during use. This will require a broader, multidisciplinary and ongoing approach in addition to the traditional review processes. This may extend throughout the weapon's life cycle to assess its ability to operate in multiple environments that require the AWS to interpret data that differs from that upon which its performance was initially reviewed. It will also address advances in AI technology that affect the AWS operation (Copeland, 2023).

Assuming the AI technology will enable changes in the AWS operation, there must be measures to ensure that the legality of such changes is assessed. In the case of small operational changes, these could be assessed by an operational or field weapons review that builds upon the weapons review conducted before introduction into service. Such weapons reviews must also be flagged with the original reviewing authority to identify whether to trigger a re-review for that particular capability. More significant changes may require specific operational limitations to be placed upon the AWS to ensure ongoing legal compliance. Specifically, an assessment of whether or not the system is capable of changing its normal or expected use will determine whether a re-review of the original assessment for legal compliance is required.

Furthermore, the review process will require more careful consideration of the expected AI technology's operating environment. The weapons review of an AWS must take into account the impact of different operating environments and operational circumstances on the AI technology. Unfamiliar conditions may risk brittleness or unpredictability in the AI technology, particularly where the data used to train the AI is focused on a particular legal regime. For example, an AI-enabled system designed to operate in an international armed conflict would require re-review if it were to be deployed in a non-international armed conflict to ensure that any rules coded into the system reflect the changed criteria for assessing targetable status.

Further, assessments of the manner in which these determinations are made by the AWS must also be undertaken. For example, in determining compliance with the law, the assumptions programmed into the AWS must also be legally compliant. For example, a fundamental IHL principle is that in cases of doubt, status is presumed to be civilian (Additional Protocol 1, 1977, arts 50(1) and 52(3)). The AWS must also therefore achieve the same kind of result in cases of doubt. In this case, the performance of the AWS in a particular context will form part of the weapons review – which is additional to the content included in a traditional weapons review.

National weapons review steps provide a basis for States to develop national mechanisms for mitigating legal, ethical and operational risks in the development and use of military AI technology. This will require multidisciplinary expertise and a policy framework that articulates both the process and standards to identify and mitigate risks. The next section proposes an approach to multirisk management through an expanded weapons review process across three broad stages during the life-cycle of military AI technology.

While there continues to be little publicly available information about the weapons review process – despite regular pledges by States undertaking to do so (Goussac et al., 2023) – a number of States have provided public versions of their weapons review process, as they relate specifically to autonomy. The US, for example, has a specific policy document, which requires a 'superior' review of any system containing autonomy, while other States have added steps to their existing weapons review processes. For example, the updated Australian approach has also created an additional step to incorporate specific considerations in review of capabilities that contain novel technologies like cyber or autonomous components.

IDENTIFICATION OF LINKAGES BETWEEN EXISTING RISK-BASED GOVERNANCE APPROACHES TO THE WEAPONS REVIEW PROCESS

Existing processes to review the legality of AWS do not specifically adopt a risk-based approach to address the broader regulatory challenge presented by autonomy. The national weapon review process may be expanded to include three broad stages of the military AI technologies lifecycle. Each stage is designed to achieve a different outcome to assist in the identification of legal, ethical, human control and operational risks. The first stage is the 'informative stage' which recognises that the design and development of AI technology are likely to occur outside the State by private defence industry, academia and research organisations. As such, much of the critical development occurs before the AI technology enters a defence acquisition process and may be done in ignorance of a State's weapons review or international law obligations. The informative stage seeks to inform those designing or developing the AI technology of the national weapons review process and its requirements through a process of self-assessment designed to identify legal risks. The second, 'determinative stage' focusses on the military acquisition process and the determination of the AI technology risks prior to its introduction into service. The determinative stage includes the traditional weapons review process. Finally, the third 'governance stage' recognises that military AI technology's functions are unlikely to be fixed and may be influenced by machine learning, environmental conditions and operational circumstances. The governance stage recognises the need for ongoing governance of the AI technology during its in-service life to ensure its use remains in compliance with the State's IHL and international law obligations (Copeland, 2023).

CONCLUDING OBSERVATIONS

Armed conflict is inherently risky. The use of AI technologies brings additional risks to the conduct of operations. These risks may be mitigated through the modification of existing risk management processes – in particular the weapons review process – to identify risk and integrate legal risk assessment across the capability life cycle from initial study and development, through acquisition during in-service life.

By utilising a mixed methodology for the regulation of AI technologies, and augmenting existing acquisition approaches – which apply across the capability' life-cycle – AI technologies can be integrated into militaries in a responsible way taking into account legal, ethical and safety assurance requirements.

The use of weapons reviews supports the identification of general compliance issues. Further, in the event that a legally binding instrument is agreed upon as a consequence of the current international debate about the regulation of LAWS, this approach also supports the identification of whether systems might incorporate the use of AI technology generally, which enables the identification of systems that 'breach the line' of what is prohibited.

This chapter has highlighted how the weapons review process can be used and adjusted to account for some of the peculiarities of AI and form part of a multifaceted governance approach that includes risk-based as well as performance-based analysis. While the weapons review itself is a risk-based process, other acquisition processes incorporate performance-based process, forming part of a regulated systems approach across the capability life cycle. Weapons review have utility in forming part of the broader regulatory framework, which itself must be multi-faceted and incorporate different regulatory methodologies in order to account for the multifaceted risk profile of using AWS within a military system. Further, we consider that through augmenting the existing process, rather than creating a bespoke and separate AI risk-mitigation process, we can enhance the governance framework already in place, while producing efficacies in acquisition that will assist militaries in retaining their technological edge.

ACKNOWLEDGEMENTS

The views in this paper represent the personal views of the authors and do not represent the views of any other organisation.

REFERENCES

American Society for Quality. (2015). *Quality management systems - requirements* (ASQ ISO 9001:2015). https://asq.org/quality-press/display-item?item=T1040

Apte, A. (2005). *Spiral development: A perspective.* Defence Acquisition Innovation Repository. https://dair.nps.edu/handle/123456789/2548

Australian Army. (2020). *Army robotic and autonomous systems strategy V2.0*. https://research-centre.army.gov.au/sites/default/files/Robotic%20and%20Autonomous%20Systems%20Strategy%20V2.0.pdf

Australian Defence Force. (2019, August 15). *Australian Defence Force Procedures 5.0.1, Joint Military Appreciation Process, Ed 2*. https://theforge.defence.gov.au/sites/default/files/adfp_5.0.1_joint_military_appreciation_process_ed2_al3_1.pdf

Australian Department of Defence (2003). *Defence procurement review (2003) ('the Kinnaird Review')*. Commonwealth of Australia.

Australian Department of Defence, Defence Science and Technology Group. (2021, February). *A method for ethical AI in defence*. (DSTG Report no DSTG-TR-3786). Commonwealth of Australia. https://www.dst.defence.gov.au/publication/ethical-ai

Australian Department of Defence, Defence Science and Technology. (2010). *Technical risk assessment handbook*. Commonwealth of Australia. https://www.dst.defence.gov.au/sites/default/files/basic_pages/documents/Technical-Risk-Assessment-Handbook_2.pdf

Australian Department of Defence. (2003). *Sovereign industrial capability priorities*. Commonwealth of Australia. https://www.defence.gov.au/business-industry/capability-plans/sovereign-industrial-capability-priorities

Australian Department of Defence. (2020). *ADF concept for robotics and autonomous systems*. Commonwealth of Australia. https://tasdcrc.com.au/wp-content/uploads/2020/12/ADF-Concept-Robotics.pdf

Australian Department of Defence. (2021). *Defence test and evaluation (T&E) strategy*. Commonwealth of Australia. https://www.defence.gov.au/about/strategic-planning/defence-test-and-evaluation-strategy

Australian Department of Industry, Science, Energy and Resources. (2019). *Australia's artificial intelligence ethics framework*. Commonwealth of Australia. https://www.industry.gov.au/data-and-publications/australias-artificial-intelligence-ethics-framework

Australian National Audit Office (ANAO). (2015, November 2). *Test and evaluation of major defence acquisitions* (Auditor-General Report No 15 of 2015-16). Commonwealth of Australia. https://www.anao.gov.au/work/performance-audit/test-and-evaluation-major-defence-equipment-acquisitions-0#footnote-099

Boothby, W. H. (2009). Weapons and the laws of armed conflict. Oxford University Press.

Boulanin, V., Bruun, L., & Goussac, N. (2021). *Key issues concerning weapons reviews and legal advice: Autonomous weapon systems and international humanitarian law* (Stockholm International Peace Research Institute Report).

Boulanin, V., Carlsson, M. P., Davison, N., & Goussac, N. (2020). *Limits on autonomy in weapon systems: Identifying practical elements of human control* (Stockholm International Peace Research Institute Report).

Bunaes, H. P. (n.d.). Bad data can kill good AI. *Global Banking and Finance Review*. https://www.globalbankingandfinance.com/bad-data-can-kill-good-ai/

Capabilities Governance and Assurance (CTO Group). (n.d.). *Governance & assurance*. https://ctogroup.com.au/capability/governance/#:~:text=Governance%20refers%20to%20the%20relationships,in%20their%20organisation's%20delivery%20capability

Cavdarski, R., Sanders L., & Liivoja, R. (2023). *Weapons reviews of autonomous weapon systems: Report on submissions to the GGE on LAWS (version 1.0)*. University of Queensland.

Certain Conventional Weapons (CCW) Convention, Group of Governmental Experts (GGE) on Emerging Technologies in the Area of Lethal Autonomous Weapons Systems (LAWS). (2016). *Advanced version of the draft report of the 2016 informal meeting of experts on lethal autonomous weapons (LAWS)*.

Certain Conventional Weapons (CCW) Convention, Group of Governmental Experts (GGE) on Emerging Technologies in the Area of Lethal Autonomous Weapons Systems (LAWS). (2017, November 20). *Report of the Main Committee II*. UN Doc CCW/GGE.1/2017/CRP.1.

Certain Conventional Weapons (CCW) Convention, Group of Governmental Experts (GGE) on Emerging Technologies in the Area of Lethal Autonomous Weapons Systems (LAWS). (2019, March 26). *Australia's system of control and applications for autonomous weapon systems.* CCW/GGE.1/2019/WP.2/Rev.1.

Certain Conventional Weapons (CCW) Convention, Group of Governmental Experts (GGE) on Emerging Technologies in the Area of Lethal Autonomous Weapons Systems (LAWS). (2021, April 19). *Working paper, chairman's summary of the 2020 session of the group of government experts on emerging technologies in the area of lethal autonomous weapons systems.* CCW/GGE.1/2020/WP/7.

Ciklum (2019, January 11). *Garbage in, garbage out: How to prepare your date set for machine learning.* https://www.ciklum.com/blog/garbage-in-garbage-out-how-to-prepare-your-data-set-for-machine-learning/

Coglianese, C. (2017). The Limits of performance-based regulation. *University of Michigan Journal of Law Reform, 50*(3), 525–563.

Copeland, D. (2023). *The Article 36 weapons review of autonomous weapons systems* [Unpublished PhD Thesis]. Australian National University.

Crootof, R. (2015). A meaningful floor for "meaningful human control. *Temple International & Comparative Law Journal, 30*(1), 53–62.

de Preux, J. (1987). Protocol I - Article 36 - New Weapons. In Y. Sandoz, C. Swinarski, & B. Zimmerman (Eds.), *Commentary on the additional protocols of 8* June 1977 *to the Geneva Conventions of 12* August 1949 (pp. 422–428). ICRC and Martinus Nijhoff.

Dempsey, J. (2022, 6 October). *Cybersecurity regulation: It's not performance based if it can't be measured.* Lawfare Blog. https://www.lawfareblog.com/cybersecurity-regulation-its-not-performance-based-if-outcomes-cant-be-measured

Enzeani, G., Koene, A., Kumar, R., Santiago, N., & Wright, D. (2021, August). *A survey of artificial intelligence risk assessment methodologies.* Ernst & Young Trilateral Research Report.

Farrant, J., & Ford, C. (2017). Autonomous weapons and weapons reviews: The UK second international weapons review forum. *International Law Studies, 93,* 389–421.

Fawcett, D. (2022, September 14). Urgent change needed in Defence's process for major acquisitions. *ASPI The Strategist.* https://www.aspistrategist.org.au/urgent-change-needed-in-defences-processes-for-major-acquisitions

Ferraris, P. (2020, February 5). *Aided detection on the future battlefield.* U.S. Army News. https://www.army.mil/article/232074

Goener, J. (2021, September 3). *Risk versus reward - understanding why we take risk and how to assess its value.* The Cove. https://cove.army.gov.au/article/risk-versus-reward-understanding-why-we-take-risk-and-how-assess-its-value.

Goussac, N., Jevglevskaja, N., Liivoja, R., & Sanders, L. (2023). *Enhancing the weapons review of autonomous weapon systems: Report of an expert meeting (Sydney, 28–30 March 2023).* University of Queensland.

Government of the Netherlands. (2023, February 16). *Call to action on responsible use of AI in the military domain.* https://www.government.nl/latest/news/2023/02/16/reaim-2023-call-to-action

Harner, I. (2020, July 3). *The 5 autonomous driving levels explained.* IOT for all. https://www.iotforall.com/5-autonomous-driving-levels-explained.

Hen, N. E. (2021, October 12). *Welcome address at the third Singapore defence technology summit.* Singapore Minister for Defence. https://www.mindef.gov.sg/web/portal/mindef/news-and-events/latest-releases/article-detail/2021/October/12oct21_speech.

Huang, C., Zeqi Zhang, Z., Mao, B., & Yao, X. (2022). An overview of artificial intelligence ethics. *IEEE Transactions on Artificial Intelligence, 4,* 799–819. https://ieeexplore.ieee.org/stamp/stamp.jsp?arnumber=9844014?

Hussain, S. M., Nurwidyantoro, W., Perera, A., Shams, H., Grundy, R., & Whittle, J. (2022). Operationalising human values in software engineering: A survey. *IEEE Access Journal*, *10*, 75269–75295. https://ieeexplore.ieee.org/document/9829732

International Committee of the Red Cross (ICRC). (2006). A guide to the weapons review of new weapons, means and methods of warfare: Measures to implement Article 36 of Additional Protocol I of 1977. https://www.icrc.org/en/publication/0902-guide-legal-review-new-weapons-means-and-methods-warfare-measures-implement-article

International Committee of the Red Cross (ICRC). (2014). Report on the ICRC expert meeting on autonomous weapon systems: Technical, military, legal and humanitarian aspects, 26–28 March 2014, Geneva. https://www.icrc.org/en/document/report-icrc-meeting-autonomous-weapon-systems-26-28-march-2014

International Committee of the Red Cross (ICRC). (2005). Customary IHL Database, Rules 17, 46-69, and 71-86. https://ihl-databases.icrc.org/customary-ihl/eng/docs/v2.

International Maritime Organisation. (2022), *Media centre, autonomous shipping.* https://www.imo.org/en/MediaCentre/HotTopics/Pages/Autonomous-shipping.aspx

International Standards Organisation. (2018). *Risk management-guidelines* (ISO Standard No. 31000:2018). https://www.iso.org/standard/65694.html

Jenkins, B. D. (2023, February 16). Under Secretary for Arms Control and International Security, Keynote Remarks by U/S Jenkins (T) to the Summit on Responsible Artificial Intelligence in the Military Domain (REAIM) Ministerial Segment. [Keynote remarks]. Responsible Artificial Intelligence in the Military Domain Summit, The Hague. https://www.State.gov/keynote-remarks-by-u-s-jenkins-t-to-the-summit-on-responsible-artificial-intelligence-in-the-military-domain-reaim-ministerial-segment/

Jevglevskaja, N. (2021). *International law and weapons review.* Cambridge University Press.

Koch, W. (2023). AI for aerospace and electronic systems: Technical dimensions of responsible design. *IEEE A&E Systems Magazine.* https://ieeexplore.ieee.org/stamp/stamp.jsp?tp=&arnumber=9980420

Lorell, M., Lowell, J., & Younossi, O. (2006). *"Evolutionary acquisition" is a promising strategy, but has been difficult to implement.* RAND Research Brief. https://www.rand.org/pubs/research_briefs/RB194.html.

Ministere Des Armees. (2019). *Report of the AI task force: Artificial intelligence in support of defence.* French Government. https://www.defense.gouv.fr/salle-de-presse/communiques/communiques-du-ministere-des-armees/communique_publication-du-rapport-du-ministere-des-armees-sur-l-intelligence-artificielle

Moy, G, et al, (2020). *Technical Report: Recent Advances in Artificial Intelligence and Their Impact on Defence* (Defence Science and Technology Group). https://www.dst.defence.gov.au/publication/recent-advances-artificial-intelligence-and-their-impact-defence

National Academies of Sciences, Engineering, and Medicine. (2018). *Designing safety regulations for high-hazard industries.* The National Academies Press. https://doi.org/10.17226/24907.

Netherlands Ministry of Foreign Affairs. (2022, November 9). *Expert workshop on responsible use of AI in military systems.* Government of the Netherlands. https://www.government.nl/latest/weblogs/reaim-2023/2022/expert-workshop-on-responsible-use-of-ai-in-military-systems

North Atlantic Treaty Organisation (NATO). (2021). *Artificial intelligence strategy.* https://www.nato.int/cps/en/natohq/official_texts_187617.htm

North Atlantic Treaty Organisation (NATO). (2022), *Data and artificial intelligence review board.* https://www.nato.int/cps/en/natohq/official_texts_208374.htm

Organisation for Economic Co-operation and Development (OECD). (2014). *The governance of regulators.* OECD Best Practice Principles for Regulatory Policy, OECD Publishing. https://doi.org/10.1787/9789264209015-en

Protocol Additional to the Geneva Conventions of 12 August 1949, and Relating to the Protection of Victims of International Armed Conflicts (AP I), June 8, 1977. https://ihl-databases.icrc.org/en/ihl-treaties/api-1977

REAIM. (2023). *REAIM 2023 call to action, responsible AI in the military domain summit co-hosted by Government of the Netherlands and Republic of Korea, The Hague, The Netherlands (16 February 2023).* https://www.government.nl/ministries/ministry-of-foreign-affairs/documents/publications/2023/02/16/reaim-2023-call-to-action

Scharre, P. (2018). *Army of none: Autonomous weapons and the future of war.* Norton & Company.

Singapore Ministry of Communications and Information. (2022). *Singapore launches world's first AI testing framework and toolkit to promote transparency; invites companies to pilot and contribute to international standards development (Media Release, 25 May 2022).* Government of Singapore. https://www.imda.gov.sg/content-and-news/press-releases-and-speeches/press-releases/2022/singapore-launches-worlds-first-ai-testing-framework-and-toolkit-to-promote-transparency-invites-companies-to-pilot-and-contribute-to-international-standards-development

Sorrell, T. (2017, March 17). *Asimov's laws of robotics aren't the moral guidelines they appear to be.* The Conversation. https://theconversation.com/asimovs-laws-of-robotics-arent-the-moral-guidelines-they-appear-to-be-74634#:~:text=But%20Asimov's%20laws%20either%20fail,way%20of%20respecting%20human%20autonomy.

Tattersall, A., & Copeland, D. (2021). Reviewing autonomous cyber capabilities. In R. Liivoja & A. Väljataga (Eds.), *Autonomous cyber capabilities under international law* (pp. 206–257). NATO CCDCOE.

U.K. Defence Science and Technology Laboratory. (2020). *Building blocks for AI and autonomy: A Dstl biscuit book.* Government of the United Kingdom. https://www.gov.uk/government/publications/buidling-blocks-for-ai-and-autonomy-a-biscuit-book.

U.K. Department for Science, Innovation and Technology. (2021). *National artificial intelligence strategy.* Government of the United Kingdom. https://www.gov.uk/government/publications/national-ai-strategy

U.S. Department of Commerce. (2022). *AI risk management framework.* National Institute for Standards and Technology. https://www.nist.gov/itl/ai-risk-management-framework

U.S. Department of Defense, Defence Innovation Unit. (2021). Responsible AI guidelines. Government of the United States. https://www.diu.mil/responsible-ai-guidelines.

U.S. Department of Defense. (2023a). *DOD Directive 3000.09 autonomy in weapon systems (25 January 2023).* U.S. Government. https://www.esd.whs.mil/portals/54/documents/dd/issuances/dodd/300009p.pdf.

U.S. Department of Defense. (2023b), *DoD announces update to DoD Directive 3000.09, 'Autonomy in weapon systems'* (Media Release, 25 January 2023). U.S. Government. https://www.defense.gov/News/Releases/Release/Article/3278076/dod-announces-update-to-dod-directive-300009-autonomy-in-weapon-systems/

U.S. Department of State, Bureau of Arms Control, Verification and Compliance. (2023). *Political declaration on responsible military use of artificial intelligence and autonomy.* U.S Government. https://www.State.gov/political-declaration-on-responsible-military-use-of-artificial-intelligence-and-autonomy/

'Killer Pays'

State Liability for the Use of Autonomous Weapons Systems in the Battlespace

8

Diego Mauri

ADDRESSING THE 'CYBERNETIC ERROR' TODAY

In his course delivered some fifty years ago at the Hague Academy of International Law – devoted to liability for ultra-hazardous activities in international law – Wilfried Jenks quickly dwelled on damages resulting from the use of cybernetic systems and thus cautioned: '[t]he question of liability for cybernetic error, or for damage resulting therefrom [...] calls for attention. These questions may at any time call for consideration on an international scale' (Wilfred Jenks, 1966, p. 169).

The moment foreshadowed by Wilfred Jenks – namely, the moment when even the international legal order must deal with 'cyber error' and its harmful consequences – seems to have arrived, as this very book and the debates from which it has derived clearly demonstrate. Addressing the topic of the military use of artificial intelligence (AI) from a multidisciplinary perspective cannot be postponed any longer. In this chapter, I will focus on a specific category of weaponry that is expected to be endowed with AI capabilities, namely Autonomous Weapons Systems (AWS): those are weapons systems that, once activated, can select and engage targets without further human intervention (Mauri, 2022, p. 14; US DoD, 2023, p. 21).

AWS do not necessarily feature AI capabilities: several weapons systems, already existing and fielded by many States' armed forces, can operate autonomously yet without resorting to AI. Examples include sensor-fused and loitering munitions (such as the Israeli Harpy and Harop), missile- and rocket-defense weapons, used for air defense of ships and ground installations (such as the US Phalanx and C-RAM). It is held that the ongoing armed conflict between the Russian Federation and Ukraine already is a laboratory for experimenting new types of AWS. Research in AI-related technologies (such as machine learning, neural networks, and evolutionary computation) will bring existing military capabilities to a higher level.

The debate around AWS – including those that will feature AI – has so far produced an impressive number of documents, ranging from official declarations and positions by States to contributions of scholarship and interventions by civil society representatives. To summarize as much as possible, at least two 'macro-strands' of discussion can be identified. These are two different but not antithetical sets of issues, as evidenced by the circumstance that most papers dealing with AWS tend to address both of them.

On the one hand, there is a concern that such systems may be employed in contravention to relevant international obligations, namely those contained in International Humanitarian Law (IHL), International Human Rights Law (IHRL), and those relating to the use of force in international relations (the so-called *jus ad bellum*) (Boothby, 2013, p. 71; Egeland, 2016, p. 89; Heyns, 2013, p. 46; McFarland, 2020; Roff, 2015, p. 37; Spagnolo, 2017; Schmitt & Thurnher, 2013). In other words, the first ground for discussing AWS is their compatibility with existing norms of international law: can they be used in accordance with relevant rules and principles applicable to the use of force against individuals and objects?

Then there is the subsequent need to properly allocate responsibility in case the use of AWS results in a violation of the mentioned obligations (Amoroso & Giordano, 2019).

To begin with, one may question whether criminal law is adequate for coping with challenges raised by AWS, in light of their complexity and the 'many hands' involved: software developers, engineers, programmers, policy-makers, military commanders, and soldiers or operators within the chains of command (Bo, 2021; McFarland & McCormack, 2014). How can criminal responsibility be properly distributed among all actors who, directly or indirectly, play a role in a specific course of action? Evidently, this issue is but magnified by the advent of AI applications to those systems.

The second issue lies in the ways to hold accountable companies (that is, legal persons) that engineer, develop, produce, and sell AWS: as a matter of fact, specific courses of action could be taken as a result of defects during the programming stage. Can those companies – typically, defense contractors – be sued before domestic courts? This issue can be seen from the perspective of corporate accountability on the international plane, which has been extensively debated by scholars in recent years under the well-known 'business and human rights' movement (Batesmith, 2014).

Third, one may investigate the last – but not least – a subject that is involved in the actual deployment of AWS, which is the State (Crootof, 2022; Hammond, 2015). No one fails to see that, at least in the near future, AWS will be developed and deployed on the basis of States' decisions, either in the battlespace or in law-enforcement operations. Certainly, one may even envisage scenarios in which AWS are hacked and employed by non-state actors, such as terrorist groups.

For the sake of the reasoning, however, I will focus exclusively on AWS operated by States during armed conflict, and thus address the issue of state responsibility, allegedly the less investigated area of responsibility deriving from AWS misdoings. Can it really be said that, according to one of the most popular theses, 'responsibility gaps' structurally (i.e., because of the very characteristics of the technology employed) result from the use of AWS? The answer to this question is determinative: according to some, the paramount reason for banning AWS is because their use is bound to generate such 'gaps'. The argument can be summarized in the words of the former UN Special Rapporteur on extrajudicial, summary, or arbitrary executions, Christof Heyns: '[i]f the nature of a weapon renders responsibility for its consequences impossible, its use should be considered as unethical and unlawful as an abhorrent weapon' (Heyns, 2013, para. 75).

Is this the case with AWS?

In the following, I will consider three different scenarios. In the first one, I will analyze future deployments of AWS intended to act in breach of applicable IHL rules and principles: this case is the one raising fewer problems (Section 8.2). I will then turn to scenarios in which AWS target unintended targets because of a malfunctioning: in this case, the international wrongdoing is the result of 'fault' on the part of the State, which does not succeed in complying with IHL due diligence obligations (Section 8.3). Last, I will focus on 'false-positives' scenarios, that is cases in which the unintended targeting cannot be traced back to any faulty conduct by the State: according to some, it is precisely because of those events that AWS would be prohibited, as it would be impossible to establish any responsibility (Section 8.4). I will address this argument by demonstrating that international law can be adapted to such 'false-positives' scenarios, both *de lege lata* and *de lege ferenda*. I eventually propose a mode of liability inspired to some that already exist in international law, which I – provokingly – name 'killer pays', with a view to demonstrating that the 'responsibility gap' argument, if used to argue for the illegality of AWS, leaves much to be desired (Section 8.5).

INTENTIONAL VIOLATIONS OF INTERNATIONAL LAW ON TARGETING

How can international responsibility be established for the intentional use of AWS in violation of applicable international norms, namely IHL? One might think of a scenario in which a State's army decides to deploy AWS, e.g., in contravention of the technical specifications provided by the manufacturing company: a system without the advanced capacity of distinguishing targets (which is apt for submarine environments) is deployed in clustered environments, such as urban guerrilla warfare. In this scenario, the selection and engagement of impermissible targets stands as a very likely, if not certain, consequence of employing AWS in a clustered environment.

The first set of IHL obligations coming to the fore are those encapsulated in the principle of distinction, which prohibits the election as targets of an attack of objects and persons protected in the context of an armed conflict. Specifically, under Article 48 of the First Additional Protocol to the Geneva Conventions of 1949, Parties to a conflict are required to distinguish 'between the civilian population and combatants and between civilian objects and military objectives' and thus to 'direct their operations only against military objectives'. Indiscriminate attacks are prohibited, which include also attacks that 'may be expected to cause incidental loss of life, injury to civilians, damage to civilian objects, or a combination thereof, which would be excessive in relation to the concrete and direct military advantage anticipated' pursuant to Article 51, para. 5, lit. b) – a provision that incorporates the so-called principle of proportionality. In addition, there are also rules imposing States a duty to take all necessary measures in order to neutralize or minimize risks for protected goods and persons, which are listed in Article 57 and can be appraised under the principle of precautions in attack.

In a scenario like the one considered here, in which the selection and engagement of an impermissible target is a result that state authorities aim to realize intentionally, I argue that it will not be hard to ascertain international responsibility and allocate it. Either by violating the negative duties listed above (i.e., the rules on distinction and proportionality) or by failing to comply with positive duties (i.e., under the rules on precautions in attack), there seems to be little to no doubt that States can be held responsible for AWS misdoings. Moreover, the nature of the weaponry employed does not affect the establishment of international responsibility: States would be held responsible as regularly happens with traditional equipment (e.g., missiles launched from a manned system).

From this perspective, it is hard to see how one could speak of 'responsibility gaps': the use of AWS (even AI-equipped systems), in lieu of less advanced, 'conventional' weaponry, has no implications whatsoever on the allocation of international responsibility. Moreover, it is worth adding that a scenario such as the one analyzed here would not raise problems even with regard to the international responsibility of the individual, namely for international crimes likely to come to the fore (primarily war crimes). In sum, international law as existing today can easily cope with such scenarios of intentional mistargeting.

FAULTY VIOLATIONS OF INTERNATIONAL LAW ON TARGETING

As is evident, the scenario addressed so far is the one raising the least difficulties. In a second scenario, state authorities intend to select and engage permissible targets but – due to inadequate planning or preparation of the operation, or system malfunction or failure that they were or should have been aware of – they end up targeting protected objects and persons. This is not the intended result of state conduct; rather, it is the result of a set of circumstances that the State failed to exercise control over.

I then need to digress a bit and address a classical question of the law of international responsibility, namely the relevance of fault (*culpa*) as a constitutive element of international responsibility of States (Gattini, 1992; Palmisano, 2007). The opposition between 'subjectivist' theories (enumerating fault among the constitutive elements of international wrongdoings) and 'objectivist' theories (which instead disregard it) seems to have been overcome, at least since the adoption of the 2001 Draft Articles on Responsibility of States for Internationally Wrongful Acts (Diggelmann, 2006).

As is known, the Draft Articles are not binding per se, but to the extent to which they largely correspond to customary law on that matter, they are of relevance to ascertaining whether *culpa* is a constitutive element of state responsibility. The answer is negative: the Draft Articles do not include fault in the elements of internationally wrongful acts. Rather, the mental element is implied in the notion of 'breach' of an international obligation. The Commentary to the Draft Article reads that '[i]n the absence of any specific requirement of a mental element in terms of the primary obligation, it is only the act of a State that matters, independently of any intention'. Put differently, the 'psychological' attitude of a State – whatever this expression may mean vis-à-vis abstract and collective entities – is as such irrelevant; conversely, what matters is if the allegedly violated primary norm encapsulates 'fault'.

It must then be asked whether international obligations coming to the fore in cases of AWS' misdoings – that is, rules and principles of IHL and IHRL dealing with targeting objects and persons – contain at least a minimum requirement of 'fault'. Once again, the analysis that I will engage with cannot be done without a discussion of specific obligations.

As far as IHL is concerned, a minimum coefficient of fault is undoubtedly present in the precautionary rules listed above, under which belligerent parties are required to exercise 'constant care' to minimize risks to the civilian population and civilian persons and property (Article 57(1)), to do 'everything feasible' to ascertain the nature of targets (Article 57(2)(a)(i)), to take 'all feasible precautions' in the choice of means and methods of conducting attacks (Article 57(2)(a)(ii)), and again to take 'all reasonable precautions' in the conduct of military operations at sea or in the air (Article 57(4)). The continuous references to such standards of feasibility and reasonableness demonstrate that fault is a constitutive element of the primary norms under scrutiny: these are, in other words, 'due diligence' obligations, as they impose a 'standard of care' that States must apply (Ollino, 2022). If States fail to exercise such standard, their conduct will be in breach of the relevant obligations. Numerous IHL provisions (e.g., in the field of the conduct of hostilities, the protection of civilians and persons *hors de combat*, the law of occupation) encapsulate that standard, and can thus be conceived as of 'due diligence' (Longobardo, 2020).

It follows that state authorities are required to take all the 'reasonable' and 'practicable' precautions that the circumstances of the case require. It is worth noting that, in practice and also in scholarship, such standards are understood in an 'objective' sense: although the state organ (e.g., the commander during a military operation) is required to decide on the basis of assessments made from the information concretely available at the time, this does not preclude a review 'from outside' and *ex post facto*. Relevant factors are, e.g., the quality and quantity of available intelligence information, the accuracy of available weapons, the urgency of the attack, and the cost/benefit assessment of additional precautionary measures (Melzer, 2016, p. 104).

One must therefore wonder whether these obligations can be considered violated in cases of erroneous targeting. At present, international jurisprudence tends to set a minimum threshold of culpability that excludes reasonable error, i.e., that error that cannot be blamed on state authorities. Some scholars speak, in these cases – and drawing largely from the parallel institute of criminal law – of 'mistake of fact', which, in order to exclude the unlawfulness of the conduct, must be 'honest and reasonable' (Milanović, 2020).

In a case regarding Ethiopia's aerial bombing of six civilian sites located in Eritrea, the Claims Commission ruled out Ethiopia's responsibility for the damage caused to protected persons and property noting how '[a]s always in aerial bombing, there were some regrettable errors of targeting and of delivery,' to be regarded obviously as 'tragic consequences of the war' but not as internationally wrongful acts as such (Eritrea-Ethiopia Claims Commission, 2005, paras. 96–97). A similar conclusion was reached by the ad hoc Committee charged with assisting the Prosecutor of the International Criminal Tribunal for the former Yugoslavia in determining whether to open an investigation regarding NATO's campaign against the former Yugoslavia. Among the various facts scrutinized by the Committee was an air attack on a railroad bridge, in the Grdelica Gorge, in which twice a NATO pilot had opened fire at the bridge, drawing in both cases a train carrying civilians. The Committee considered that such a mistake did not warrant the opening of a criminal investigation against the pilot (Committee Established to Review the NATO Bombing Campaign Against the Federal Republic of Yugoslavia, 2000). Although this example concerned a different model of responsibility, namely, the international criminal responsibility of the individual, this circumstance confirms what has been noted above in terms of state responsibility: an attack involving damage to protected objects and persons is not to be considered unlawful if such an attack is not due to fault at the very least, namely if the factual error turns out to be 'honest and reasonable'.

State practice goes in the same direction. Indeed, in those cases where such damage results from attacks committed during armed conflicts (or in similar contexts), States tend not to admit to wrongdoing, while paying *ex gratia* sums of money to the victims' families (Crootof, 2022, p. 1098; Ronen, 2009). The US, for instance, has made *ex gratia* payments on a regular basis, on the basis of the National Defense Authorization Act (Lattimer, 2022). This was the case in the August 2021 drone strike in Kabul during the US retreat from the country, which resulted in the unintentional killing of ten civilians (including seven minors). The Pentagon spokesperson justified that engagement as an 'honest mistake', affirming that '[e]xecution errors combined with confirmation bias and communication breakdowns led to regrettable civilian casualties' (Borger, 2021).

Again, this is a matter of precautionary rules, which, as mentioned above, must be considered in tandem with those of distinction and proportionality. If, at the time of engagement, state authorities can rely on an 'honest and reasonable' belief, it is because precautionary obligations have been properly discharged. Thus if, in fact, the operation was planned taking all practicable and reasonable measures depending on the circumstances of the case and, in spite of this, the unintended outcome – the selection and engagement of impermissible targets – nevertheless occurred, this state of affairs cannot be blamed on the state authorities and thus on the State as a whole: technically speaking, no breach of IHL has occurred.

Ongoing discussions on AWS take into account the need that States must abide by due diligence obligations contained in IHL. According to the Guiding Principles affirmed by the Group of Governmental Experts on AWS, States are required to carry out 'risk assessments' and adopt 'mitigation measures' from the initial stages of weapon system development (including the development of an AI system) to operational deployment (GGE, 2019). Furthermore, measures to protect the physical and non-physical security of the weaponry should be taken at the development and acquisition stages, also to avert the risk of these components falling into the hands of terrorist groups and triggering a potentially dangerous proliferation process (GGE, 2019). This set of precautionary measures adds to those related to the choice of armaments and planning of operations, in order to minimize risks to protected individuals and objects.

If, therefore, AWS end up selecting and engaging an impermissible target, and this is attributable to insufficient planning of the operation as a whole (e.g., for choosing a weapon system that has not been adequately tested for use in such contexts, or for failing to properly supervise the system during deployment), it will be possible to assert international state responsibility on the basis of existing rules. No 'responsibility gaps' would arise.

Having said this, I have to indulge further in a point of utmost relevance. One could well argue how difficult – if not impossible – it is to reason in terms of 'honest' or 'honest and reasonable belief' when it is a weapon system, not a human decision-maker, that performs critical functions (target selection and engagement). As a matter of fact, existing standards have been developed starting from individuals: how can compliance be assessed in such cases? It is vital to 'interrogate' the system, which will have to be able to provide an intelligible explanation to the human operator so as to check whether the erroneous engagement is due to a malfunction of the system or is to be considered as a 'false positive'. In the former case, the system 'erred' and the human operator could (and should) have known; in the latter case, the system 'erred' without anyone being in a position to reasonably prevent the harmful course of action of the machine.

Let me now draw some conclusions. Leaving aside scenarios of intentional use of AWS to attack impermissible targets, the 'honest belief' standard and the due diligence nature of precautionary rules both demonstrate the existence of a minimum coefficient of culpability as a constituent element of the primary obligations of IHL. This leads to the key question: what happens if the error of the weapon system – which selected and engaged an impermissible target – is not due to a defect in the planning of the operation, nor to a malfunction of the machine, of which the state authorities should have had knowledge (and, therefore, mitigated)? If the minimum coefficient of culpability is not demonstrated in the actual case, neither can the violation of an international obligation be said to qualify, which precludes international responsibility of the State.

This is exactly the scenario in which a 'responsibility gap' can be said to arise. It is left for us to understand whether such a gap is intrinsically deemed to arise and impossible to fill, as some have argued.

THE 'FALSE POSITIVE' QUANDARY

In my opinion, the question that closes the previous section is in the negative. Put differently, I argue that, as things stand, the 'responsibility gap' in cases of mistargeting that is not due to fault on the part of state authorities (which I refer to as 'false positive') can be filled by international law, and thus banning AWS on this sole basis is ultimately unwarranted.

To demonstrate this, I will now proceed by illustrating two sets of legal toolkits that could fill such a gap.

The Teleological Interpretation of Primary Obligations (*de lege lata* Solution)

The first toolkit is a *de lege lata* means, in the sense that no new law is required to fill the 'responsibility gap'; rather, existing law can be interpreted in a way that makes it possible to address 'false positives'.

I am referring to the teleologically oriented interpretation of relevant obligations of IHL and IHRL, typically on a case-by-case basis by monitoring bodies and international courts. Such an 'evolutionary' interpretation could rely on hermeneutic canons that are rooted in international jurisprudence and that have been refined over the decades: technological evolution has always informed the interpretation of relevant obligations in an attempt to 'update' them (Bjorge, 2014).

In this sense, the field of human rights is rich in examples. The IHRL concept of due diligence has been progressively expanded so as to apply to scenarios in which it is hard to find fault on the part of state authorities. For instance, in the *Kotilainen and Others v Finland* case, the European Court of Human Rights extended the duty to adopt necessary measures to protect lives to cover killings that occurred at the hands of an individual who had broken into a school and opened fire on ten people (App. No. 62439/12, 17 September 2020). The Court concluded that there had been a violation of the right to life despite the following circumstances: (i) there were no particular deficiencies in the domestic rules on the use of firearms; (ii) the weapon was duly possessed by the subject, nor were there any detectable procedural deficiencies; (iii) there was no real and immediate risk to the lives of the victims prior to the attack, which should have been known to the state authorities. The reason that led to the finding of liability lay in the breach of a 'special duty of diligence' (*Kotilainen and Others v Finland*, para 89) arising from the general obligation to protect every individual from the use of firearms, a duty that should have led to the confiscation of the firearm.

This judgment signals a remarkable extension of due diligence obligations in the field of the right to life. On close inspection, the culpability of state authorities is held to exist *in re ipsa*: nothing more is required than the mere causal link between state conduct and the event in order to prove fault. This hermeneutical approach – which ultimately nullifies fault – may be extended to different scenarios, such as those involving the use of

AWS in operational contexts. The 'lowering' of the threshold of culpability would make it possible to establish state responsibility and fill the 'responsibility gap' in cases where, based on the interpretation prevailing to date, it is not easy to prove any fault.

Admittedly, such a *de lege lata* instrument, though useful (especially in the early days of employing AWS), does not seem satisfactory for a number of reasons. First, it presents a partial solution, in that it applies only to those cases in which the respondent State is party to a IHRL and IHL instrument establishing a monitoring body. The US, Israel, China, and Russia – to mention only a few – will thus avoid the scrutiny of any such mechanisms, merely because they have not accepted the jurisdiction of any such mechanisms. Second, it is by no means certain that international bodies will adopt this hermeneutical tool in cases of employment of AWS, particularly those featuring AI capabilities, with regard to which no precedent can be found. Thirdly – but no less importantly – on a theoretical level, such an interpretive stance clearly stretches existing law: although justifiable in the name of filling 'responsibility gaps,' the risk of compromising the already fragile IHRL and IHL edifice is material.

The Adoption of a Liability Instrument on AWS (*de lege ferenda* Solution)

I argue that a more convincing solution than forcing existing law by way of interpretation can be found if only one approached the issue of AWS from a slightly different standpoint. The basic need that arises from any impermissible targeting is that the victims of such an attack – family members of those who lost their lives, owners of destroyed property, or the targets themselves, if the attack was non-lethal – have access to some form of compensation for their losses.

A topic that has recently begun to be discussed in the field of AI application is the possible resort to forms of 'strict' or 'absolute' liability, that is liability that stems from the commission of an act that is not prohibited per se, but which is likely to cause damage to persons and objects. This model of responsibility *sine delicto* has been the subject of extensive scholarship in international law (Barboza, 1994; Montoje, 2010). The first authors began to engage with this topic around the 1960s, focusing on activities that, being made possible by technological progress, were likely to cause damage to persons and objects (including the environment), but were still not prohibited (Jenks, 1966; Dupuy, 1976). The International Law Commission, namely the then Special Rapporteur Mr. Roberto Ago, decided to dedicate a specific topic to that issue, removing it from the more general topic of state responsibility.

Back then, the common expression was 'international liability for injurious consequences arising out of acts not prohibited by international law'. There were two key elements of this model of liability: the 'hazardous' character of the activity in question, which had to lead at the very least to damage of a certain importance, and the 'lawful' character thereof (Montoje, 2010, p. 508). Curiously, the ILC did not make it to adopt a single text, but again divided its work into two texts, the former devoted to the prevention of transboundary harm resulting from dangerous activities (ILC, 2001), and the latter to the allocation of the ensuing losses (ILC, 2006).

Those texts contained general rules and principles aiming to combat the harmful effects of a specific set of activities, namely industrial activities (conducted as a rule, though not exclusively, by private individuals), on the environment (especially that over which a state having jurisdiction bordering on that in which the activity in question takes place). The insertion of such rules in non-binding acts, and with limited impact on state practice, finds its justification in States' reluctance to charge themselves of damages produced by private entities.

It is commonly held that those rules, in particular those of 2006, must be traced back to a well-known principle of international environmental law, i.e., the 'polluter pays' principle as established, for the first time, in the 1992 Rio Declaration (UNGA, 1992). The content of such primary norm – whose correspondence to customary law is debated – reflects the two elements of strict or absolute liability as illustrated above, namely the 'hazardous' and 'lawful' characters of the activity at hand (Boyle, 2009; Gervasi, 2021, p. 348).

It must be said, however, that until now this model of liability has had limited success in international law. Leaving aside treaties that require State parties to introduce forms of strict (civil) liability into their legal systems, which would go beyond the scope of our analysis, the only area of international law in which strict or absolute liability has been adopted is the law of outer space.

The Treaty on Principles Governing the Activities of States in the Exploration and Use of Outer Space, including the Moon and Other Celestial Bodies (OST) establishes that the State that launches, procures the launching of or from the territory of which the launching of a space object takes place is 'internationally liable' for damage caused to other States, individuals and legal entities (Article VII). Analogously, the later 1972 Convention on International Liability for Damage Caused by Space Objects (hereinafter also '1972 Convention') stipulates that the launching State 'shall be absolutely liable to pay compensation for damage caused by its space object on the surface of the Earth or to aircraft in flight' (Article II). On the contrary, if damage occurs elsewhere and is suffered by another space object or by persons or property aboard it, the fault liability regime revives (Article III).

Theoretically speaking, the dichotomy between the traditional regime of responsibility and the 'strict' or 'absolute' liability one, can be justified on the basis of the diversity of the overall activity under consideration. As is clear, in the first case the injured party too engages in dangerous activity (i.e., the launching of space objects), so that in the 1970s – but the same holds today – it seemed more correct, in case of damage, to allocate losses on both the parties involved in the same, dangerous activity. On the contrary, the 'strict' or 'absolute' liability regime – which thus imposes a mere duty to 'compensate' losses, without any proof of fault – better fits in those 'asymmetrical' cases, in which only one of the parties conducts an activity likely to generate damage (Condorelli, 1990; Pedrazzi, 2008; Wilfred Jenks, 1966, p. 153). As a confirmation of the rationale behind this dichotomy, the 1972 Convention enshrines, as cause for exoneration from absolute liability and thus from the duty to compensate only 'gross negligence' and 'act or omission done with intent to cause damage' (willful conduct) on the part of the injured (Article VI).

It is worth noting that the 1972 Convention also establishes a mechanism for compensation of damage (Articles X–XXI). This mechanism can be activated within one year from the accident, through diplomatic channels, and in case of failure each State can request the establishment of a Claims Commission that will settle the dispute by binding award. Such procedure is characterized by a non-jurisdictional character and does not contemplate the participation of private actors (whether natural or legal persons), i.e., typically, the very victims of the damages caused by space objects. Admittedly, private actors will always be able to assert their arguments before domestic courts, making use of the applicable rules of civil liability (Schmalenbach, 2022). However, one must not forget that domestic litigation comes with a plethora of obstacles when international conduct is at stake: to name one, the international rules on state immunity – which are of a customary nature – prohibit States from adjudicating the conduct of other States that constitutes the exercise of sovereign powers (*par in parem non habet jurisdictionem*). All those limits, coupled with the complicatedness of the procedure as a whole, allegedly constitute the 'main flaw' of the entire Convention (Pedrazzi, 2008, para. 15).

In light of all this, it is not surprising that practice regarding this mode of liability is almost nonexistent. The most quoted case regards the 1978 incident involving the USSR and Canada (often referred to as the Cosmos 954 case), which was settled by diplomatic means without the USSR acknowledging expressly its liability under the 1972 Convention. The USSR eventually paid Canada half the amount claimed, without any reaction from the damaged State. More recent collision cases, even in space (i.e., in those cases in which the traditional regime of fault liability as per the 1972 Convention would apply), have never led to the activation of the diplomatic means envisaged (Schmalenbach, 2022, p. 535).

Turning now to AWS, the establishment of an international regime of absolute or strict liability for 'false positives' generated by the use of AI systems in the military field would fill the sole 'responsibility gap' opening up as the result of such technology. Again, this sort of 'gap' has made the object of extensive discussions among scholars, yet in contexts that do not feature resort to AI systems. Any conduct that is not attributable at least to fault on the part of state authorities (either in the planning of the operation or in the concrete act of targeting) and that ends up producing damage to persons and property is not subject to a general duty to compensate losses, to the point that it is believed that such gap 'for civilian harm is built into the structure of the law of armed conflict. A structural change is needed to close it' (Crootof, 2022, p. 1070; Ronen, 2009). On the contrary, the practice of *ex gratia* payments also in those instances in which the lawfulness of the impugned state conduct can be reasonably doubted confirms the lack of instruments to properly ensure compensation in favor of the victims.

In sum, it seems that the gist of the matter is greater than the specific AI-related issue. However, turning back to discussing the need for appropriate tools for compensating victims is momentous now, in times when developments in the AI field in the military and in law enforcement could but magnify the lack of such tools. If not now, when?

While the 'social' need is more imperative than ever, the question remains as to how to correct existing law. From a *de lege ferenda* perspective, I argue that the best way for ensuring victims of IHL violations – which in the near future could be perpetrated through AWS – proper redress is to establish a right to compensation in a legally binding instrument, as States were able to do back in the 1970s in the matter of liability for space activities.

A future treaty should cover, in my view, the following aspects.

First, in line with the 1972 Convention, it should establish two different regimes of state responsibility. On the one hand, the 'fault-based' regime for violations of IHL due to malfunctioning that the State should have foreseen and neutralized before fielding AWS, and more generally for any violations that can be traced back to primary rules regarding distinction, proportionality, and precautions in attack. For instance, the treaty should strive to articulate a strict regime of test & evaluation, validation & verification (TEVV) for AWS, so as to set the bar high. On the other hand, for those scenarios that cannot be included in the former, that is for 'false positive' scenarios, a 'strict' or 'absolute' liability regime should be introduced as a residual form of indemnification of victims.

Second, so as to confer granularity to the former regime, it should outline, as much as feasible, specific duties regarding the prevention of AWS misdoings, and – at the very least – provide mechanisms for the exchange of best practices among States. Granted, this obligation would reasonably assume a quite generic character, due to the reluctance of States to share key military knowledge among themselves.

Third, in order to ensure the protection of victims of 'false positives,' it should provide forms of compensation for damage to property and persons resulting from the use of AWS (again, along the lines of the aforementioned 1972 Convention). As is the case of space activities, state conduct is not prohibited by international law, but is hazardous to the extent that it is likely to generate catastrophic damage: the liability model is, therefore, best suited to the case.

Fourth, and so as to correct the limitations of existing mechanisms, the future treaty should contain adequate means of guaranteeing victims effective access to justice, either through the establishment of an *ad hoc* supervisory body (for instance, a Claims Commission to which individuals can apply directly) or through the provision of obligations to be incorporated into domestic legal orders.

I do not claim that such an instrument would not be ambitious: one may even wonder whether the proposed content for a treaty on AWS risks having a chilling effect toward States, which would have quite a hard time signing it. This could be troublesome, particularly if one considers that such a treaty could not be joined by those States that will be the first to develop and deploy AWS in operational scenarios.

On a more realistic note, those objections stand and, albeit itchy for those who profess 'idealism' in international law, must be taken seriously. This notwithstanding, the purpose of the present contribution has been reached: I demonstrated that the 'responsibility gap' that AWS are bound to open up can be filled, both on the basis of existing law (yet with some difficulties) and by adopting new law (yet with a considerable amount of ambition).

If AWS are to be regarded as 'abhorrent', this is not due to responsibility-related reasons.

TOWARD A 'KILLER PAYS' PRINCIPLE

The 'cybernetic error' foreshadowed some fifty years ago by Wilfred Jenks has not only become a reality, but, in the near future and as a result of the proliferation of AWS, it is likely to generate damage to property and persons that existing rules of international state responsibility are able to regulate to a discrete extent, but not entirely. The advent of AI capabilities will magnify even more those troubling damages.

The first set of issues derives from cases of intentional mistargeting by AWS, that in those scenarios in which it is possible to detect an intention to violate IHL and IHRL rules and principles on the part of state authorities. The intentional selection and engagement of impermissible targets – either objects or persons – gives rise to the responsibility of States in accordance with existing rules, not to mention, in certain cases, also the international criminal responsibility of the individuals involved. Put differently, the intentional use of technology that, however advanced, does not guarantee adequate levels of operability in certain contexts is conduct that is already 'covered' by existing norms: AWS' peculiar characteristics – namely, the possibility to select and engage targets without further human intervention – are relevant only to a very limited extent.

The case of unintended attacks on impermissible targets is different. As the primary rules of IHL and IHRL are held to incorporate a minimum coefficient of culpability (that is, fault), if the misdoing is attributable to a malfunctioning of the system that state authorities were in a position to prevent, neutralize, and in any case minimize, and if they fail to do so, no 'responsibility gap' arises. The standard against which this fault is to be assessed is the one encapsulated in the 'honest and reasonable belief' formula, as crafted in both IHL and IHRL. Due diligence obligations impose a standard of care on States when they develop and deploy AWS.

The last set of cases that have been analyzed in the present chapter deals with the unintended attack on persons and objects and is due neither to the 'bad intention' of state authorities nor to 'fault' on their part; rather, the selection and engagement of impermissible targets depends on the unpredictable way in which AWS operate in contact with the real world. If no blame can be placed on state authorities, since there is no primary norm actually violated and therefore no internationally wrongful act, it logically follows that neither can one reason in terms of responsibility, due to the defect of one of the two essential elements of the internationally wrongful act, that is the breach of an international obligation incumbent on the State.

By saying this, I do not argue that States will be released from the respect of the due diligence obligations illustrated above, quite the contrary: the liability regime would operate only for those instances where no fault whatsoever can be traced back to the State, which implies that State's performance of due diligence obligations (including those related to TEVV) will be closely scrutinized. Simply skipping those obligations by internalizing indemnification costs will not work, as in those case the existing rules of international responsibility will kick in.

To conclude that AWS should be (or, according to some, are) banned is however unwarranted: it has been demonstrated that this argument, captured by the 'responsibility gap' expression, proves too much. More specifically, it fails to take into account a twofold set of tools that can be used to 'fill the gap': on the one hand, interpreting existing law so as to cover those 'false positive' scenarios (*de lege lata*); on the other hand, adopting new law to establish a 'strict' or 'absolute' regime of liability in the absence of an internationally wrongful act, which does have precedents in the international legal order.

In this paper, I argued in favor of the latter option: the best way to fill the responsibility gap is by way of a treaty establishing a form of responsibility without internationally wrongful acts. For the sake of clarity, it must be stressed that this term should not mislead: it is, in fact, a primary norm that would impose a duty to compensate victims on the State. The prerequisites of this regime, as conceived in the international legal system, are the lawful character of the state conduct in question and its structural hazardousness for objects and, maybe more worryingly, persons. This regime resonates clearly with the rationale of a cornerstone of international environmental law, that is the 'polluter pays' principle (Boyle, 2009). Suffice to say that this principle was conceived as a tool for guaranteeing that victims of activities which were not prohibited under international law, but which were of an hazardous or ultra-hazardous nature, be able to obtain prompt and adequate compensation. As provocative as it may sound, one could thus speak of a 'killer pays' principle that the international community should develop to properly address cases of 'responsibility gaps' arising from the incremental use of AI in the military field.

Of course, what has been said so far does not imply that there are no other reasons to ban, or at least strictly regulate, AWS. It could, for example, be argued that the use of force, especially against persons, in the absence of specific deliberation by a human operator, is contrary to the human dignity of the target, and thus contravenes fundamental principles of both IHL and IHRL (Amoroso, 2020; Tamburrini, 2016). However, this is evidently a different set of arguments, which cannot be addressed here. It sufficed to show that the argument focusing on the irremediable opening of 'responsibility gaps' as a result of the use of AWS is, on closer inspection, unconvincing.

REFERENCES

Amoroso, D. (2020). *Autonomous weapons systems and international law. A study of human-machine interactions in ethically and legally sensitive domains.* Edizioni Scientifiche Italiane.

Amoroso, D., & Giordano, B. (2019). Who is to blame for autonomous weapons systems' misdoings? In E. Carpanelli & N. Lazzerini (Eds.), *Use and misuse of new technologies. Contemporary challenges in international and European law* (pp. 211–232). Springer.

Barboza, J. (1994). International liability for the injurious consequences of acts not prohibited by international law and protection of the environment. *Collected Courses of the Hague Academy of International Law, 247*(3), 291–406.

Batesmith, A. (2014). Corporate criminal responsibility for war crimes and other violations of international humanitarian law: The impact of the business and human rights movement. In C. Harvey, J. Summers, & N. White (Eds.), *Contemporary challenges to the laws of war. Essays in honour of Professor Peter Rowe* (pp. 285–312). Cambridge University Press.

Bjorge, E. (2014). *The evolutionary interpretation of treaties.* Cambridge University Press.

Bo, M. (2021). Autonomous weapons and the responsibility gap in light of the *mens rea* of the war crime of attacking civilians in the ICC statute. *Journal of International Criminal Justice, 19*(2), 275–299.

Boothby, W. (2013). Autonomous attack - Opportunity or spectre? *Yearbook of International Humanitarian Law, 16*, 71–88.

Borger, J. (2021). 'Honest mistake': US strike that killed Afghan civilian was legal. *The Guardian.*

Boyle, A. (2009). Polluter pays. *Max Planck Encyclopedia of Public International Law.*

Committee Established to Review the NATO Bombing Campaign against the Federal Republic of Yugoslavia (2000). Final report to the Prosecutor.

Condorelli, L. (1990). La réparation des dommages catastrophiques causés par les activités spatiales. In *La réparation des dommages catastrophiques. Travaux des XIIIe Journées d'études juridiques Jean Dabin* (pp. 263–300). Bruylant.

Crootof, R. (2022). War torts. *New York University Law Review, 97*(4), 1063–1142.

Diggelmann, O. (2006). Fault in the law of state responsibility - Pragmatism ad infinitum? *German Yearbook of International Law, 49*, 293–306.

Dupuy, P. M. (1976). *La responsabilité internationale des États pour les dommages d'origine technologique et industrielle*. Pedone.

Egeland, K. (2016). Lethal autonomous weapon systems under international humanitarian law. *Nordic Journal of International Law, 85*(2), 89–118.

Eritrea-Ethiopia Claims Commission. (2005). Partial Award. Western Front, Aerial Bombardment and Related Claims.

Gattini, A. (1992). La notion de faute à la lumière du projet de convention de la Commission du Droit International sur la responsabilité internationale. *European Journal of International Law, 3*(2), 253–186.

Gervasi, M. (2021). *Prevention of environmental harm under general international law. An alternative reconstruction*. Edizioni Scientifiche Italiane.

GGE. (2019). Guiding principles affirmed by the group of governmental experts on emerging technologies in the area of lethal autonomous weapons system, *Annex III*. CCW/MSP/2019/9.

Hammond, D. (2015). Autonomous weapons and the problem of state accountability. *Chicago Journal of International Law, 15*(2), 652–687.

Heyns, C. (2013). *Report of the United Nations Special Rapporteur on extrajudicial, summary or arbitrary executions* (A/HRC/23/47).

ILC. (2001). Draft articles on prevention of transboundary harm from hazardous activities, with commentaries. In *Yearbook of the International Law Commission* (pp. 148–170).

ILC. (2006). Draft principles on the allocation of loss in the case of transboundary harm arising out of hazardous activities, with commentaries. In *Yearbook of the International Law Commission* (pp. 59–90).

Lattimer, M. (2022). Civil liability for violations of IHL: Are the US and UK moving in opposite directions? *EJIL:Talk!*.

Longobardo, M. (2020). The relevance of the concept of due diligence for international humanitarian law. *Wisconsin International Law Journal, 37*(1), 44–87.

Mauri, D. (2022). *Autonomous weapons systems and the protection of the human person. An international law analysis*. Edward Elgar Publishing.

McFarland, T. (2020). *Autonomous weapon systems and the law of armed conflict: Compatibility with international humanitarian law*. Cambridge University Press.

McFarland, T., & McCormack, T. (2014). 'Mind the gap': Can developers of autonomous weapons systems be liable for war crimes? *International Law Studies, 90*, 361–385.

Melzer, N. (2016). *International humanitarian law: A comprehensive introduction*. ICRC Publications.

Milanović, M. (2020). Mistakes of fact when using lethal force in international law: Part I". *EJIL:Talk!*. https://www.ejiltalk.org/mistakes-of-fact-when-using-lethal-force-in-international-law-part-i/

Montoje, M. (2010). The concept of liability in the absence of an internationally wrongful act. In J. Crawford, A. Pellet, S. Olleson, & K. Parlett (Eds.), *The law of international responsibility* (pp. 503–513). Oxford University Press.

Ollino, A. (2022). *Due diligence obligations in international law: A theoretical study*. Cambridge University Press.

Palmisano, G. (2007). Fault. *Max Planck Encyclopedia of Public International Law.*

Pedrazzi, M. (2008). Outer space, liability for damage. *Max Planck Encyclopedia of Public International Law.*

Roff, H. (2015). Lethal autonomous weapons and jus ad bellum proportionality. *Case West Reserve Journal of International Law, 47*(1), 37–52.

Ronen, Y. (2009). Avoid or compensate? Liability for incidental injury to civilians inflicted during armed conflict. *Vanderbilt Journal of Transnational Law, 42*(1), 181–225.

Schmalenbach, K. (2022). Convention on international liability for damage caused by space objects. In P. Gailhofer, D. Krebs, K. Schmalenbach, & R. Verheyen (Eds.), *Corporate liability for transboundary environmental harm* (pp. 523–536). Springer.

Schmitt, M., & Thurnher, J. (2013). "Out of the loop": Autonomous weapon systems and the law of armed conflict. *Harvard National Security Journal, 4*, 231–281.

Spagnolo, A. (2017). Human rights implications of autonomous weapon systems in domestic law enforcement: Sci-fi reflections on a lo-fi reality. *Questions of International Law, 43*, 33–58.

Tamburrini, G. (2016). On banning autonomous weapons systems: From deontological to wide consequentialist reasons. In N. Bhuta, S. Beck, R. Geiss, H. Liu, & C. Kress (Eds.), *Autonomous weapons systems: Law, ethics, policy* (pp. 112–142). Cambridge University Press.

UNGA. (1992). *Report of the United Nations conference on environment and development. Annex I.* Rio Declaration on Environment and Development, A/CONF.151/26.

US DoD. (2023). *Directive 3000.09. Autonomy in weapons systems* (pp. 1–24).

Wilfried Jenks, W.C. (1966). Liability for ultra-hazardous activities in international law. *Collected Courses of the Hague Academy of International Law, 117*(1), 105–196.

Military AI and Accountability of Individuals and States for War Crimes in the Ukraine

9

Dan Saxon

INTRODUCTION

On February 16, 2023, the United States Department of State issued a "Political Declaration on Responsible Military Use of Artificial Intelligence and Autonomy" (Political Declaration). The Political Declaration provided definitions of artificial intelligence (AI) and autonomy that I adopt for this chapter:

> For the purposes of this Declaration, artificial intelligence may be understood to refer to the ability of machines to perform tasks that would otherwise require human intelligence — for example, recognizing patterns, learning from experience, drawing conclusions, making predictions, or taking action — whether digitally or as the smart software behind autonomous physical systems. Similarly, autonomy may be understood to involve a system operating without further human intervention after activation.

The preamble to the Political Declaration expressed several principles including that the use of "AI in armed conflict must be in accord with applicable international humanitarian law, including its fundamental principles. Military use of AI capabilities needs to be accountable, including through such use during military operations within a responsible human chain of command and control."

The ongoing armed conflict between Russia and Ukraine is "an unprecedented testing ground for AI" (Fontes & Kamminga, 2023). This chapter attempts to unpack some of the challenges to determine the responsibility of individuals and states for the (mis) use of military AI during the war. To preserve accountability for events that occur during armed conflict, soldiers and commanders must conduct combat according to norms entrenched in both international and domestic law, so that military activity does not take place in a normative void (Beinisch, The Public Committee Against Torture in Israel v. The Government of Israel, 2005). I use the phrases "international humanitarian law (IHL)", "the law of armed conflict", and "the law of war" synonymously in this chapter to describe the legal framework that codifies these norms.

In the interest of brevity, rather than provide a country-wide survey of alleged crimes committed in Ukraine with weapon systems that use AI, this chapter focuses on the more limited context of Russia's 2022–2023 aerial campaign to destroy Ukrainian energy infrastructure. First, I review the facts known about these attacks and the technology operating one of the primary weapons used by the Russian armed forces to carry them out – the Iranian-made Shahed drone. Next, I review the basic principles of IHL, in particular the rules of targeting, which are particularly relevant to the use of military AI. The remainder of the chapter examines how Russia's operation of the Shahed weapon system in the context of repeated targeting of Ukraine energy installations likely constitutes war crimes and the possibilities of holding persons and States (e.g. Russia and Iran) accountable for these offenses.

RUSSIAN ATTACKS ON UKRAINE'S ELECTRICAL INFRASTRUCTURE USING MILITARY AI

During the autumn of 2022 and the early months of 2023, Russian armed forces launched multiple, widespread attacks against electrical power stations and related infrastructure in Ukraine (Hutch, 2023; Santora, 2023; Schwirtz & Mpoke Bigg, 2022). Between October 10, 2022 and February 1, 2023 alone, Russia launched at least 13 waves of attacks using hundreds of long-range missiles and drones carrying explosives (Report of the Independent International Commission of Inquiry on Ukraine [COI], 2023). The attacks affected 20 of Ukraine's 24 regions and systematically targeted powerplants and other installations crucial for the transmission of electricity and the generation of heat across the country. The strikes damaged or destroyed a large portion of Ukraine's energy production and distribution facilities just as cold weather descended on the region (Hutch, 2023; Schwirtz & Mpoke Bigg, 2022). Even half of Moldova lost power as a result of one attack, as its energy grid is tied to Ukraine's (Santora & Gibbons-Neff, 2022).

By December 2022, every one of Ukraine's thermal and hydroelectric power installations had been damaged by Russian strikes (Lander et al., 2022). Entire regions and millions of people were left for periods without electricity or heat during the winter and, consequently, with reduced access to water, sanitation, medical treatment, and education (COI, 2023). During one bombardment, all of Ukraine's nuclear power plants – which provide fifty percent of the country's energy supply – "went into blackout" (Santora, 2023). Fortunately, meltdowns of the nuclear cores were avoided and the reactors went offline.

Although public information about civilian harm was available after the first few attacks, Russia continued to target energy infrastructure. By April 2023, Russia had launched more than 1,200 missiles and drones against Ukraine's power installations (Moloney, 2023). Whilst some of these attacks may have targeted lawful military objects that supported the Ukrainian military, the sheer scale of these assaults and their effects indicates that many of them were disproportionate, indiscriminate, and intended to instill terror in the civilian population. If true, such operations would be violations of the law of armed conflict, i.e. war crimes.

Since the start of the armed conflict in February 2022, Russian forces have deployed several kinds of drones in the Ukraine that operate with military AI. For example, the KUB-BLA and Lancet drones are relatively short-distance, loitering munitions that use cameras and algorithms to detect and identify military objects by class and type in real-time (Automated Decision Research, 2023). The AI technology expands the area monitored during a single flight by 60 times and increases the drone's lethality and autonomy.

The repeated operations against electrical plants and networks relied heavily on a particular type of "Kamikaze" drone called Shahed (Hambling, 2023). Purchased from Iran in 2022, the different versions of Shahed drones are longer-range UAVs that, prior to launch, are programmed to strike and destroy targets by diving into them and exploding (Chulov et al., 2023). Often launched in "swarms" of five to a dozen drones or more, Shahed proved effective against Ukraine energy installations, in spite of Ukraine military's air defenses. They are comparatively small and fly at low altitudes, which makes radar detection difficult until the vehicle is close to the target (Hutch, 2023). Even when only a fraction of a swarm reaches its objective, the drones' warhead (bearing 30–50 kg of explosives depending on the design), inflicts substantial damage (Brennan, 2022).

Shahed drones can be programmed with algorithms to perform various tasks such as surveillance and attacks (Army Recognition, 2023). In addition, when satellite-linked guidance systems fail, Shahed's guidance system can switch to an inertial navigation system, whereby raw sensor data about dynamic states such as angular velocity and acceleration is processed by a computer. The software, by means of different fusion algorithms, can estimate attitude, position, and velocity, allowing for course corrections. (OE Data Integration Network, 2023). The system also detects when sensors malfunction and then discards the input from the affected sensors and compensates for the loss with other available sensors. This makes the system robust against sensor failures and the drone's accuracy is described as "spectacular" and "uncanny" (Rubin, 2023).

The use of military AI to assist Russian attacks on Ukraine's electrical assets raises questions about the ability to hold drone operators and commanders accountable for war crimes that occur during these operations. Nonetheless, it would be premature to examine issues of individual criminal responsibility prior to determining whether these strikes are lawful, or violations of IHL. The next section, therefore, explains the important principles and rules of the law of armed conflict and assesses the legality of Russia's concerted assaults on Ukraine's energy system.

BASIC PRINCIPLES AND RULES OF IHL

The application of modern IHL is an attempt to achieve an equitable balance between humanitarian requirements and the demands of armed conflict (ICRC Commentary to Art. 57, Additional Protocol I to the Geneva Conventions of 12 August 1949 (API) [ICRC Commentary], 1987; May & Newton, 2014), e.g. between the principles of humanity and military necessity. The principle of "humanity" – the heart of IHL (International Court of Justice, Legality of the Threat or Use of Nuclear Weapons [Nuclear Weapons], 1996) – prohibits the infliction of suffering, injury or destruction not actually necessary for the accomplishment of a legitimate military purpose. This tenet is based on the concept that once a military purpose has been achieved, the further infliction of suffering is unnecessary (U.K., The Joint Service Manual of the Law of Armed Conflict [JSP 383], 2004). The humanitarian character of the principles of the law of armed conflict applies to all forms of warfare and all kinds of weapons, including future weapons.

Francis Lieber (1863) defined "military necessity" as "the necessity of those measures which are indispensable for securing the ends of the war, and which are lawful according to the modern law and usages of war." The U.K. armed forces use a more nuanced definition that mirrors the principle of humanity:

> [m]ilitary necessity is now defined as "the principle whereby a belligerent has the right to apply any measures which are required to bring about the successful conclusion of a military operation and which are not forbidden by the laws of war." Put another way a state engaged in an armed conflict may use that degree and kind of force, not otherwise prohibited by the law of armed conflict, that is required in order to achieve the legitimate purpose of the conflict, namely the complete or partial submission of the enemy at the earliest possible moment with the minimum expenditure of life and resources

> *(JSP 383, Amendment 3, 2010, para. 2.2).*

In addition to humanity and military necessity, two "crucial" principles determine the effectiveness of modern IHL (Nuclear Weapons, paras. 77–78). First, the principle of distinction establishes that belligerents must always distinguish between enemy combatants and civilians and never intentionally target civilians or civilian objects (Art. 48, API, 1977). Consequently, indiscriminate attacks, i.e. those that are of a nature to strike military objectives and civilians without distinction, as well as the use of weapons that are indiscriminate, are unlawful (Art. 51(4), API). Second, belligerent parties may not employ means and methods of warfare in a manner that causes superfluous injury or unnecessary suffering (Art. 35(2), API). The phrase "means of combat" generally refers to the weapons used while "methods of combat" generally refers to the way in which weapons are used (ICRC Commentary, 1987). This constraint reflects a "fundamental customary principle" of the law relating to the conduct of hostilities; that the right of belligerents to adopt means of injuring the enemy, including the choice of weapons, is not unlimited (Roberts & Guelff, 2000).

The Law of Targeting: The Use of Force During Armed Conflict. In order to understand how the employment of military AI impacts the exercise of force, it is necessary to review the process(es) professional armed forces undertake to plan and execute attacks. In modern warfare, the process of selecting and engaging targets can be extraordinarily complex, involving multiple stakeholders, interests, and values and includes a mix of human thinking, automation, and autonomy. Essentially, the targeting process identifies resources that the enemy can least afford to lose or that provide her with the greatest advantage. Subsequently, targeters identify the subset of those targets that must be neutralized to achieve success. According to U.S. military doctrine, valid targets are those that have been vetted and those that "meet the objectives and criteria outlined in the commander's guidance and ensures compliance with the law of armed conflict and rules of engagement" (U.S., Joint Publications 3-60, "Joint Targeting," [JP 3-60], 2013).

Four general principles guide the targeting process. First, it should be focused, i.e. every target proposed for engagement should contribute to attaining the objectives of the mission. Second, targeting should be "effects-based", i.e. it attempts to produce desired effects with the least risk and least expenditure of resources (Anderson & Waxman, 2017). Third, it is interdisciplinary in that targeting entails participation from commanders and their staffs, military lawyers, analysts, weaponeers, other agencies, organizations, and multinational partners. Finally, targeting should be systematic; a rational process that methodically analyses, prioritizes, and assigns assets against targets. A single target may be significant because of its particular characteristics. The target's real importance, however, "lies in its relationship to other targets within the operational system" of the adversary (JP-360, para. 1.2.1, 2013).

There are two general categories of targeting: deliberate and dynamic. Deliberate targeting is more strategic; it shapes the battlespace and addresses planned targets and efforts, i.e. beyond the next twenty-four hours. Systematic attacks over months on Ukraine's power plants and related facilities would be an example of deliberate, strategic targeting. Dynamic targeting manages the battlespace and refers to decisions requiring more immediate responses, usually within the current twenty-four hour period (JP-360, 2013).

Targets have temporal characteristics in that their vulnerability to detection, attack, or other engagement varies in relation to the time available to engage them. Targets that are especially time-sensitive present the greatest challenges to targeting personnel who must compress their normal decision cycles into much shorter periods. As all or most of the energy infrastructure attacked by Russian forces since October 2022 was stationary, it is unlikely that it represented a particularly "time-sensitive" target in the usual sense of the term (although Ukraine's increasing capacity to improve its air defenses might have raised the temporal urgency of some of the attacks).

As mentioned above, targeting decisions must satisfy the law of war obligations. In this context, targeting personnel bears three essential responsibilities. First, they must positively identify and accurately locate targets that comport with military objectives and rules of engagement. Second, they must identify possible concerns regarding civilian injury or damage to civilian objects in the vicinity of the target (U.S., "No-Strike and the Collateral Damage Estimation Methodology" [NSCDEM], 2012). Finally, they must conduct incidental damage estimates with due diligence and within a framework of the accomplishment of mission objectives, force protection, and collateral damage mitigation (Australia, "Targeting" [ADDP], 2009).

After targets are engaged, commanders must assess the effectiveness of the engagement. "Direct" effects are the immediate consequences of military action whilst "indirect" effects are the delayed and/or displaced second, third, or higher-order consequences, resulting from intervening events or mechanisms. Importantly for this chapter, effects can "cascade", i.e. ripple through a targeted system and affect other systems (ADDP, 2009). The assessment process is continuous and helps commanders adjust operations as necessary and make other decisions designed to ensure the success of the mission (NSCDEM, 2012).

Finally, the work of targeting is increasingly an automated (if not autonomous) process. Computer applications speed the accurate development and use of information that matches objectives with targeting and facilitates the assessment of effects. Nonetheless, U.S. military doctrine holds that, whilst automation increases the speed of the targeting process, "it is not a replacement for human thinking or proactive communications" and personnel must "fully comprehend foundational targeting concepts" (NSCDEM, 2012). The next section describes the most important targeting rules of IHL with respect to the use of Shahed drones to attack Ukraine's electrical infrastructure.

Applicable Rules of Targeting in IHL

The IHL provisions prescribing how belligerents should conduct targeting – i.e. Articles 48–59 of API – integrate the principles of military necessity and humanity. "The question who, or what, is a legitimate target is arguably the most important question in the law of war ...." (Waxman, 2008). The targeting rules attempt to delineate the parameters for the use of force during armed conflict and therefore are the most relevant to a discussion of the use of AI in weapon systems.

Articles 48 and 52 enshrine the customary law duty of parties to an armed conflict to distinguish between the civilian population and combatants and between civilian objects and military objectives, and thus direct attacks only against combatants and/or other military objectives such as enemy installations, equipment, and transport (Waxman, 2008) Article 52 defines "military objectives" as "those objects which by their nature, location, purpose or use make an effective contribution to military action and whose total or partial destruction, capture or neutralization, in the circumstances ruling at the time, offers a definite military advantage" (API, 1977). The principle of military necessity does not provide a basis for derogation from the prohibition on attacking civilians and civilian objects (Prosecutor v. Stanislav Galić, IT-98-29-A; Galić, 2006).

In parallel, Article 51 (4) expresses the customary law prohibition of indiscriminate attacks, which include:

- Those which are not directed at a specific military objective
- Those which employ a method or means of combat which cannot be directed at a specific military objective; or

- Those which employ a method or means of combat the effects of which cannot be limited as required by API

Attacks that employ means of combat which cannot discriminate between civilians and civilian objects and military objectives are "tantamount to direct targeting of civilians" (Prosecutor v. Pavle Strugar, IT-01–42-A, note 689; Strugar, 2008). Similarly, the encouragement of soldiers to fire weapons for which they lack training may be indicative of the indiscriminate nature of an attack. Furthermore, the indiscriminate nature of an attack may be circumstantial evidence that the attack actually was directed against the civilian population (Galić, 2006).

Precautions in Attacks

Article 57 of API addresses the precautions that "those who plan or decide upon" an attack must exercise to avoid or minimize civilian casualties. Planners and executors of attacks must do everything feasible to verify that the target of the attack is a military objective and the provisions of API do not forbid the operation. Furthermore, belligerent forces must "take all feasible precautions in the choice of means and methods of attack" to avoid and minimize incidental injury to civilians and damage to civilian objects (Art. 57, API, 1977). "Feasible precautions" are precautions that are practicable or practically possible considering all circumstances ruling at the time, including humanitarian and military considerations (Henckaerts & Doswald-Beck, 2005). Thus, feasibility determinations depend on diverse factors such as access to intelligence concerning the target and the target area, availability of weapons, personnel, and different means of attack, control (if any) over the area to be attacked, the urgency of the attack and "additional security risks which precautionary measures may entail for the attacking forces or the civilian population" (Wright, 2012, p. 827). As technology develops, however, the scope of what is "practicable", and therefore legally necessary, may expand accordingly (Beard, 2009).

The rule of proportionality, expressed in Articles 51 (5) (b) and 57 (2) (a) (iii), is the most challenging obligation within the realm of "precautions-in-attack." This rule requires parties to armed conflict to "refrain from deciding to launch any attack which may be expected to cause incidental loss of civilian life, injury to civilians, damage to civilian objects, or a combination thereof, which would be excessive in relation to the concrete and direct military advantage anticipated" (Art. 57 (2) (a) (iii)). This duty requires consideration and balancing of at least three abstract values: *"excessive* incidental injury to civilians and/or damage to civilian objects," "concrete and direct" and "military advantage."

The adjective "excessive" is important because, as Dinstein observes, incidental civilian damage during armed conflict is inevitable due to the impossibility of keeping all civilians and civilian objects "away from the circle of fire in wartime" (Dinstein, 2012). However, the term does not lend itself to empirical calculations as it is impossible to prove, for example, that a particular factory is worth X number of civilians (Rogers, 2004). A variety of relevant military, moral, and legal concerns may inform an assessment of what is "excessive" such as the importance of the military objective, the number of civilians at risk from the attack, as well as the risks to friendly forces and civilians if the attack does not occur (U.S. Department of Defense "Law of War Manual," 2015). Furthermore, calculations of *expected* incidental damage to civilians (whether excessive or not) will always be approximations "to help inform a commander's decision making" (NSCDEM, 2012). Accordingly, military commanders must use their common sense and good faith when they weigh up the humanitarian and military interests at stake.

The requirements of the Article 57 rules concerning precautions-in-attack (as well as the other targeting rules codified in API) reflect elementary considerations of humanity and the IHL principle that civilians and civilian objects shall be spared, as much as possible, from the effects of hostilities (Galić, 2006). Similarly, these rules speak to military necessity and the need of armed forces for disciplined soldiers who will fight effectively and facilitate the re-establishment of peace. "[I]t is clear that no responsible military commander would wish to attack objectives which were of no military interest. In this respect humanitarian interests and military interests coincide" (ICRC Commentary, 1987). Thus, this dual proscriptive and permissive approach runs through the laws and customs of war from the writings of Grotius, Vattel and their contemporaries to modern-day treaties and customary IHL.

Military AI in Weapon Systems and Compliance with the Laws of Targeting

As the laws of targeting are effects-based, nothing in IHL per se makes the application of these targeting rules using weapons operating with military AI unlawful, *provided* that the weapons system utilizing AI is capable of compliance with the rule(s). The current legal standard for weapon systems, including those employing machine learning and other forms of AI, is whether or not that system can be used in compliance with the traditional principles of the law of armed conflict, including minimizing death and injury to civilians and damage to civilian objects (Jensen, 2020).

Accordingly, professional armies must "expect military commanders employing a system with [military AI] to engage in the decision-making process that is required by international humanitarian law" (Jackson, 2014). Logically, it is impossible for commanders to *direct* weapons at specific military objectives, as required by Article 51 (4) (b) of API, without a proper understanding of the weapon. Thus, in many jurisdictions deployment of weapons utilizing military AI without a proper understanding of how the system works will constitute an indiscriminate attack and be subject to criminal sanction (M Schmitt, personal communication, March 15, 2014).

Moreover, prior to deploying a weapon system using, the superior must ensure one of two criteria: (i) once programmed, the AI software controlling the weapon system has the robust capacity to comply with Article 57, or (ii) deployment of the weapon system is itself an expression of a "feasible precaution in the choice of means and methods of attack" within the meaning and spirit of the law (Jackson, 2014). Depending on the degree of autonomy provided by the AI software, if the commander deploys an autonomous weapon platform, she may lose her ability to take additional feasible precautions as well as make proportionality judgments.

ARE RUSSIA'S DRONE ATTACKS ON ENERGY INFRASTRUCTURE LAWFUL?

During an armed conflict, an attack on the enemy's energy infrastructure can be lawful only if the item targeted is a military objective as described in Article 52(2) of API. Power infrastructure that supports military facilities, equipment, or activities qualifies as a military objective so long as it "makes an effective contribution to enemy military action and neutralizing it will yield a military advantage to the attacker," in this case, Russian armed forces (Schmitt, 2022). This is so even when the power installation is a "dual-use" object that also supports the civilian population. Those portions of a power grid, however, upon which the military does not rely and that can be struck separately to retain their civilian character, are not military objectives and should not be attacked. Russia's repeated use of Shahed drones to carry out attacks on Ukraine's energy installations, therefore, warrants review under the precautions-in-attack and proportionality rules of IHL.

Precautions in Russia's Attacks

As discussed above, Article 57 of API obliges belligerent forces to exercise all feasible precautions in the choice of means and methods of attack to avoid and minimize incidental injury to civilians and damage to civilian objects. The person launching the attack(s) must endeavor to spare the civilian population as much as possible (ICRC Commentary, 1987). In the context of Russia's widespread and systematic attacks against Ukraine's energy infrastructure, it is difficult to believe that the Russian military has taken all feasible precautions to minimize harm to civilians and damage to civilian objects when: (i) the repeated attacks occurred for months throughout the autumn and winter of 2022/2023 (Picheta, 2023); (ii) the attacks struck energy installations across most of Ukraine territory; (iii) the strikes continued during the coldest months of the year when loss of energy has the greatest impact; (iv) the attacks affected millions of civilians; and (v) the effects of the attacks on the civilian population were public and common knowledge around the world (Lander et al., 2022; Schwirtz & Mpoke Biggs, 2022).

In November 2022, Vasily Nebenzya, Russia's ambassador to the United Nations Security Council, defended his armed forces' assault on Ukraine's energy system: "To weaken and destroy the military potential of our opponents, we are conducting strikes with precision weapons against energy and other infrastructure, which is used for the purpose of military supplies to Ukrainian units" (Santora & Gibbons-Neff, 2022).

The extraordinary impact of these attacks on the civilian population, however, suggests that Russian armed forces took little or no precautions to limit harm to civilians and civilian objects. On the contrary, the use of "precision weapons" to attack a broad spectrum of energy targets across Ukraine, with no apparent effort to differentiate between power systems and services that support the Ukrainian armed forces and those that supply the civilian population, indicates an *intent to maximize* civilian suffering. The use of high-precision weapons allows the completion of operations with less waste of ammunition and human suffering (Kostenko, 2022). Nebenzya's statement – as well as the systematic damage and destruction of Ukraine's energy facilities – indicate that Russian forces carried out these attacks with knowledge and understanding of the capabilities of these weapons. The availability of accurate, long-range weapons, including the Shahed drone, gave the Russian military the ability to focus their attacks on the energy infrastructure that sustains Ukraine's army whilst avoiding, as much as possible, harm to civilians. This it failed to do.

Proportionality in Russia's Attacks

As discussed above the rule of proportionality, expressed in Articles 51 and 57 of API, is the most challenging duty within the scope of "precautions-in-attack." Proportionality assessments require commanders and targeters to make a careful judgment, using common sense and good faith, that balances the foreseeable military advantage resulting from the attack against the expected harm to civilians (ICRC Commentary, 1987). Art. 8 (2) (b) (iv) of the Rome Statute of the International Criminal Court prohibits attacks where the anticipated civilian injury and damage is "clearly excessive" to the expected military advantage. No similar provision exists in treaty or customary law that criminalizes failures to take feasible precautions under Arts. 57 (2) (a) (i) or (ii) (R. Geiss, personal communication).

Whilst proportionality judgments can be complex depending on the circumstances of the combat domain, the facts known about Russia's attacks on Ukraine's energy infrastructure indicate that many of the attacks were disproportionate (COI, 2023). After just the first week of attacks in October 2022, thirty percent of Ukraine's power plants were damaged, causing electrical blackouts across the country (Schwirtz & Mpoke Bigg, 2022). Thus, substantial harm to civilian objects and the civilian population was foreseeable by that time, and, due to the approaching winter weather, more severe harm caused by additional attacks was predictable. At that time, under IHL, Russia should have suspended or canceled these strikes (Art. 57(2)(b), API, 1977). Nonetheless, Russian attacks on the power system continued, sometimes involving dozens or up to a hundred missiles and drones at one time, disrupting the energy supply to millions of persons.

Moreover, if by mid-to-late autumn 2022, it was foreseeable that continued attacks would result in the suffering of millions of Ukrainian civilians, to constitute proportionate attacks the anticipated military advantage for Russian forces should have been extraordinarily high. The waves of strikes, however, have had little impact on ongoing Ukrainian military operations. On the contrary, over time, they exhausted Russia's supply of missiles and drones (Schmitt, 2022a). It is difficult to imagine, therefore, that as the attacks occurred over months, proportionality assessments met the legal standard.

Taken together, these circumstances point to a broader conclusion about Russia's conduct of attacks against Ukraine's energy infrastructure: at least some of them were indiscriminate, either because they were not directed at a specific military objective, because they were disproportionate, or both (Art. 51(4)(a) & 5(b), API, 1977). The evidence suggests that, when conducting these attacks, Russian armed forces made no effort to find an equitable balance between the principles of military necessity and humanity. Attempts to find and maintain this balance form the basis of the entire law of armed conflict (ICRC Commentary, 1987). By launching and continuing indiscriminate attacks, Russian soldiers and commanders violated IIIL.

The Intent to Cause Terror

In late November 2022, one highly respected commentator suggested that, as Russian attacks on Ukrainian energy infrastructure "have gone on for so long, are so widespread, and are so intense, that it is difficult to attribute any purpose to them other than terrorizing the civilian population" (Schmitt, 2022b). The law of armed conflict prohibits attacks or threats of violence the primary purpose of which is to spread terror among the civilian population (Art. 51(2), API, 1977). Terror can be defined as "extreme fear" (Prosecutor v. Ratko Mladić, MICT-13-56-A, 2021, para. 315; Mladić, 2021). Attacks intended to terrorize civilians may include indiscriminate and disproportionate attacks and such acts may contribute to establishing the intent to terrorize (Mladić, 2021).

Acts or threats of violence that cause death or serious injury to body or health are only one possible mode of commission of terror (Mladić, 2021). Whilst victims must suffer grave consequences as a result of the attacks or threats of violence, "grave consequences" can include, but are not limited to death or serious physical or mental injury Prosecutor v. Dragomir Milosević, IT-98-29/1-A, 2009, para. 33; Milosević, 2009). Importantly, the *actual* infliction of terror on civilians is not a requirement of this offense, although evidence of actual terrorization can contribute, for example, to establishing that the infliction of terror was the primary purpose of the attacks. Causing terror must be the *primary* purpose of the attack or threats of violence, but it need not be the only one (Milosević, 2009]. The intent to terrorize may be inferred from a number of factors including the nature, manner, timing, and duration of the acts or threats.

It is to be expected that a civilian population will be afraid during the chaotic times of war (Prosecutor v. Radomir Karadžić, IT-95-5/18-T, 2016). However, the situation of civilians in Ukraine during the winter of 2022–2023 was unique. The intent to terrorize can be inferred by the decision of the Russian military to launch repeated attacks against energy infrastructure serving the civilian population just at the start of the coldest period of the year. The use of swarms of dozens of missiles and drones in these attacks across broad regions of Ukraine also suggests an intent to cause fear in the civilian population. Continuing the attacks for months whilst millions of Ukrainian civilians were without power in freezing temperatures could only contribute to a sense of fear and helplessness and Russian leaders must have been aware of this dynamic. Considered as a whole, these circumstances suggest, at a minimum, that an important purpose of the attacks on Ukrainian power installations was to instill terror in the civilian population.

THEORIES OF INDIVIDUAL CRIMINAL RESPONSIBILITY FOR UNLAWFUL ATTACKS WITH WEAPONS THAT USE AI

Theories of Individual Criminal Responsibility for Unlawful Attacks with LAWS

Preliminarily, two general kinds of individual criminal responsibility may arise when soldiers and/or their commanders violate IHL. First, "direct" responsibility arises from an individual's acts or omissions that contribute to the commission of crimes (Prosecutor v. Stanislav Galić, IT-98-29-T, 2003, para. 169). Second, "superior" or "command" responsibility emanates from the failure of military or civilian superiors to perform their duty to prevent their subordinates from committing such crimes, and/or the failure to fulfill the obligation to punish the perpetrators thereafter (Prosecutor v. Momčilo Perišić, IT-04-81-A, 2013, paras. 86–87). The latter form of liability, therefore, implies criminal responsibility by omission.

Moreover, each theory of individual criminal liability contains objective and subjective elements: the actus reus – the physical act necessary for the offense – and the mens rea – the necessary mental element (Prosecutor v. Zejnil Delalić, et al., IT-96-21-T, 1998; Delalic, 1998). The principle of individual guilt requires that an accused can be convicted of a crime only if her mens rea comprises the actus reus of the crime (Prosecutor v. Mladen Naletilić, a.k.a. "Tuta" & Vinko Martinović, a.k.a. "Štela," IT-98-34-A, 2006). A conviction absent mens rea would violate the presumption of innocence. Thus, to convict an accused of a crime, she must, at a minimum, have had knowledge of the facts that made her conduct criminal.

Similarly, under the Rome Statute of the International Criminal Court (ICC) conviction can occur "only if the material elements are committed with intent and knowledge" (Art 30, 1998 [Rome Statute, 1988]) This conjunctive approach requires the accused to possess a volitional element encompassing two possible situations: 1) she knows that her actions or omissions will bring about the objective elements of the crimes, and she undertakes such actions or omissions with the express intent to bring about the objective elements of the crime, or 2) although she does not have the intent to accomplish the objective elements of the crime, she is nonetheless aware that the consequence will occur in the ordinary course of events (The Prosecutor v. Abdallah Banda & Saleh Jerbo, Corrigendum to Decision on Confirmation of Charges, ICC-02/05-03/09, 2011).

Theories of Direct Responsibility

For the sake of brevity, this chapter focuses on two possible modes of the direct responsibility of Russian commanders for their employment of Shahed drones to attack Ukraine's energy infrastructure: commission and ordering.

Individual "commission" of a crime entails the physical perpetration of a crime or engendering a culpable omission in violation of criminal law (*Prosecutor v. Fatmir Limaj*, IT-03-66-T, 2005, para. 509 [Limaj, 2009]). The actus reus of this mode of criminal liability is that the accused participated, physically or otherwise directly, in the material elements of a crime, through positive acts or omissions, whether individually or jointly with others. In the case law of the ad hoc tribunals for the former Yugoslavia and Rwanda, the requisite mens rea for commission is that the perpetrator acted with the intent to commit the crime, or with an awareness of the probability, in the sense of the substantial likelihood, that the crime would occur as a consequence of his/her conduct (Limaj, 2005). The Rome Statute of the ICC, however, excludes the application of the dolus eventualis standard, as well as the mens rea of recklessness. Instead, the criminal mens rea exists if the accused means to commit the crime, or, she is aware that by her actions or omissions, the crime will occur in the ordinary course of events. (Prosecutor v. Thomas Lubanga Dyilo, Judgment Pursuant to Article 74 of the Statute, ICC-01/04-02/06, 2012).

In addition, at the ICC, criminal responsibility may accrue when the accused makes an essential contribution to a plurality of persons acting with a common criminal purpose. The accused must be aware of her essential contribution, and must act with the intention that the crime occur, or with the awareness that by implementing the common plan, the crime "will occur in the ordinary course of events" (Art. 25(3) (d), Rome Statute, 1998). In the case law of the ad-hoc tribunals for Rwanda and the former Yugoslavia, culpable participation in a common criminal purpose is referred to as "joint criminal enterprise" and requires a significant contribution to the realization of the crime (Prosecutor v. Vujadin Popović, et. al., IT-85-88-T, 2010, para. 1027; Popović, 2010).

Responsibility under the mode of "ordering" ensues when a person in a position of authority orders an act or omission with the awareness of the substantial likelihood that a crime will be committed in the execution of that order, and, if the person receiving the order subsequently commits the crime (Prosecutor v. Boškoski & Tarčulovski, IT-04-02-A, 2010, para. 160; Bokoški & Tarčulovski, 2010). Orders need not take a particular form and the existence of orders may be established using circumstantial evidence. Liability ensues if the evidence demonstrates that the order substantially contributed to the perpetrator's criminal conduct (Boškoski & Tarčulovski, 2010).

The Theory of Superior Responsibility

When crimes occur due to the misuse of weapons systems operating with AI, the theory of superior responsibility also may be appropriate to hold commanders accountable. The superior-subordinate relationship lies at the heart of the doctrine of a commander's liability for the crimes committed by her subordinates. During armed conflict, the role of commanders is decisive (ICRC Commentary, 1987) and it is the position of command over subordinates and the power to control their actions (and comply with international law) that form the legal basis for the superior's duty to act, and for her corollary liability for a failure to do so (Limaj, 2009).

In general terms, pursuant to the statute and jurisprudence of the ad-hoc tribunals, a military or civilian superior may be held accountable if the superior knew or had reason to know that her subordinates were committing or about to commit criminal acts and failed to take necessary and reasonable measures to prevent the crimes and/or punish the perpetrators (Statute of the International Residual Mechanism for Criminal Tribunals, 2010). The Rome Statute of the ICC alters the evidentiary thresholds for holding civilian and military commanders accountable under the theory of superior responsibility. In addition to the three elements found in the law of the ad-hoc tribunals, prosecutors at the ICC must establish that the crimes committed by subordinates occurred as a result of the superior's "failure to exercise control properly over such forces" (Rome Statute, 1998). In short, it is necessary to prove that the superior's omission increased the risk of the commission of the crimes charged (*The Prosecutor v. Jean Pierre Bemba Gombo*, Decision Pursuant to Article 61 (7) (a) and (b) of the Rome Statute on the Charges of the Prosecutor v. Jean-Pierre Bemba Gombo, 01/05-01/08, 2009).

The superior/subordinate relationship

A superior-subordinate relationship exists when a superior exercises effective control over her subordinates, i.e. when she has the material ability to prevent or punish their acts. Factors indicative of an accused's position of authority and effective control include the official position she held, her capacity to issue orders, whether *de jure* or *de facto*, the procedure for appointment, the position of the accused within the military or political structure and the actual tasks that she performed. The indicators of effective control are more a matter of evidence than of substantive law and depend on the specific circumstances of each case. The concept of superior is broader than immediate and direct command "and should be seen in terms of a hierarchy encompassing the concept of control" (ICRC Commentary to Art. 86, 1987, para. 3544). Thus, more than one superior may be held responsible for her failure to prevent or punish crimes committed by a subordinate, regardless of whether the subordinate is immediately answerable to the superior or more distantly under her command (Prosecutor v. Sefer Halilović, IT-01-48-T, 2005).

The superior's knowledge of the criminal acts of her subordinates

A superior's mens rea, i.e. her knowledge that her subordinates were about to commit or had committed crimes may be actual knowledge or the availability of "sufficiently alarming" information that would put her on notice of these events (Strugar, 2008, paras. 297–304). Such knowledge may be presumed if the superior had the means to obtain the knowledge but deliberately refrained from doing so (Delalić, 1998).

At the ICC, instead of requiring proof that the superior "had reason to know" that her forces were committing or had committed crimes, the court's "knowledge" standard for military commanders compels prosecutors to establish that she "should have known" about such crimes (Art. 28(1)(a), Rome Statute, 1998). This standard requires the commander "to ha[ve] merely been negligent in failing to acquire knowledge" of her subordinates' unlawful conduct (Strugar, 2008, paras. 297–304). The "knowledge" requirement for demonstrating the liability of civilian superiors is higher: "the superior either knew, *or consciously disregarded* information which clearly indicated that the subordinates were committing or about to commit such crimes" (Art. 28(2)(a), Rome Statute, 1988).

Necessary and reasonable measures to prevent the crimes and/or punish the perpetrators

"Necessary" measures are the measures appropriate for the superior to discharge her obligation (showing that she genuinely tried to prevent or punish) and "reasonable" measures are those reasonably falling within the material powers of the superior (Halilović, 2005). A superior will be held responsible if she fails to take such measures that are within her material ability and the superior's explicit legal capacity to do so is irrelevant provided that she has the material ability to act (Limaj, 2005).

"The determination of what constitutes 'necessary and reasonable measures' is not a matter of substantive law but of fact, to be determined on a case-by-case basis" (Popović, 2010). This assessment depends upon the superior's level of effective control over her subordinate(s). Depending upon the circumstances of the case, "necessary and reasonable" measures can include carrying out an investigation, providing information in a superior's possession to the proper administrative or prosecutorial authorities, issuing orders aimed at bringing unlawful conduct of subordinates in compliance with IHL and securing the implementation of these orders, expressing criticism of criminal activity, imposing disciplinary measures against the commission of crimes, reporting the matter to the competent authorities, and/or insisting before superior authorities that immediate action be taken (Popović, 2010).

APPLICATION OF THE THEORIES OF INDIVIDUAL CRIMINAL RESPONSIBILITY TO THE USE OF SHAHED DRONES TO ATTACK ENERGY INFRASTRUCTURE IN UKRAINE

Application of Theories of Direct Responsibility

In cases involving deliberate, unlawful attacks with weapons utilizing AI, proof of a commander's individual criminal responsibility under the direct modes of commission and ordering will be relatively simple. For example, if, during armed conflict, a commander *intentionally* employs a weapon system in circumstances where the system's capabilities for compliance with IHL are inadequate (such as a notoriously inaccurate weapon within a densely-populated urban area where civilians are known to be present), and death and injuries to civilians occur, that commander is culpable for the commission of a war crime (Schmitt, 2014).

In the context of Ukraine, it is possible that some of the apparently disproportionate attacks perpetrated with Shahed drones occurred as a result of computational errors at the programming stage of the targeting process. Nonetheless, given Shahed's reputation as a precise and accurate weapon, the repeated launch over time of so many disproportionate attacks on energy infrastructure suggests that the relevant personnel acted with the criminal intent to commit these offenses.

In addition, as mentioned above, at the ICC, under the mode of "commission," criminal responsibility may accrue when the accused makes an essential contribution to a plurality of persons acting with a common criminal purpose. The repeated waves of Shahed drone attacks on Ukraine energy installations between October 2022 and early 2023 required the coordinated planning and efforts of many Russian commanders, intelligence and targeting analysts, logistical personnel, engineers/programmers, drone operators, etc. An undetermined number of Russian commanders and their subordinates who planned, coordinated, conducted, and monitored the effects of these attacks made essential contributions to a common criminal design to launch attacks on energy infrastructure in violation of IHL. The identities of some of these personnel are available in the public domain (Grozev, 2022).

Similarly, in professional armed forces, the strategic and repeated use of important and expensive resources, such as Shahed drones, over many months, only occurs under the direction and orders of high-level commanders. Assuming sufficient evidence exists of orders given by particular individuals, those commanders (including civilian superiors such as President Putin) can be held accountable under the theory of liability of ordering for these unlawful attacks.

Application of Superior Responsibility

In certain scenarios, the theory of superior responsibility may be appropriate to hold military commanders and/or civilian superiors responsible for failing to prevent and/or punish crimes perpetrated with military AI-driven weapons systems. For example, if a commander at the operational level becomes aware that a subordinate officer at the tactical level is using such weapons to perpetrate unlawful attacks, the operational commander has a duty to prevent further misconduct and punish his subordinate (Arts. 86 and 87, API, 1977). As long as evidence exists demonstrating the three essential elements of a commander's effective control over subordinates who commit crimes, the commander's knowledge, and a failure to take necessary and reasonable measures to prevent further crimes and punish the perpetrators, criminal liability should ensue.

In the case of Russia's drone attacks on Ukrainian energy installations, the existence of commanders' effective control over subordinates who actually launched the attacks is evidenced by the planned and coordinated nature of the attacks, repeated over months. The pattern of these attacks and the sophisticated weapons used indicate that they occurred under the direction and control of superiors.

With respect to the "knowledge" element, commanders of Russian forces operating the Shahed drone will not easily convince a court that they were unaware that their subordinates operated these weapons unlawfully during 2022–2023. This premise should hold true regardless of whether courts apply the "had reason to know" standard of the ad-hoc tribunals or the "should have known" standard of the ICC. The attacks on Ukraine's energy infrastructure were public knowledge. Moreover, any state or organized armed group with the resources and ability to employ such weapons – such as Russia – will also have the means and the communications technology to monitor how these weapons are used. Any competent commander utilizes all possible methods to observe the progress and operations of her subordinate units (NATO, Allied Joint Doctrine, 2019).

The availability of electronic records also will minimize the challenge that physical and/or temporal "remoteness" poses to the accountability of superiors. Physical "remoteness" in this context refers to the geographical distance between the acts or omissions of a superior and the location of the criminal conduct. Temporal "remoteness" refers to the time elapsed between the accused's acts or omissions and the execution of the crimes. Modern communications increasingly provide superiors with real-or-nearly-real-time access to circumstances and events, including combat occurring far from command centers and headquarters. Furthermore, the internet and social-media technology create virtual links between front-line areas and all parts of the world. These connections, combined with electronic records of commanders' decision-making processes, reduce the physical and temporal distances between a superior and events in the battlespace and reveal much about a superior's mental state.

With respect to the last prong for establishing criminal liability under the theory of superior responsibility, the scope of "necessary and reasonable measures" to prevent further crimes and punish the subordinates involved in misconduct may vary when weapons operating with AI are used to carry out unlawful attacks. For example, the use of swarm technology will undoubtedly increase the tempo of military engagements (Fiddian, 2012). The faster pace of combat – and unlawful conduct – will reduce a superior's opportunities to prevent crimes.

Nonetheless, these concerns do not apply to Russia's attacks on Ukraine's power facilities using Shahed drones. Whilst Russia utilized swarms of these drones during these assaults, days or weeks passed between the waves of strikes, giving Russian commanders ample time to attempt measures to avert future unlawful conduct by their subordinates. Orders could have been issued to suspend or modify attacks in accordance with Article 57(2)(b) of API, reports to higher superiors could have been sent, and, in a last resort, commanders could have resigned. Absent evidence that Russian commanders took such measures, it appears that each of the three elements necessary to establish superior responsibility for these offenses is met.

The Responsibility of Russia and Iran for Attacks on Ukraine's Energy Infrastructure

States incur international responsibility by acts imputable to them that violate a rule or rules of international law (Cheng, 1953). The clarity and power of rules of state responsibility, therefore, are necessary to complement the processes of individual criminal responsibility and, hopefully, to set standards for accountability that reduce the likelihood of violations of international law. For example, when states deliberately employ weapon systems in the commission of serious violations of international law, they will be in affirmative breach of their international legal obligations. A customary rule of international law provides that the conduct of any organ of a State – such as the behavior of Russian forces in Ukraine – must be regarded as an act of that State (Armed Activities on the Territory of the Congo, [Congo v. Uganda], 2005). Accordingly, the conduct of individual Russian soldiers and commanders must be considered as the conduct of a Russian state organ. Consequently, as a party to the armed conflict Russia is responsible for all acts by individuals forming part of its armed forces (Art. 91, API, 1977).

In addition to deliberate violations of the law, international legal decisions have (implicitly or explicitly) recognized a duty of states to exercise due diligence and prevent harm with respect to the design, manufacture, and use of weapons. For example, in 1996, the United States Government agreed to pay nearly 132 million U.S. dollars to the Government of Iran as compensation for the 1988 shoot-down of an Iranian passenger plane by the USS Vincennes, a U.S. warship operating in the Strait of Hormuz. During a highly fluid situation involving multiple surface and air vessels from the U.S. and Iranian armed forces, the ship's defense system correctly indicated that the Iranian plane was ascending. Human error, however, contributed to the mistaken belief on the part of the Vincennes' Captain that the civilian airliner was actually an Iranian fighter plane preparing to attack the ship (Schmitt, 2013).

In the context of the Russian military's recent (and apparently deliberate) widespread damage and destruction of much of Ukraine's energy network, the determination of Russia's responsibility as a state should mirror many of the legal and evidentiary questions discussed vis a vis its armed forces' compliance with the law of armed conflict (Congo v. Uganda, 1996). The existence of prolonged pain and suffering resulting from the loss of electricity, gas, water, etc, will be relevant to the State's obligation of reparation (International Law Commission, Draft Articles on Responsibility of States for Internationally Wrongful Acts with Commentaries [Draft Articles], 2001).

The evaluation of *Iran's* possible State responsibility for the disruption and destruction of Ukraine's energy grid, however, requires a different analysis. In situations where one State voluntarily provides aid or assistance to another with a view to facilitate the commission of internationally wrongful acts by the receiving State, the assisting State will be responsible to the extent that its own conduct contributed to the wrongful act(s). State responsibility for this kind of assistance occurs when three criteria are satisfied: (i) the assisting State must be aware of the circumstances making the conduct of the assisted State internationally wrongful; (ii) the help or support must be given with a view to facilitate the commission of the wrongful act, and must actually do so; and (iii) the act must be such that it would have been wrongful had it been committed by the assisting State itself (Draft Articles, 2001).

With respect to Russia's use of Shahed drones to destroy Ukraine's energy infrastructure, Iran initially exported 46 UAVs, including Shahed, to Russia in early August 2022 and provided relevant training on the technology. In mid-October, Iran sent dozens of Revolutionary Guard specialists to eastern and southern Ukraine to train members of the Russian military to operate drones. On 21 October, Ukraine reported that it had killed ten Iranian trainers in two separate strikes. In November Russia and Iran reportedly made an agreement to manufacture at least 6,000 drones, including an advanced version of the Shahed-136 model, in a new factory to be constructed in the Russian town of Yelabuga. The drone shipments continued into the winter and in late December 2022, Ukraine reported that Iran had provided 1,700 Shahed drones to Russia (United States Institute for Peace, "Timeline: Iran Russia Collaboration on Drones," 2023).

The evidence available in the public domain strongly indicates that the three threshold criteria required to establish Iran's responsibility for assisting Russia's disproportionate and indiscriminate attacks on Ukraine's power installations are met. First, given the public nature of the civilian suffering caused by Russia's use of Shahed drones to target these facilities, the Iran government's awareness of these circumstances is beyond doubt. Second, given Iran's continuing supply of Shahed drones (and training on how to use them) to Russia during the months when the attacks occurred, it is reasonable to infer that Iran purposely provided this assistance to facilitate these (unlawful) Russian operations. Iran's commitment to assist Russian efforts to manufacture these drones on Russian territory in the future only reinforces this inference. The fact that Ukraine expended its own resources to target and kill Iranian trainers demonstrates that Iran's assistance actually facilitated Russia's unlawful attacks.

Lastly, whilst Iran has signed API, it has not ratified the Protocol. Nonetheless, the prohibitions on launching disproportionate and indiscriminate attacks (as well as attacks whose primary purpose is to terrorize the civilian population) are part of customary IHL and consequently binding upon Iran as well as Russia. Therefore, had Iran's military launched disproportionate and indiscriminate assaults directly on Ukraine's energy infrastructure, such attacks would have constituted internationally wrongful acts.

CONCLUSIONS

Russia's use of weapons systems operating with AI software to target Ukraine's power installations and grid during 2022–2023 raises three significant issues. First, in the execution of these attacks, did the Russian armed forces violate their duties under the rules of IHL? Second, is it possible to hold individual soldiers and commanders accountable for crimes that occurred as a result of the use of the Shahed drone in these assaults? Lastly, do Russia and Iran, as States, incur responsibility for breaches of their international legal obligations? I have tried to demonstrate that most likely the answer to each question is "yes." Indeed, Russia's use of AI technology that increases the accuracy of its attacks illustrates the greater likelihood that breaches of IHL occurred.

As weapons system technology – including AI technology – advances, however, the resolution of these issues during current and future conflicts will become increasingly complex. As an "AI laboratory," the Russia/Ukraine war is a "major stepping stone toward the networked battlefield and the AI wars of the future" (Fontes & Kamminga, 2023). Due to advances in AI, the autonomy and speed of weapons systems will increase, making it more difficult for human agents to control their behavior and the outcomes of even lawful targeting decisions.

Indeed, the use of swarms of programmed Shahed drones to attack Ukrainian energy targets portends future assaults with large numbers of more autonomous UAVs, and not only by Russia. Ukrainian AI experts consider "multi-agent autonomous robot swarms technology, including UAVs" to be a principal technology for AI application (Shevchenko et al., 2022). Not surprisingly given its current predicament, Ukraine hopes to develop "completely new technologies in the field of AI" between now and 2030 (Shevchenko et al., 2022).

The sheer speed and complexity of increasingly sophisticated AI-driven weapons can blur the causal links between human targeting decisions and violations of the laws of war. When apparent violations of IHL occur using weapon systems bearing more advanced AI, the establishment of human and state responsibility will be much more challenging.

REFERENCES

Anderson, K & Waxman, M. (2017). Debating autonomous weapon systems, their ethics, and their regulation under international law. In R. Brownsword, E. Scotford, & K. Yeung (Eds.), *The Oxford handbook of law, regulation, and technology* (pp. 1097 & 1114). Oxford University Press.

Armed Activities on the Territory of the Congo (Democratic Republic of the Congo v. Uganda), Judgment, paras. 213, 220 & 245. (2005). https://www.icj-cij.org/sites/default/files/case-rel ated/116/116-20051219-JUD-01-00-EN.pdf

Army Recognition. (2023, July 2023). *Shahed 136.* https://www.armyrecognition.com/iran_ unmanned_ground_aerial_vehicles_systems/shahed-136_loitering_munition_kami-kaze-suicide_drone_iran_data.html

Automated Decision Research. (2023). *Weapons systems with autonomous functions used in Ukraine.* Retrieved September 1, 2023, from https://automatedresearch.org/news/weapons-systems-with-autonomous-functions-used-in-ukraine/

Beard, J. (2009). Law and war in the virtual era. *American Journal of International Law, 103*(3), 409–445. https://www.jstor.org/stable/40283651

Beinisch, D. (2005). *The Public Committee against Torture in Israel v. The Government of Israel,* HCJ 769/02. https://www.haguejusticeportal.net/Docs/NLP/Israel/Targetted_Killings_ Supreme_Court_13-12-2006.pdf.

Brennan, D. (2022, December 30). Shahed 136: The Iranian Drones Aiding Russia's Assault on Ukraine. *Newsweek.*

Cheng, B. (1953). *General principles of law as applied by international courts and tribunals.* Stevens & Sons Limited.

Chulov, M, Sabbagh, D., & Mando, N. (2023, February 12). *Iran smuggled drones into Russia using boats and state airline, Sources Reveal.* The Guardian. https://www.theguardian. com/world/2023/feb/12/iran-uses-boats-state-airline-smuggle-drones-into-russia

Crowe, W. (1988). Letter attached to Investigation Report: Formal Investigation into the Circumstances Surrounding the Downing of Iran Air Flight 655, para. 9.

Dinstein, Y. (2012). The principle of distinction and cyber war in international armed conflicts. *Journal of Conflict & Security Law, 17*(2), 261–277. https://www.jstor.org/stable/26296230

Draft Articles on Responsibility of States for Internationally Wrongful Acts with Commentaries. (2001). https://legal.un.org/ilc/texts/instruments/english/commentaries/9_6_2001.pdf.

Fiddian, P. (2012). UAV swarm technology trial success. *Armed Forces International News.* https://www.armedforces-int.com/news/uav-swarm-technology-trial-success.html

Fontes, R., & Kamminga, J. (2023, March 24). Ukraine: A living lab for AI warfare. *National Defence.* https://www.nationaldefensemagazine.org/articles/2023/3/24/ukraine-a-living-lab-for-ai-warfare#

Grozev, C. (2022). *The remote control killers behind Russia's cruise missile strikes on Ukraine.* Bellingcat. https://www.bellingcat.com/news/uk-and-europe/2022/10/24/the-remote-control-killers-behind-russias-cruise-missile-strikes-on-ukraine/

Hambling, D. (2023, April 20). The secret behind Russia's swarms of deadly drones. *Popular Mechanics.* https://www.popularmechanics.com/military/weapons/a43519073/iranian-drones-ukraine/

Henckaerts, J. & Doswald-Beck, L. (2005). *Customary International Law, Vol. 1: Rules*, (p. 54). Cambridge University Press.

Hutch, T. (2023, January 13). How cheap drones replaced fighter jets in the battle for Ukraine skies. *Popular Mechanics.* https://www.popularmechanics.com/military/weapons/a42478097/how-cheap-drones-replaced-fighter-jets-in-ukraine/

ICRC Commentary to Art. 57, Additional Protocol I to the Geneva Conventions of 12 August 1949 (API) (1987).

Jackson, R. (2014, April 10). Autonomous weaponry and armed conflict [Conference Session]. Annual Meeting of American Society of International Law, Washington DC, United States. https://www.asil.org/resources/video/2014-annual-meeting

Joint Publications 3-60, "Joint Targeting," U.S. Department of Defence, (January 31, 2013). https://www.justsecurity.org/wp-content/uploads/2015/06/Joint_Chiefs-Joint_Targeting_20130131.pdf

JP 3-60, Joint Targeting, p. II–5; "Operations Series, ADDP 3.14," (2009). *Targeting* (2nd ed.). Australia Department of Defence. https://www.defence.gov.au/foi/docs/disclosures/021_1112_Document_ADDP_3_14_Targeting.pdf

JSP 383, The Joint Service Manual of the Law of Armed Conflict. (2004). https://www.gov.uk/government/publications/jsp-383-the-joint-service-manual-of-the-law-of-armed-conflict-2004-edition.

JSP 383, The Manual of the Law of Armed Conflict, U.K. Ministry of Defence, Amendment 3, (2010). https://assets.publishing.service.gov.uk/government/uploads/system/uploads/attachment_data/file/27871/20100929JSP383Amendment3NoterUpChs14Sep10.pdf

Kostenko, O. (2022). The probability of military aggression of autonomous AI: Assumptions or imminent reality (analyzing the facts of Russian war against Ukraine). *Analytical and Comparative Jurisprudence*, 179–183. https://app-journal.in.ua/wp-content/uploads/2022/05/35.pdf

Lander, M., Stevens, M., & Mpoke, M. (2022, December 11). Ukraine faces more outages and strikes russian-controlled melitopol. *New York Times.* https://www.nytimes.com/2022/12/11/world/europe/ukraine-odesa-outages-melitopol.html?searchResultPosition=3

Legality of the Threat or Use of Nuclear Weapons. (1996). I.C.J. Reports. paras. 77 & 78.

May, L., & Newton, M. (2014). *Proportionality in International Law.* Oxford University Press.

Moloney, M. (2023, April 8). Ukraine to export energy again after months of Russian attacks. *BBC.* https://www.bbc.com/news/world-europe-65220003

NATO Allied Joint Doctrine for the Conduct of Operations. (2019, February). https://www.coemed.org/files/stanags/01_AJP/AJP-3_EDC_V1_E_2490.pdf

No-Strike and the Collateral Damage Estimation Methodology (October 12, 2012). Chairman of the Joint Chiefs of Staff Instruction, U.S. Department of Defense. https://int.nyt.com/data/documenttools/no-strike-collateral-damage-estimation/6632f2785aff5bba/full.pdf

OE Data Integration Network. (2023). *Shahed-136 Iranian Loitering Munition Unmanned Aerial Vehicle (UAV).* Retrieved September 1, 2023, from https://odin.tradoc.army.mil/Search/All/Shahed

Picheta, R. (2023, March 10). Russia tests Ukraine's defenses with a rarely-used missile. *CNN*. https://edition.cnn.com/2023/03/10/europe/ukraine-russia-missiles-air-defenses-explainer-intl/index.html

Prosecutor v. Abdallah Banda & Saleh Jerbo, Corrigendum to decision on confirmation of charges, ICC-02/05-03/09, Pre-Trial Chamber I, para. 153, (2011). https://www.icc-cpi.int/sites/default/files/CourtRecords/CR2011_02580.PDF

Prosecutor v. Boškoski & Tarčulovski, Judgment, IT-04-82-A, Appeals Chamber, para. 160, (2010). https://www.refworld.org/cases,ICTY,4bff84af2.html

Prosecutor v. Dragomir Milosević, Judgment, IT-98-29/1-A Appeals Chamber, para. 33, (2009). https://www.icty.org/x/cases/dragomir_milosevic/acjug/en/091112.pdf

Prosecutor v. Fatmir Limaj, Judgment, IT-03-66-T, para. 509. (2005). https://www.refworld.org/cases,ICTY,48ac17cc2.html

Prosecutor v. Mladen Naletilić, a.k.a. "Tuta" & Vinko Martinović, a.k.a. "Štela," Judgment, IT-98-34-A, Appeals Chamber, para. 114. (2006). https://cld.irmct.org/assets/filings/Judgement-Naletilic.pdf

Prosecutor v. Momčilo Perišić, Judgment, IT-04-81-A, Appeals Chamber, paras 86-87. (2013). https://www.icty.org/x/cases/perisic/acjug/en/130228_judgement.pdf

Prosecutor v. Pavle Strugar, Judgment, IT-01-42-A, Appeals Chamber, paras. 297–304 & note 689. (2008). https://www.icty.org/x/cases/strugar/acjug/en/080717.pdf

Prosecutor v. Radomir Karadžić, Judgment, Case No. IT-95-5/18-T, para. 4599. (2016). https://www.icty.org/x/cases/karadzic/tjug/en/160324_judgement.pdf

Prosecutor v. Ratko Mladić, Judgment, Case No. IT-95-5/18-T, paras. 460 & 466. (2016). https://www.icty.org/x/cases/karadzic/tjug/en/160324_judgement.pdf

Prosecutor v. Ratko Mladić, Judgment, Case No. MICT-13-56-A, Appeals Chamber, para. 315. (2021). https://ucr.irmct.org/LegalRef/CMSDocStore/Public/English/Judgement/NotIndexable/MICT-13-56-A/JUD285R0000638396.pdf

Prosecutor v. Sefer Halilović, Judgment, IT-01-48-T, paras. 63, 73–74. (2005). https://www.refworld.org/cases,ICTY,48ac3a6c2.html

Prosecutor v. Stanislav Galić, Judgment, IT-98-29-A, Appeals Chamber, para. 130. (2006). https://www.refworld.org/cases,ICTY,47fdfb565.html

Prosecutor v. Stanislav Galić, Judgment, IT-98-29-T, para. 169. (2003).

Prosecutor v. Vujadin Popović et al., Judgment, IT-85-88-T, para. 1027. (2010).

Prosecutor v. Zejnil Delalić et al., Judgment, IT-96-21-T, paras. 424 and 425. (1998).

Protocol Additional to the Geneva Conventions of 12 August 1949, and Relating to the Protection of Victims of International Armed Conflicts ("API"), June 8, 1977 (n.d.). https://ihl-databases.icrc.org/en/ihl-treaties/api-1977

Report of the Independent International Commission of Inquiry on Ukraine (COI). (March 15, 2023). A/HC/52/62. https://www.ohchr.org/sites/default/files/documents/hrbodies/hrcouncil/coiukraine/A_HRC_52_62_AUV_EN.pdf

Roberts, A., & Guelff, R. (2000). *Documents on the law of war* (3rd ed.). Oxford University Press.

Rogers, A. (2004). *Law on the battlefield* (2nd ed.). Manchester University Press.

Rome Statute of International Criminal Court. (1998, July 17). https://www.icc-cpi.int/sites/default/files/RS-Eng.pdf

Rubin, U. (2023). *Russia's Iranian-made UAVs: A technical profile*. Royal United Services Institute. https://rusi.org/explore-our-research/publications/commentary/russias-iranian-made-uavs-technical-profile

Santora, M. (2023, April 11). How Ukraine's power grid survived so many Russian bombings. *New York Times*. https://www.nytimes.com/2023/04/11/world/europe/ukraine-war-infrastructure.html?searchResultPosition=6

Santora, M., & Gibbons-Neff, T. (2022, November 23). Russian missile barrage cuts power and water across Ukraine. *New York Times.* https://www.nytimes.com/2022/11/23/world/europe/russia-ukraine-missiles-power.html?referringSource=articleShare&smid=nytcore-ios-share&login=email&auth=login-email

Schmitt, M. & Thurnher J. (2013). Out of the loop: autonomous weapon systems and the law of armed conflict, *Harvard International Law Journal.* https://harvardnsj.org/2013/05/22/out-of-the-loop-autonomous-weapon-systems-and-the-law-of-armed-conflict

Schmitt, M. (2014, February 24). The international legal context. In *Autonomous military technologies: policy and governance for next generation defense systems* [symposium] Chatham House, London, United Kingdom.

Schmitt, M. (2022a, November 25). Ukraine symposium - Further thoughts on Russia's campaign against Ukraine's power infrastructure. *Articles of War.* https://lieber.westpoint.edu/further-thoughts-russias-campaign-against-ukraines-power-infrastructure/

Schmitt, M. (2022, October 22). Ukraine symposium - Attacking power infrastructure under international humanitarian law. *Articles of War.* https://lieber.westpoint.edu/attacking-power-infrastructure-under-international-humanitarian-law/

Schwirtz, M., & Mpoke Bigg, M. (2022, October 22). Russia hits Ukraine's power infrastructure with some of the biggest strikes in recent weeks. *New York Times.* https://www.nytimes.com/2022/10/22/world/europe/russia-ukraine-war-strikes.html?searchResultPosition=5

Shawcross, H. (1946). Closing Argument, *The Trial of German Major War Criminals,* vol. 19 (HM Stationery Office 1948), 430.

Shevchenko, A. et al. (2022). The Ukrainian AI strategy: premises and outlooks. *Proceedings of the 2022 12th International Conference on Advanced Computer Information Technologies (ACIT).* 511–515.

Statute of the International Residual Mechanism for Criminal Tribunals S/RES/1966 (2010). https://www.irmct.org/sites/default/files/documents/101222_sc_res1966_statute_en_0.pdf

Talbot Jensen, E. (2020). The (erroneous) requirement for human judgment (and error) in the law of armed conflict, *International Law Studies,* 96, 28–29.

The Lieber Code. (1863). https://ihl-databases.icrc.org/en/ihl-treaties/liebercode-1863?activeTab=undefined

The Prosecutor v. Jean Pierre Bemba Gombo, Decision Pursuant to Article 61 (7) (a) and (b) of the Rome Statute on the Charges of the Prosecutor v. Jean-Pierre Bemba Gombo, ("Decision on Confirmation of Charges"), ICC 01/05-01/08, para. 425. (2009). https://www.icc-cpi.int/sites/default/files/CourtRecords/CR2016_18527.PDF

U.S. Department of Defense "Law of War Manual". (2015). https://dod.defense.gov/Portals/1/Documents/pubs/DoD%20Law%20of%20War%20Manual%20-%20June%202015%20Updated%20Dec%202016.pdf?ver=2016-12-13-172036-190

United States Institute for Peace. (2023). *Timeline: Iran-Russia collaboration on drones.* https://iranprimer.usip.org/blog/2023/mar/01/timeline-iran-russia-collaboration-drones

Waxman, M. (2008). Detention as targeting: Standards of certainty and detention of suspected terrorists. *Columbia Law Review, 108*(6), 1365–1430. https://www.jstor.org/stable/40041788

Wright, J. (2012). Excessive ambiguity: Analysing and refining the proportionality standard. 94 *International Review of the Red Cross, 94*(866), 819–854. 10.1017/S1816383113000143

Scapegoats! Assessing the Liability of Programmers and Designers for Autonomous Weapons Systems

10

Afonso Seixas Nunes, SJ

INTRODUCTION

Autonomous Weapon Systems (AWS) as emergent technologies of warfare are source opposing views regarding their opportunity (Kalpouzos, 2020, pp. 289–291). The plurality of opinions creates a feeling of bewilderment when one must answer specific questions that AWS raise. Who is accountable for IHL violations caused by Autonomous Weapons Systems (AWS)? The risks of the employment of AI systems, namely the absence of direct human control and unpredictability have been highlighted by many States and scholars (*2023 Austria Working Paper. Group of Governmental Experts on Emerging Technologies in the Area of Lethal Autonomous Weapons System Geneva, 6-10 March, and 15-19 May 2023 Item 5 of the Provisional Agenda*, 2023; (*2023 Australia, Canada,*

192   DOI: 10.1201/9781003410379-12 **CC BY-ND - Attribution-NoDerivs**

Japan, Poland, The Republic of Korea, The United Kingdom, and the United States, 2023, p. Article 3(2); Acquaviva, 2023, pp. 6–11; Bostrom, 2014, pp. 144–145). However, little attention has been given to the obligation of 'constant care' required by the rules of precaution enshrined in Article 57 API. Nor has much attention been paid to the possibility of applying Article 28 (a)(i) of the Rome Statute to designers and programmers. This chapter argues that whenever a designer or programmer is entrusted by a military commander to program a system for a mission and he/her foresees the possibility of an AWS causing a violation of International Humanitarian Law (IHL) (prohibited act – *actus reus*) and is willing to take that risk (guilty mind – *mens rea*), he/she violates the obligation of 'constant care' and should be found liable whenever that risk leads to an actual violation of IHL.

The chapter proceeds in four parts. Part one will look to the specificity of the rules of precaution (Article 57, 1977 Additional Protocol I to the 1949 Geneva Conventions (API) and to the definition of AWS, highlighting 'adaptability' as the feature that truly distinguishes AWS from previous weapon systems. Parts 2 and 3 will look at two sets of problems: first, the distinction between 'errors' and 'accidents' and second, the possibility of liability for 'accidents' whenever the human operator fails to represent properly the risk of violations of IHL caused by the deployment of an AWS. Finally, Part 4 will argue that, in order to find designers and programmers liable for subjective recklessness the category of *dolus eventualis* should be restored as a requisite guilty state of mind in international criminal law (ICL).

Understanding the Rules of Precaution

Article 57 API falls under the heading of Precautions in Attack and is a fundamental principle of international customary law, applicable both to International and Non-International Armed Conflicts (International Committee of the Red Cross, 1987, p. 2191; International Law Association Study Group on the Conduct of Hostilities in the 21st Century, 2017, p. 372). The principle of precaution demands that States 'in the conduct of military operations take "constant care" to spare civilian life, prevent civilian injury, and preserve civilian objects' (Article 57 (1) API; Article 48 and 51 API). While neither the 1949 Geneva Conventions nor the respective Additional Protocols define what 'constant care' means (Jenks & Liivoja, 2018), the Tallinn Manual 2.0 on cyber operations can be of help. Per that Manual, 'the law admits of no situation in which, or time when, individuals involved in the planning and execution process may ignore the effects of their operations on civilians or civilian objects. In the cyber context, this requires situational awareness at all times, and not merely at the preparatory stage of an operation' (Schmitt, Michael N. & Vihul, 2017, p. 477). There is nothing preventing the same interpretation in the context of kinetic autonomous military operations (Jensen, 2020a, p. 587). Indeed, the increasing distance between human operators and AWS outcomes should demand more attentive care from those who design and program an AWS. Therefore, the obligation of 'constant care' should not be taken generically but actually as an operational principle that shall guide every military operation, that is, 'any movements, maneuvers and other activities whatsoever carried out by the armed forces with a view to combat' (International Committee of the Red Cross, 1987, p. 2191; Queguiner, 2006, p. 797). The newness of AWS is not a limitation for any legal regulation because IHL general rules are applicable in the absence of specific conventional or customary rules for AWS (McFarland, 2022, p. 395). As the International Court of Justice (ICJ) stated 'the intrinsically humanitarian character of the legal principles (…) permeates the entire law of armed conflict and applies to all forms of warfare and to all kinds of weapons, those of the past, those of the present and those of the future' (*ICJ, Advisory Opinion on the Legality of the Threat and Use of Nuclear Weapons, Judgment*, 1996, Para 85).

A second obligation concerns 'those who decide or plan upon an attack'. Such persons must take 'all feasible precautions in the choice of means and methods of attack with a view to avoiding, and in any event to minimizing, incidental loss of civilian life, injury to civilians and damage to civilian objects' (Article 57(2)(a)(ii) API). As the Commentary to the Additional Protocols has the opportunity to clarify, 'the words "everything feasible" means everything that was practicable or practically possible, taking into account all the circumstances at the time of the attack' (International Committee of the Red Cross, 1987, p. 681; Para 2198). Once again the Tallin Manual is of help when it states that 'those charged with approving cyber operations, mission planners should, where feasible, have technical experts available to assist them in determining whether appropriate precautionary measures have been taken' (Schmitt, Michael N. & Vihul, 2017, p. 477). In the context of AWS, Winter goes further by highlighting the importance of software developers because 'autonomous weapons are, more than anything else, a product of code designed by an array of military and civilian programmers' and furthermore 'a software developer could be prosecuted on the basis of individual accountability in the event that they programmed an autonomous weapon, intentionally or recklessly, in such a way that it would violate IHL' (Winter, 2021, p. 53;56). Though Winter limits his analysis to programmers involved in a Joint Criminal Enterprise (JCE), that is, an engagement where all parties share a 'common purpose' to pursue a criminal conduct (Article 28(3)(d) ICC Statute), the same can, and should, be said for rank and file programmers involved in any military action.

The question posed in this chapter is rather different: should programmers be found liable for violations of IHL before the deployment of an AWS? The question is not an easy one. In 2004, Andreas Matthias wrote an article with the enigmatic title, "*The responsibility gap: Ascribing responsibility for the actions of learning automata*". In it, Matthias highlighted that 'learning automata' could do things that were neither predictable nor reasonably foreseeable by their human overseers, and no one could be found liable for those unlawful outcomes (Matthias, 2004). In 2015, Human Rights Watch on the same line of thought published the report *Mind the Gap. The Lack of Accountability for Killer Robots* (Human Rights Watch & International Human Rights Clinic, 2015). The topic also received attention from legal scholars Tim McFarland, Tim McCormack (McFarland & McCormack, 2014) and Thomas Chengeta (Chengeta, 2016), but the possibility of *dolus eventualis* for programmers has not ever been considered.

The answer to the problems raised above imply, first and foremost, some attention to particular features of AWS.

Understanding AWS

In 2014, State parties to the *Convention on Prohibitions or Restrictions on the Use of Certain Conventional Weapons which may be Deemed to be Excessively Injurious or to Have Indiscriminate Effects* (CCW) got together to debate the future of AWS (Bo et al., 2022; Solovyeva & Hynek, 2023). Little consensus has been achieved among States' delegates regarding the definition of an autonomous weapon system (Cath et al., 2017, p. 2). However, most scholars accept without any contestation the definition presented by the USA DoD 3000.09 (US Department of Defense, 2012). According to the directives an AWS is:

> a weapon system that, once activated, can select and engage targets without further intervention by an operator. This includes, but it is not limited to, operator-supervised autonomous weapon systems that are designed to allow operators to override operation of the weapon system, but can select and engage targets without further operator input after activation
>
> *(US Department of Defense, 2012, 2023a).*

This definition, however, should not be taken without a pinch of salt. First, it does not distinguish AWS from current weapon systems already in use like Phalanx, or Harpy, which are not able to identify targets on their own, let alone adapt to unpredictable new circumstances on the battlefield. In all those systems, the target is pre-programmed and the machine follows a pre-determined rigid set of rules that will not allow any deviation from the order received by the operator. It is logical then to conclude that the autonomy of these types of systems is rather limited.

Autonomous systems are different from the systems mentioned above. As Schmitt observed 'the crux of autonomy is a capability to identify, target and attack a person or object without human interference' (Schmitt, 2013, p. 4). The distinctive element is then the 'autonomous' capability to *identify what/who is a target without human intervention*. Put simply, this is the 'ability of a system to behave in a desired manner, or achieve the goals previously imparted to it by its operator, without needing to receive the necessary instructions from outside itself on an ongoing basis' (McFarland, 2021, p. 6).[1] Regarding the capability of a weapon system to identify targets, two possible situations arise. First, the deployment of 'automated systems' that are programmed to identify a target based on a particular radar emission profile but are not able to understand or adapt to the surrounding environment (Suchman & Weber, 2016, p. 76). Second, systems that determine independently what/who would constitute a military target, including systems that use Human-Based Intelligence to interact with their environment and create their own set of values. Such systems are 'original', in the sense that they only rely indirectly on human programming to provide the tools for the system itself to obtain an outcome (Arkoudas & Bringsjord, 2014, pp. 35–36; Bostrom, 2014, p. 22).

In light of the above, systems that can adapt to the mutable circumstances of the battlefield raise the most concerns. As the US DoD highlights, those systems can lead to unlawful and unpredictable outcomes. This element has received little attention from States and scholars, but it remains one of the elements, if not *the* element, that deserves the most attention for understanding the level of care that an AWS's deployment will demand. Once an AWS is deployed, the individual weapon will be called not only to assess the situation (observation and collection of new data), but also to process all the information collected (orientation) in order to select an outcome from possible alternatives (decision) and engage a military target (action) without human intervention. That is to say, the selection-engaging process will not have or allow the presence of a human operator. The *algorithm for the mission*, will interact and adapt to multiple and unpredictable circumstances of the battlefield and will remain unknown to the programmers and designers of the *algorithm for the mission* (Seixas-Nunes, SJ, 2022a, pp. 429–439).

Adaptability, however, does not stand alone. The system will also require 'assertiveness', that is, 'an adaptive system needs to have a feedback mechanism, which keeps the system focuses[d] on its objective by changing its internal state as the environment changes' (Sartor & Omicini, 2016, p. 49). An autonomous system 'does not necessitate direct oversight by a (human) commander for every decision made' but requires asking 'whether the system is fulfilling the intent of the commander' (Liivoja et al., 2022, p. 642). As the American military puts it, 'the system is designed to complete engagements within a timeframe and geographic area, as well as other applicable environmental and operational parameters, consistent with the commander and operator intentions', adding that if the system is 'unable to do so, the system will terminate engagements or obtain additional operator input before continuing the engagement' (US Department of Defense, 2023, Section 4; 4.1).

In light of the considerations made above, Taddeo and Blanchard (2022) have suggested that an AWS is:

> an artificial agent which, at the very minimum, is able to change its own internal states to achieve a given goal, or a set of goals, within its dynamic operating environment and without the direct intervention of another agent, and which is deployed with the purposes of exerting kinetic force against a physical entity (whether an object or a human being) and to this end is able to identify, select or attack the target without the intervention of another agent
>
> *(Taddeo & Blanchard, 2022, p. 15).*

Although this definition has the merit of highlighting the adaptability dimension of AWS, it qualifies an AWS as an 'artificial agent'. The qualification of AWS as an 'artificial agent' may raise the eyebrows of some, but the term does not necessarily imply an 'ontological' move, comparing humans agents with artificial agents as some scholars fear (ICRC – International Committee of the Red Cross, 2019, p. 31; Liu, 2016, p. 327). The term chosen, however, raises the question of whether it is the best term for legal purposes. The agency is inherently linked to the notion of liability, and not even adaptive AWS can or should be considered moral agents. Thus, for the purposes of this chapter, an autonomous weapon system will be defined as a *weapon system designed and programmed for a mission to be able to be adaptive, and to identify, select and engage military targets, without human intervention* (Seixas-Nunes, SJ, 2022b, pp. 82–89). This definition reinforces the instrumental side of AWS, and for the purposes of this chapter accentuates the responsibility that designers and programmers have toward the mission that will be entrusted to the system.

'ERRORS' AND 'ACCIDENTS' AND AWS

It is undeniable that any battlefield is a source of unpredictable situations. Most of the time, soldiers are required to answer fast, flawless, and lawfully in very narrow time frames. AWS represent, therefore, a kind of long-expected 'technical messiah' for human imperfection, because 'humans are extremely bad at making the kind of rational judgments that complying with IHL requires – particularly when they find themselves in dangerous and uncertain situation like combat. It is precisely human as they are (…) that explains why combat using machines are likely to be far more "humane" than combat with human soldiers' (Heller, 2023, pp. 17–18; Trabucco & Heller, 2022, pp. 20–23). Indeed, there is no reason to doubt that machine-learning technology and AWS will improve massive data analysis in volume, variety, and velocity, making 'weapon systems capable of speed, accuracy, and precision' (Jensen, 2020b, p. 46). However, the challenge of acquiring accurate data or the inability to produce datasets that replicate combat conditions might well condemn AWS to failure (Atherton, 2022).

In spite of all the advantages that AWS may bring to the battlefield, machines can never be considered moral actors. Thus, it is legitimate to ask whether AWS could introduce a new type of culpability to combat. Arguments such as 'human soldiers are not necessarily better' than AWS (Trabucco & Heller, 2022, p. 26) or to postulate that IHL demands the 'best application possible' and not necessarily the 'best human application possible' (Jensen, 2020b, pp. 54–55) do not prevent us from considering whether programmers and designers should be found liable for IHL violations.

As computer scientists note, 'unobserved regions of the AI decision-making process are prone to normal accident – a type of "inevitable" accident that emerges in situations where the components are densely connected, tightly coupled, and opaque in their processing' (Feldman et al., 2019). Indeed, AI systems struggle to adapt to changing conditions. Even methods such *Transfer Learning* that seek to increase adaptability, show some weaknesses as vulnerability to adversary attacks (Donges, 2022). McFarland (2020) also argues that 'the ability to predict how an AWS will behave in a given situation is limited by the complexity arising from various sources, and that is that it is almost impossible to avoid software errors' (McFarland, 2020, p. 63). The risk of unlawful unpredictable outcomes begs for a legal answer. Scholars have been acknowledging the situation of 'errors of software', postulating that in those types of situations an AWS will 'fall outside of the commander's command and effective control' and individual accountability cannot rise (Acquaviva, 2023, p. 2; Arkin, 2009, p. 138; Buchan & Tsagourias, 2020; Seixas-Nunes, SJ, 2022a). However, IHL provides a legal framework for these types of situations. The rules of precaution and mechanisms of legal reviews for new weapons imposed on States obligations of due diligence (Article 57 API and Article 36 API). Military designers and programmers must consider the prospect of unpredictable hazards, and the texture of risk, associated with the principles of IHL compliance. It will be up to those human operators to differentiate between acceptable and unacceptable risks.

It is the position of this chapter that a proper assessment of incorrect decisions leading to unacceptable risks cannot escape the scrutiny of ICL for violations of IHL caused by AWS poor programming or design (Bhuta & Pantazopoulos, 2016, p. 294; Winter, 2021, p. 53). As Longobardo rightly argues 'the principle of precaution (…) is an autonomous source of obligations under international humanitarian law, which has the peculiar capacity to allow a scrutiny of the preparatory conduct that anticipates actual hostilities' (Longobardo, 2019, p. 81). If it is true that the mere acceptance of risk is not in itself unlawful under IHL, the perception of possible war crimes is a substantial risk that cannot be ignored by International Law.

ACCOUNTABILITY FOR COMMANDERS

Since 2014, State parties to the CCW have reaffirmed the need for mechanisms of individual accountability for violations of IHL caused by AWS on the battlefield (*2019 Guiding Principles Affirmed by the Group of Governmental Experts on Emerging Technologies in the Area of Lethal Autonomous Weapons System*, 2019). In the most recent US Department of State *Political Declaration on Responsible Military Use of Artificial Intelligence and Autonomy* it also provides that 'States should adopt, publish, and implement principles for the responsible design, development, deployment, and use of AI capabilities by their military organizations' (US Department of State, 2023). The problem remains, however, for how to establish those mechanisms for 'crimes' committed by AWS in which the unlawful outcome was foreseen by human operators.

ICL provides some guidance. For example, it provides a spectrum of international crimes that *in theory* could be committed by AWS. Serious violations of IHL amounting to war crimes are codified and defined in international instruments such as the 1949 Geneva Conventions, Additional Protocol I, and the Rome Statute of the International Criminal Court (Rome Statute) as well as in national laws on war crimes and customary international law. The fact that AWS will allow human operators to be removed from the selection-engagement process, begs the urgent question as to who, and in what circumstances, will be held accountable for violations that might occur (Jensen, 2020b, p. 37). As Acquaviva explains 'a link of causation between (or causality) between a human act and the harm(…) is a requirement almost invariably built into criminal law systems'(Acquaviva, 2023, p. 5).

Autonomous technologies change, however, the traditional role of human operators. There is an increasing process of dissociation of communication between humans and autonomous technologies. In spite of the multiple AI successes, one cannot obviate the complexities inherent to any process of communication, because it is at this level where the double process of scapegoating finds its roots. Whenever an error occurs one of two possibilities can easily be chosen: the military commander is found liable for everything that happens on the battlefield (Sassoli, 2014, p. 328) or, instead, the error was committed by the machine, and therefore no one can be held accountable. It is the machine's fault! (Buchan & Tsagourias, 2020, p. 658). While the former raises the question of how commanders might be liable for unforeseeable consequences that result from software errors, the latter calls into question whether IHL accepts that level of decriminalization of AWS. Thus, scapegoating commanders or machines cannot be the legal path that one should go down when it remains possible to look more closely at the criminal culpability of programmers and designers.

Chiodo (2022) argues that with the progress and achievements of AI humans are abandoning autonomy and taking on some kind of automation and vice-versa. Technology becomes a sort of scapegoat to free human operators from the burden of individual responsibility for the results produced by autonomous technology. As the author argues, humans do not simply become 'absent at all from decision-making processes: they keep participating in them (…) but their role significantly changes – their role moves from bearing individual responsibility for the decision-making to notify that the decision-making process is automated' (Chiodo, 2022, p. 43).

One should ask whether Chiodo description entails not so much a 'responsibility gap' as many scholars would argue, but rather a process of decriminalization of ICL. As strange as this phenomenon may sound, Heller (2023) accepts this legal path. The possibility of AWS causing violations of IHL and creating an 'accountability gap' is not new because 'other situations aside from AWS create accountability gaps (Heller, 2023, pp. 62–63). However, one may wonder if the existence of possible 'accountability gaps' is the desirable legal answer for weapon systems that raise concerns of predictability. For example, a situation in which the AWS causes an indiscriminate attack that turned out to 'be excessive in relation to the concrete and direct military advantage anticipated' by the military commander who activated the system (Article 51(5)(b) API). Since AWS operate autonomously, it is possible that the commander will have no information or time to cancel or suspend the attack as required by law (Article 57(2)(b) API). Should this situation be accepted as another legitimate 'accountability gap' parallel to others? No. Designers and programmers should also be considered within the liability chain. The relationship between AWS and the operator is an entirely new one, comparable with no other, and raises original and new legal concerns that deserve an answer and not an analogy between humans and machines (Kraska, 2021, p. 414).

The rules for military commanders are well- and long-established in military doctrine (*Australian Defence Doctrine Publication 06.4*, 2006; *UK - The Joint Service Manual of the Law of Armed Conflict*, 2004, p. 5.6.1; US Department of Defense, 2015, p. 191) and in International ICL (Article 28 Rome Statute; Article 85 API), according to which military commanders are considered the 'guarantor' of IHL and its fundamental principles (Buchan & Tsagourias, 2020, p. 651). This fundamental role of the military commanders is expressed in the rules of Precaution enshrined in Article 57 API. *The International Law Association Study Group on the Conduct of Hostilities in the 21st Century*, considered this concept as important as the Principle of Distinction (Article 48; Article 50 API) and the Principle of Proportionality (Article 51(5)(b); Article 57(2)(b) API); the 'issue of precautions has remained under-researched and problematically under-emphasized', in spite of the customary nature of such norm (Henckaerts & Doswald-Beck, 2009, Rule 15; International Law Association Study Group on the Conduct of Hostilities in the 21st Century, 2017, p. 93; *Prosecutor v Zoran KrupresKic et al*, 2000, Para 524). Indeed, commanders are responsible to take 'constant care' for all operational judgments inherent to warfighting (Article 57(1) API). According to the rule of precaution 'those who plan or decide upon an attack shall take all feasible precautions to verify that the principle of distinction is respected (Article 57 (2)(a)(i) API); to take all feasible precaution in choice of means and methods of attack with a view to avoiding, and in any event to minimizing, incidental loss of civilian life, injury to civilians and damage to civilian objects' (Article 57(2)(a)(ii) API); and refrain from deciding to launch any attack which may be expected to cause incidental loss of civilian life, injury to civilians, damage to civilian objects or a combination thereof, which would be excessive in relation to the concrete and direct military advantage anticipated (Article 57(2)(a)(iii) API). These obligations have received attention from scholars. For example, Jensen considers that IHL only demands the 'best application' and not the 'best human possible application', (Jensen, 2020a). It will always be up to the commander to decide whether an AWS is the best suitable weapon for the specific context of the battlefield 'based upon a consolidated assessment [and] such assessment must be made for every attack' (McFarland, 2022, p. 398; Sassoli, 2014, p. 321).

In light of the above, two aspects deserve consideration: first, the possible of 'direct responsibility' for military commanders for unforeseeable unlawful consequences on the battlefield caused by AWS; second, what level of understanding is required of the military commanders regarding the AWS' operating technicalities.

Direct Responsibility

It is certainly compelling to advocate 'direct responsibility' for unforeseen unlawful consequences on the battlefield caused by AWS. On this point, Kraska (2021) argues, similarly to Sassoli (2014), that 'in the event an AWS proves to be indiscriminate the commander would be held accountable. Every weapon system in the combat zone and every method of training (…) falls within the remit of the commander's direct or individual accountability' and 'although it may seem "unfair" to impose liability on commanders for incidents occurring beyond their immediate control (…) the armed forces routinely do just that' (Kraska, 2021, pp. 438–439). This understanding may sound like a panacea for all accountability ills, but fails to convince.

Buchan and Tsagourias (2020) raise two different types of situations. The first arises when an autonomous system 'acts differently than instructed (…) but remains "within the parameters of command and control'. The second situation arises when the 'effective control would be lacking' because the system would act 'on its own initiative in order to pursue a goal outside of the framework of commander's command and effective control and command responsibility cannot arise' (Buchan & Tsagourias, 2020, p. 658). The second type of situation will always be complex to prove. However, as the ICJ had the opportunity to argue in the *Nicaragua Case*, it is the obligation of States to respect and ensure that the IHL 'in all circumstances' derives 'not only from the Conventions themselves, but from the general principles of humanitarian law to which the Conventions merely give specific expression' (*Military and Paramilitary Activities in and against Nicaragua (Nicaragua v. United States of America)*, 1986, Para 220). The duty of States to ensure that IHL 'in all circumstances' is in accordance with obligations of due diligence and the doctrine of State responsibility whatever is the system the State is deploying (Seixas-Nunes, SJ, 2022a, p. 455).

The second type of situation is, however, the most interesting one: situations in which the system executes the order differently than the initially programmed and happens to commit a war crime. The difference here is that the human operators represented the risk of IHL violation before activating the system.

Military Commanders 'Understanding' AWS

But what of accidents? 'Accidents' tend to occur due to poor design and programming of the system. The question that emerges then is whether military commanders should be familiar with all the intricacies of an AWS in order to make a better judgment of if, when, and how to deploy an AWS? Should that knowledge impact the responsibility of military commanders whenever the system causes accidents?

This question has not been ignored by scholars. Kraska (2021), for example, argues that commanders 'have an obligation to understand what AWS can do in a particular environment in which it is used, like any other weapon' but 'they are not required to understand the intricacies of how the weapon works' (Kraska, 2021, p. 438). The position, however, deserves further development.

Attention, first, should be given to Article 36 API. Article 36 API demands that 'in the study, development, acquisition or adoption of a new weapon, means and methods of warfare, a High Contracting party is under the obligation to determine whether its employment would in some or all circumstances, be prohibited by this protocol or by any other rule of international law applicable to the High Contracting Party'. A weapon's review cannot be a prerequisite only for the deploying State, but also should serve to instruct military commanders on how the system operates, its strengths and weaknesses, and when its deployment could become a potential source of violations of IHL (Copeland, 2014, pp. 12–14; Copeland et al., 2022, pp. 6–8).

Another argument, Article 1, Article 6 and Article 83 API impose obligations on States 'to train', 'to disseminate' and 'to include the study' of IHL in the programs of military instruction. In terms of autonomous technologies, the disinterest for how a weapon will perform on the battlefield can have severe consequences – namely accidents that could be easily avoided if the armed forces were duly instructed on IHL and the technicalities of the systems (Longobardo, 2019, p. 78; US Department of State, 2023).

But what of commanders, and even States, that lack a basic understanding of technology? The Eritrea-Ethiopia Claims Commission (EECC) provides an example. Specifically, it took into consideration the fact that the State of Eritrea pilots were 'utterly inexperienced' in the use of AWS and committed violations of IHL as a result. In such an instance, culpability should be modified accordingly:

> 109. The Commission must also take into account the evidence that Eritrea had little experience with these weapons and that the individual programmers and pilots were utterly inexperienced, and it recognizes the possibility that, in the confusion and excitement of June 5, both computers could have been loaded with the same inaccurate targeting data(…)

And

> 110. The Commission believes that the governing legal standard for this claim is best set forth in Article 57 of Protocol I, the essence of which is that all feasible precautions to prevent unintended injury to protected persons must be taken in choosing targets, in the choice of means and methods of attack and in the actual conduct of operations.

The Ethiopia-Eritrea case raises the possibility that accidents could become the norm when operators lack the education necessary to comprehend AWS. In such instances, programmers and designers might be even more culpable than their superiors, for it is they who best know the dangers inherent in the systems that they build. If the deployment of AWS by advanced nations raises concerns, in other words, then the use of AWS by developing nations should spark real fear. Accidents, for such parties, are likely to spiral completely out of control. The position of the EECC asserts the responsibility of the military commander to take 'constant care' and 'feasible precautions' to train and make his/her subordinates acquainted with the specificities of new technologies of warfare in general. There is no reason why the same reasoning cannot be applied to AWS.

In light of the above, it is now possible to distinguish, firstly, the legitimacy of holding military commanders responsible for the violations of the rules of precaution and, secondly, the question of holding designers and programmers accountable for accidents on the battlefield.

Military Commanders and the 'Accidents'

The Trial Chamber in the Oric´ Case upheld the claim:

> The capacity to sign orders is an indicator of effective control, provided that the signature on a document is not purely formal or merely aimed at implementing a decision made by others, but that the indicated power is supported by the substance of the document

> *(Prosecutor v. Naser Oric, 2006, p. Para 312).*

From this statement it can be concluded that the orders given by the commander circumscribe his/her responsibility. Although the capacity to sign orders is addressed in Oric´ as a matter of command responsibility, it can also provide information as to what the precautionary measures taken by the commander were before deploying the AWS, which contributes to the matter of the commander's *mens rea* since orders might well reflect the commander's intentionality.

The problem emerges, however, in cases where harm, or the risk of harm, to protected people or objects, is caused by an AWS user's insufficient care, control, or diligence. According to an interpretation given by the ICTY, customary international law allows a presumption of negligent lack of knowledge if the commander had information that 'put him on notice of offences committed by subordinates'; while under the Rome Statute it is crucial to determine whether the commander would, in the exercise of their duties, have gained knowledge of the commission of the crime by their subordinates.

But what, specifically, does a commander's proactive duty to seek and scrutinize information concerning an attack with an AWS entail? Looking at what the EECC concluded, the military commander, under the rules of precaution, would have to be aware of every single technological update.

However, the consideration made above is not without problems. The question of whether omissions, such as a failure to suspend an attack with AWS expected to be unlawful (Article 57(2)(b) API), is a mode of commission of war crime remains unsettled. Some national laws provide a legal basis for the criminalization of war crimes committed by omissions; Article 86 of AP I requires states to 'repress grave breaches which result from a failure to act when under a duty to do so'. However, whether there is a rule of customary international law on 'commission by omission' is subject to debate (Ambos, 2002, p. 1002). Some omissions, such as failing to gather or use available information to verify targets, could amount to violations of the IHL primary rule on the duty to take precautions. However, war crime provisions do not criminalize failures to take precautions. Failures to take precautions could, in principle, be taken into account as contextual elements to prove that a commander had the intent to target civilians (Article 25 and Article 30 Rome Statute). Even so, there are still questions as to what, in concrete terms, is demanded by the rules of precaution on the part of the commander, especially what information the precautionary rule on target verification requires in the use of AWS. Particularly acute in the deployment of AWS would be to distinguish what kind of omissions by military commanders could indeed become a source of individual criminal responsibility, under the Rome Statute.

The second element for the attribution of individual criminal responsibility is the mental element, or 'guilty mind' (*mens rea*). Like the material element, a fundamental difficulty for the establishment of the mental element of a war crime, involving AWS or not, is that the element is codified differently in API and the Rome Statute. Under Article 85(3) API, war crimes stemming from violations of the rule of distinction and proportionality must be committed 'willfully'. In contrast, paragraphs (i) and (iv) of Article 8(2)(b) Rome Statute require that the prohibited attack is executed 'intentionally'; there is some debate as to whether intentionality under Article 8 coincides with the general *mens rea* requirement of 'intent and knowledge' under Article 30 of the Rome Statute.

In the context of AWS, these differences in codification do not raise as much concern where direct intent (*dolus directus*) can be established. An example might be when the user deliberately programs and launches an AWS to attack the civilian population. If the user's intent is established, then it is undisputedly covered by the mental elements of '*willfulness*' under API and 'intentionality' under the Rome Statute. Moreover, situations of indirect intent (*dolus indirectus*), where the user launches an attack using an AWS and is 'practically' or 'virtually' certain that the attack will be directed against civilians or result in civilian death or injuries are covered by these mental elements. However, it could be problematic where the use of an AWS enables the user to shield their 'intent' or knowledge.

Difficulties emerge in cases where harm, or the risk of harm, to protected people or objects, is caused by an AWS user's insufficient care, control, or diligence. In such cases, the mental element that needs to be established is recklessness, *dolus eventualis* (a special kind of intent involving foreseeing and accepting the consequences and risks of actions), or *negligence*.

Making Designers, Programmers, and Manufacturers AccountableThe capability of AWS to operate without the need for human intervention can create a challenge because some mechanical actions do not fit into any form of international accountability. This problem has been addressed before, but it is useful to recall some factors that are frequently forgotten. First, machine learning algorithms are, by definition, less predictable; certainly, they import new risks into the battlespace. However, States at the CCW have always insisted on the importance of legal reviews, and of being kept fully informed as to how AWS will operate. Unpredictability can cause violations of IHL that may put the interests of the home state in jeopardy, an eventuality that prompts states to rigorously test the systems.

Second, some may argue that many of the personnel involved with manufacturing and programming AWS do not belong to the armed forces of a state and are, therefore, not subject to International Law but 'to domestic tort law'. Apart from this, it is already difficult to make manufacturers and programmers accountable even in a state's internal forum (Cass, 2015). Notwithstanding, each AWS, apart from its general design, will be programmed for a specific mission that a commander entrusts to the system. As is clear, and as Arkin (2009) underlines, it is inconceivable that an AWS could be deployed without specific ROE and other constraints designed and programmed into it to equip it for a particular mission. Thus designers, programmers, manufacturers, and technicians should indeed be seen as part of the military (Winter, 2021, p. 65). In this situation, they would be obliged to conduct themselves in accordance with International Law, as would any other participant in the hostilities, as well as being subject to domestic law. It is true, however, that very well-established military manufacturers are private, can work for several different states, and can hardly be considered part of any one State's armed forces. This fact, however, should not be seen as an obstacle that excludes designers, programmers, and manufacturers from compliance with international values.

Aside this argument, the ICC argued that the principle of 'unity of command' or 'singleness of command' implies that, "[f]or the proper functioning of an army, there can be only one individual in command of any particular unit at one time". However, the determination of whether a person has effective authority and control rests on that person's material power to prevent or repress the commission of crimes or to submit the matter to a competent authority' (*Situation in the Central African Republic in the Case of the Prosecutor v. Jean-Pierre Bemba Gombo*, 2009, Para 698). Thus, in matters of autonomous technology whoever has the 'authority', 'knowledge' of its operations can be included in the chain of command as a military commander *de facto* (Buchan & Tsagourias, 2020, pp. 658–658). The power to design and program the algorithm for the mission can be understood as a *commander de facto*.

The ICC Statute establishes the principle of complementarity between international and domestic jurisdiction regarding the international criminalization of the individual. Thus, there is nothing precluding designers, manufacturers, and programmers from being prosecuted at a national level by the state deploying the AWS 'unless the State is unwilling or unable genuinely to carry out the investigation or prosecution' (Article 17(1)(a) ICC Statute) and the jurisdiction of the ICC comes into force. As an example, the *Final Report to the Prosecutor by the Committee Established to Review the NATO*

Bombing Campaign Against the Federal Republic of Yugoslavia states that the misidentification of certain civilian objects was "due to the inadequacy of the supporting data bases and the mistaken assumption the information they contained would necessarily be accurate", and as consequence "the CIA has also dismissed one intelligence officer and reprimanded six senior managers"(*ICTY Final Report to the Prosecutor by the Committee Established to Review the NATO Bombing Campaign against the Federal Republic of Yugoslavia*, 2000, p. Para 83–84).

It is true, however, that the category of 'risk' is not recognized by ICL. The principle of legality can, indeed, be an obstacle. But, as Winter explains, designers and programmers responsible for 'critical functions', displacing the decision-making power from commanders, should not lead to an accountability gap but to a process reinforcing IHL (Winter, 2021, pp. 77–80). Designers and programmers are the bearers of legal but also moral responsibility for the operationality of AWS.

Finally, a third and more contentious issue is that some authors discredit 'negligent behavior' as an appropriate legal category in cases involving AWS. Crootof (2016, p 1383) argues that although lethal accidents do happen in the battlespace, 'if an individual could be held criminally liable for negligent actions in war and if her commander would be indirectly liable for negligence, every commander would be a war criminal'. Moreover, admitting negligence would undermine the moral legitimacy of IHL because it would be immoral to hold someone accountable 'for [the] independent and unpredictable actions of autonomous weapons systems' (Crootof, 2016, p. 1384).

It is important to prevent the introduction of any 'scapegoating mechanism' that might be applied to AWS operatives, individuals who could not have been in any position to know of potentially illegal outcomes. However, it would be illogical, legally, for there not to be a mechanism to hold individuals accountable for proven conscious behavior in the design, manufacture, and programming of an AWS; second, the standard to be applied to AWS' operators should not be that of 'negligence', since such is excluded from Article 30 ICC Statute, but rather *dolus eventualis*.

Crootof (2016) argues that it is proper that ICL should prohibit certain behaviors and, therefore, that it does not apply in situations involving AWS since they have no moral agency. Instead, she suggests Tort Law (which is focused 'on wrong-doing, fault, and regulation of valuable but sometimes dangerous activities') would facilitate state responsibility by 'delimitating what violations are sufficiently serious to require reparation' and make the state liable for full reparation in cases of injury caused by AWS (Crootof, 2016, p. 1384). What is suggested in this chapter is to a certain extent parallel. Intentional unlawful conducts involving substantial risk should be forbidden because of the dependence of AWS on their programmers.

It is true that the 'subjective element' will always be an object of debate and reasoning for the International Court, and it will not always be a straightforward interpretation task but the Lubanga Case has opened a door for *dolus eventualis* to be considered (*Prosecutor v. Thomas Lubanga Dylo*, 2009, para 927).

The foregoing analysis and inclusion of the category of *dolus eventualis* cannot proceed without snags. The possible requirements of measuring and quantifying the level of risk are currently the subject of lively debate (Summers, 2014, pp. 680–681). Notwithstanding, if a human operator is aware and conscious of a substantial risk of IHL violation, yet disregards that possibility, he/she should be held accountable because the consequences were so predictable; this was pointed out by the Pre-trial Chamber in the *Lubanga Case*. In the final analysis, an important philosophical question should be put to International Criminal Lawyers: taking into account the nature and features of AWS, should the failure to acknowledge a substantial risk be as blameworthy as the decision to commit a crime? The two situations obviously invoke different levels of culpability, but is it not the case that an outcome caused by *dolus eventualis* has the same causal link as an outcome caused by *dolus directus* conduct, that is, a disregard for the most important international values? Designers and programmers are the closest link of causation between an AWS and harm, especially when aware of the possibility of an unlawful consequence. Finally, it is important to acknowledge that the substantial risk posed by AWS comes from the fact that they are a technology not yet mastered and might require more restrictive criminal measures. In 1938, for example, in the advent of aerial warfare, the *Resolution of the League of Nation Assembly Concerning Protection of Civilians against Bombing from the air in case of War* provided that "any attack on legitimate military objects must be carried out in such a way that the civilian population (…) are not bombed through negligence" (Cohen & Zlotogorski, 2021, p. 26). Although this proposal was not accepted it is indicative that the emergence of new technologies suggests accentuated forms of liability whenever the risk associated to them is still in a fog of uncertainty.

CONCLUSION

Autonomous technologies call the attention of scholars to their newness. The physical and cognitive distance from the battlefield that such systems create is not without risks. One of them hotly debated is the unpredictability associated to machine learning. Military commanders, under the rules of precaution, are responsible for constantly care for the principle of distinction and rules of proportionality. Such commitment may involve the delegation of the algorithm for the mission to designers and programmers. These human operators will be called to make assessments of the risk inherent to any military mission making them truly commanders *de facto*. This level of responsibility cannot be hidden between two different processes of scapegoating whether by making military command-ers accountable for unlawful unforeseeable outcomes or by blaming the machines for accidents. This chapter posits that *dolus eventualis* should be considered for the purpose of holding designers/programmers responsible for war crimes caused by AWS.

NOTE

1 However, the understanding of 'autonomy' determines precisely the position of States on matters of legitimacy of AWS, namely if it will be ever possible to accept AWS on the battlefield. (Jensen, 2020a, p. 579).

REFERENCES

2019 Guiding Principles affirmed by the Group of Governmental Experts on Emerging Technologies in the Area of Lethal Autonomous Weapons System (CCW/MSP/2019/9). (2019). CCW. https://documents-dds-ny.un.org/doc/UNDOC/GEN/G19/343/64/PDF/G1934364.pdf?OpenElement

2023 Australia, Canada, Japan, Poland, The Republic of Korea, The United Kingdom, and the United States. (2023). https://docs-library.unoda.org/Convention_on_Certain_Conventional_Weapons_-Group_of_Governmental_Experts_on_Lethal_Autonomous_Weapons_Systems_(2023)/CCW_GGE1_2023_WP.4_US_Rev2.pdf

2023 Austria Working Paper. Group of Governmental Experts on Emerging Technologies in the Area of Lethal Autonomous Weapons System Geneva, 6-10 March, and 15-19 May 2023 Item 5 of the Provisional agenda. (2023). https://docs-library.unoda.org/Convention_on_Certain_Conventional_Weapons_-Group_of_Governmental_Experts_on_Lethal_Autonomous_Weapons_Systems_(2023)/CCW_GGE1_2023_WP.1_Rev.1.pdf

Acquaviva, G. (2023). Crimes without Humanity? Artificial intelligence, meaningful human control and international criminal law. *Journal of International Criminal Justice, 21*(2), 1–26.

Ambos, K. (2002). Superior responsibility. In A. Cassese, P. Gaeta, & J. R. W. D. Jones (Eds.), *The Rome statute of the international criminal court: A commentary: Vol. III* (pp. 823–872). Oxford University Press. https://papers.ssrn.com/sol3/papers.cfm?abstract_id=1972189

Arkin, R. C. (2009). *Governing lethal behavior in autonomous robots.* Chapman & Hall Book.

Arkoudas, K., & Bringsjord, S. (2014). Philosophical foundations. In K. Frankish & W. M. Ramsey (Eds.), *The Cambridge handbook of artificial intelligence* (pp. 34–58). Cambridge University Press.

Atherton, K. (2022, May). *Understading the errors introduced by military AI applications.* Brookings. https://www.brookings.edu/articles/understanding-the-errors-introduced-by-military-ai-applications/

Australian Defence Doctrine Publication 06.4. (2006). Defence Publishing Service Department of Defence.

Bhuta, N., & Pantazopoulos, S.-E. (2016). Autonomy and uncertainty: Increasingly autonomous weapon systems and the international legal regulation of risk. In N. Bhuta, S. Beck, R. Geis, H.-Y. Liu, & C. Kreis (Eds.), *Autonomous Weapons Systems. Law, Ethics, Policy* (pp. 284–300). Cambridge University Press.

Bo, M., Bruun, L., & Boulanin, V. (2022). *Retaining human responsibility in the development and use of autonomous weapon systems. On accountability for violations of international humanitarian law involving AWS.* Sipri - Stockholm International Peace Institute. https://www.sipri.org/publications/2022/other-publications/retaining-human-responsibility-development-and-use-autonomous-weapon-systems-accountability

Bostrom, N. (2014). *Superintelligence: Paths, dangers, strategies.* Oxford University Press.

Buchan, R., & Tsagourias, N. (2020). Autonomous cyber weapons and command responsibility. *International Law Studies, 96*(20), 645–673. https://digital-commons.usnwc.edu/ils/vol96/iss1/21/

Cass, K. (2015). Autonomous weapons and accountability: Seeking solutions in the law of war. *Loyola of Los Angeles Law Review, 48*(Spring), 1017–1067.

Cath, C., Wachter, S., Mittelstadt, B., Taddeo, M., & Floridi, L. (2017). Artificial intelligence and the "good society": The US, EU, and the UK approach. *Science and Engineering Ethics, 24*(2), 505–528.

Chengeta, T. (2016). Accountability gap: Autonomous weapon systems and modes of responsibility in international law. *Denver Journal of International Law and Policy, 45*(1), 1–50.

Chiodo, S. (2022). Human autonomy, technological automation (and reverse). *AI & Society, 37,* 39–48. https://link.springer.com/article/10.1007/s00146-021-01149-5

Cohen, A., & Zlotogorski, D. (2021). *Proportionality in international humanitarian law. Consequences, precautions and procedures (Vol. 6).* Oxford University Press.

Copeland, D. P. (2014). Legal review of new technology weapons. In H. Nasu & R. McLaughlin (Eds.), *New technologies and the law of armed conflict.* Asser Press, Springer.

Copeland, D. P., Liivoja, R., & Sanders, L. (2022). The utility of weapons reviews in addressing concerns raised by autonomous weapons systems. *Journal of Conflict & Security Law, 28*(2), 285–316. https://doi.org/10.1093/jcsl/krac035

Crootof, R. (2016). War torts: Accountability for autonomous weapons. *University of Pennsylvania Law Review, 164*(6), 1347–1402.

Donges, N. (2022, August 25). *What is transfer learning? Exploring the popular deep learning approach.* Builtin. https://builtin.com/data-science/transfer-learning

Feldman, P., Dant, A., & Massey, A. (2019). *Integrating artificial intelligence into weapon systems.* Arxiv Cornell University. https://arxiv.org/abs/1905.03899

Heller, K. J. (2023). The concept of "the human" in the critique of autonomous weapons. *Harvard National Security Journal,* 1–74. https://dx.doi.org/10.2139/ssrn.4342529

Henckaerts, J.-M., & Doswald-Beck, L. (2009). *Customary international humanitarian law. Volume I: Rules.* Cambridge University Press.

Human Rights Watch, & International Human Rights Clinic. (2015). *Mind the gap. The lack of accountability for killer robots* (p. 39). https://www.hrw.org/report/2015/04/09/mind-gap/lack-accountability-killer-robots

ICRC - International Committee of the Red Cross. (2019). *International humanitarian law and the challenges of contemporary armed conflicts. Recommitting to protection in armed conflict on the 70th anniversary of the Geneva conventions* (pp. 5–81). ICRC. https://www.icrc.org/sites/default/files/document/file_list/challenges-report_new-technologies-of-warfare.pdf

ICTY Final Report to the Prosecutor by the Committee established to Review the NATO Bombing Campaign against the Federal Republic of Yugoslavia. (2000). https://www.icty.org/en/press/final-report-prosecutor-committee-established-review-nato-bombing-campaign-against-federal

International Committee of the Red Cross. (1987). *Commentary on the additional protocols of 8 June 1977 to the Geneva conventions of 12 August 1949.* International Committee of the Red Cross.

ICJ, Advisory Opinion on the Legality of the Threat and Use of Nuclear Weapons, Judgment. (July 8, 1996).

International Law Association Study Group on the Conduct of Hostilities in the 21st Century. (2017). The conduct of hostilities and international humanitarian law: Challenges of 21st century warfare. *International Law Studies, 93,* 322–388. https://www.ila-hq.org/index.php/study-groups?study-groupsID=58

Jenks, C., & Liivoja, R. (2018, December 11). Machine autonomy and the constant care obligation. *ICRC Law and Policy.* https://blogs.icrc.org/law-and-policy/2018/12/11/machine-autonomy-constant-care-obligation/?utm_source=ICRC+Law+%26+Policy+Forum+Contacts&utm_campaign=e9de618589-LP_EMAIL_BLOG_POST_2018_12_11_03_37&utm_medium=email&utm_term=0_8eeeebc66b-e9de618589-79024717&mc_cid=e9de618589&mc_eid=b8997e9f4e

Jensen, E. T. (2020a). Autonomy and precautions in the law of armed conflict. *International Law Studies, 96*(1), 577–602. https://digital-commons.usnwc.edu/ils/vol96/iss1/19/

Jensen, E. T. (2020b). The (erroneous) requirement of human judgment (and error) in the law of armed conflict. *International Law Studies, 96,* 26–57. https://papers.ssrn.com/sol3/papers.cfm?abstract_id=3548314

Kalpouzos, I. (2020). Double elevation: Autonomous weapons and the search for an irreducible law of war. *Leiden Journal of International Law, 33*, 289–312.

Kraska, J. (2021). Command responsibility for AI weapon systems in the law of armed conflict. *International Law Studies, 97*, 407–447. https://digital-commons.usnwc.edu/ils/vol97/iss1/22/

Liivoja, R., Massingham, E., & McKenzie, S. (2022). The legal requirement for command and the future of autonomous military platforms. *International Law Studies, 99*, 640–675.

Liu, H.-Y. (2016). Refining responsibility: Differentiating two types of responsibility issues raised by autonomous weapons systems. In N. Bhuta, S. Beck, R. Geis, H.-Y. Liu, & C. Kreis (Eds.), *Autonomous weapons systems. Law, ethics, policy* (pp. 325–344). Cambridge University Press.

Longobardo, M. (2019). Training and education of armed forces in the age of high-tech hostilities. In E. Carpanelli, & N. Lazzerini (Eds.), *Use and misuse of new technologies: contemporary challenges in international and European law*. Springer.

Matthias, A. (2004). The responsibility gap: Ascribing responsibility for the actions of learning automata. *Ethics and Information Technology, 6*(3), 175–183.

McFarland, T. (2020). *Autonomous weapon systems and the law of armed conflict: Compatibility with international humanitarian law*. Cambridge University Press.

McFarland, T. (2021). The concept of autonomy. UQ eSpace - Research Paper, 1–20. https://espace.library.uq.edu.au/view/UQ:d2aac6d

McFarland, T. (2022). Minimum levels of human intervention in autonomous attacks. *Journal of Conflict & Security Law, 27*(3), 387–409.

McFarland, T., & McCormack, T. (2014). Mind the gap: Can developers of autonomous weapons systems be liable for war crimes? *International Law Studies, 90*, 361–385.

Military and Paramilitary Activities in and against Nicaragua (Nicaragua v. United States of America). (June 27, 1986). ICJ (Judgement).

Prosecutor v. Naser Oric, IT-03-68-T (ICTY June 30, 2006).

Prosecutor v. Thomas Lubanga Dylo, ICC-01/04-01/06-803, Decision on the Confirmation of Charges (ICC - Pre-Trial Chamber I January 29, 2009).

Prosecutor v Zoran KrupresKic et al, IT-95-16-T (ICTY - Trial Chamber January 14, 2000).

Queguiner, J.-F. (2006). Precautions under the law governing the conduct of hostilities. *International Review of the Red Cross, 88*(864), 793–821.

Sartor, G., & Omicini, A. (2016). The autonomy of technological systems and responsibilities for their use. In N. Bhuta, S. Beck, R. Geis, H.-Y. Liu, & C. Kreis (Eds.), *Autonomous weapons systems: Law, ethics, policy* (pp. 39–73). Cambridge University Press.

Sassoli, M. (2014). Autonomous weapons and international humanitarian law: Advantages, open technical questions and legal issues to be clarified. *International Legal Studies, 90*, 308–340.

Schmitt, M. N. (2013). Autonomous weapon systems and international humanitarian law: A reply to the critics. *Harvard National Security Journal Feature*, 1–37. https://papers.ssrn.com/sol3/papers.cfm?abstract_id=2184826

Schmitt, M. N., & Vihul, L. (Eds.). (2017). *Tallinn manual 2.0 on the international law applicable to cyber operations*. NATO Cooperative Cyber Defence Centre of Excellence.

Seixas-Nunes, SJ, A. (2022a). Autonomous weapons systems and the procedural accountability gap. *Brooklyn Journal of International Law, 46*(2), 421–478. https://brooklynworks.brooklaw.edu/cgi/viewcontent.cgi?article=1973&context=bjil

Seixas-Nunes, SJ, A. (2022b). *The legality and accountability for autonomous weapon systems. A humanitarian law perspective*. Cambridge University Press.

Situation in the Central African Republic in the Case of the Prosecutor v. Jean-Pierre Bemba Gombo, ICC-01/05-01/08 (ICC June 15, 2009).

Solovyeva, A., & Hynek, N. (2023). When stigmatization does not work: Over-secularization in efforts of the campaign to stop killer robots. *AI & Society*. https://link.springer.com/article/10.1007/s00146-022-01613-w

Suchman, L., & Weber, J. (2016). Human-machine autonomies. In N. Bhuta, S. Beck, R. Geis, H.-Y. Liu, & C. Kreis (Eds.), *Autonomous weapons systems. Law, ethics, policy* (pp. 75–102). Cambridge University Press.

Summers, M. A. (2014). The problem of risk in international criminal law. *Washington University Global Studies Law Review, 13*(4), 667–697.

Taddeo, M., & Blanchard, A. (2022). A comparative analysis of the definitions of autonomous weapons systems. *Science and Engineering Ethics, 28*(5), 1–22. https://link.springer.com/article/10.1007/s11948-022-00392-3

The Netherlands. (2017). 2017 *Netherlands Group of Governmental experts of the high contracting parties to the convention on prohibitions or restrictions on the use of certain conventional weapons which may be deemed to be excessively injurious or to have indiscriminate effects.* https://reachingcriticalwill.org/images/documents/Disarmament-fora/ccw/2017/gge/documents/WP2.pdf

Trabucco, L., & Heller, K. J. (2022). Beyond the ban: Comparing the ability of autonomous weapon systems and human soldiers to comply with IHL. *Fletcher Forum of World Affairs, 46*(2), 15–31.

UK - The Joint Service Manual of the Law of Armed Conflict. (2004). UK Ministry of Defense. chrome-extension://efaidnbmnnnibpcajpcglclefindmkaj/https://assets.publishing.service.gov.uk/media/5a7952bfe5274a2acd18bda5/JSP3832004Edition.pdf

US Department of Defense. (2012). *DoD 3000.09. Autonomy in weapon systems.* https://www.hsdl.org/?abstract&did=726163

US Department of Defense. (2015). *Department of defense law of war manual.* General Council of the Department of Defense.

US Department of Defense. (2023). DoD Directive 3000.09 - Autonomy in weapon system (update 2023). https://www.defense.gov/News/Releases/Release/Article/3278076/dod-announces-update-to-dod-directive-300009-autonomy-in-weapon-systems/

US Department of State. (2023, February 16). *Political declaration on responsible military use of artificial intelligence and autonomy.* United States Government Website. https://www.state.gov/political-declaration-on-responsible-military-use-of-artificial-intelligence-and-autonomy/

Winter, E. (2021). The accountability of software developers for war crimes involving autonomous weapons: The role of joint criminal enterprise doctrine. *University of Pittsburgh Law Review, 83*(1), 51–86.

SECTION III

Human Control in Human–AI Military Teams

Rethinking 'Meaningful Human Control'

11

Linda Eggert

INTRODUCTION

The debate about autonomous weapon systems (AWS) is marred by definitional issues. A recent report found that, between 2012 and 2020, states and major international organizations proposed 12 different definitions of AWS (Taddeo & Blanchard, 2022). One common denominator is the assumption that AWS, once activated, have the capacity to select and engage targets, including human persons, without further human intervention. For example, France mentions 'a total absence of human supervision, meaning there is absolutely no link (communication or control) with the military chain of command' (France, 2016; Taddeo & Blanchard, 2022); Germany defines AWS as 'weapons systems that completely exclude the human factor from decisions about their employment' (Germany, n.d.; Taddeo & Blanchard, 2022); Switzerland mentions the capacity of 'carrying out tasks governed by IHL in partial or full replacement of a human in the use of force, notably in the targeting cycle' (Switzerland, 2016; Taddeo & Blanchard, 2022); Canada's Department of National Defence mentions the 'capability to independently compose and select among various courses of action' (Department of National Defense, 2018); and the ICRC broadly defines AWS as 'weapons that select and apply force to targets without human intervention' (ICRC, 2020).

In line with the mainstream, I will take diminished or absent human involvement and control after activation to be a distinguishing feature of AWS. The notion of autonomy that applies to weapon systems differs from autonomy in the conventional philosophical sense, according to which autonomy essentially involves acting for one's own reasons. In its application to AWS, the label 'autonomous' typically captures a

DOI: 10.1201/9781003410379-14

machine's capacity to carry out tasks independently of a human person. Given a target list or description, provided by a human, autonomous systems are able to identify, track, and destroy targets without further human guidance.

This possibility has given rise to several ethical, political, and legal concerns, especially regarding *lethal* AWS, which will be the subject of this chapter. Prominent worries concern the moral permissibility of 'killing by algorithm', feared violations of human dignity, the potential unavoidability of 'responsibility gaps', the potential avoidance of public scrutiny over governments' use of force, and the adequacy of international humanitarian law (IHL) and international human rights law (IHRL) at governing the use of AWS. Additional fears concern the risk of proliferation, asymmetric warfare, and risks to international security. After all, the option of outsourcing warfighting to robots, as well as AWS' constant combat readiness, may lower the barrier for entry and lead to more wars (Leveringhaus, 2022).

A common claim, in response to many of these concerns, is that AWS must remain under 'meaningful human control' (MHC) (Amoroso & Tamburrini, 2020; Article 36, 2017; HLS International Human Rights Clinic, 2020; Horowitz & Scharre, 2015; Moyes, 2016; Roff & Moyes, 2016). As Taddeo and Blanchard observe in their report of discussions by the Group of Governmental Experts held at the Convention on Certain Conventional Weapons, 'Many interventions stressed that the notion of meaningful human control could be useful to address the question of autonomy' (Taddeo & Blanchard, 2022, p. 4). To name but one example, in its 2016 Memorandum to CCW Delegates, Human Rights Watch (HRW) stated that 'Mandating human control would resolve many of the moral and legal concerns that fully autonomous weapons raise' (Human Rights Watch, 2016). It also states that 'Such a requirement would protect the dignity of human life, facilitate compliance with international humanitarian and human rights law, and promote accountability for unlawful acts'.

The aim of this chapter is to challenge this widespread faith in MHC. MHC is a prominent and popular notion especially in policy debates about the ethics of AWS. However, it is not a particularly lucid notion. Indeed, a common complaint in the academic literature is that, though MHC is frequently invoked, it is neither clear what it means nor what it requires (Amoroso & Tamburrini, 2021; Ekelhof, 2019; Horowitz & Scharre, 2015). How we should define MHC is an important question, especially if MHC is to be a necessary condition for the permissible deployment of certain weapons. The focus of this chapter is on an even more foundational, surprisingly seldom examined question – namely, what, from a *moral* standpoint, the notion is meant to achieve. By working out precisely how MHC is meant to dispel specific ethical concerns about AWS, this chapter seeks to help us arrive at a deeper and more comprehensive understanding of the idea. Clarifying why we may want MHC in the first place will also help us better understand how we should define it.

More specifically, what this chapter illuminates are the *normative limits* of MHC. The possibility this chapter asks us to consider is that the justificatory force of MHC is significantly less transformative than both policy and scholarly discussions suggest. By way of illustration, I will focus on three principal problems, in response to which mainstream debates commonly appeal to MHC as the go-to solution. All three concerns were highlighted by Christof Heyns, then-UN special rapporteur on extrajudicial, summary, or arbitrary executions, in his influential 2013 report which outlined a series of issues arising from 'lethal autonomous robotics and the protection of life' (Heyns, 2013).[1]

The first problem is what I will call the *Compliance Problem*. This is the concern that AWS may not be able to comply with the laws of war (IHL). Existing robots lack the cognitive capacities, including contextual understanding, necessary to comply with IHL's key principles of distinction, proportionality, and precaution (Anderson & Waxman, 2017; Geiss & Lahmann, 2017; Krupiy, 2015; Sassóli, 2014; Schmitt & Thurnher, 2013). The second problem is what I will call the *Dignity Problem*. This assumes that delegating life-and-death decisions to algorithms is inherently wrong and thus constitutes an indefensible affront to human dignity (Asaro, 2012; Birnbacher, 2016; Leveringhaus, 2016; Heyns, 2017; ICRC, 2018). The third is the *Responsibility Problem*. This is the prominent worry that autonomy in weapon systems may render it impossible to ascribe responsibility for harms wrongfully caused (Heyns, 2013).[2]

My claim is that MHC does considerably less to address the Compliance Problem, the Dignity Problem, and the Responsibility Problem than typically assumed. First, it is an open question what the relevant principles are that AWS would need to comply with. Besides, the degree of control necessary to ensure compliance with whatever principles are applicable may ultimately be incompatible with autonomy in weapon systems. Second, MHC is no guarantee that human dignity will not be violated. Third, MHC may help provide a basis for responsibility, but, I argue, primarily by rendering harms caused by AWS instances of wrongful omissions by human agents who failed to intervene and prevent those harms.

Ultimately, we face an awkward tension between human control and autonomy in machines. If the very attributes of AI, such as superhuman speed, that make the use of AI systems desirable have as their flipside a lack of controllability, MHC seems incompatible with autonomy in weapon systems. To be sure, there may be no loss of human control *pre* activation. But that may not address concerns of a lack of human control *post* activation. If autonomy in weapon systems captures the capacity to identify and attack targets without human intervention *post* activation, then arguments for MHC over life-and-death decisions are ultimately arguments for banning AWS. By illuminating complexities that have so far remained obscure and paving avenues for further enquiry, this chapter offers a starting point for rethinking the normative role of 'meaningful human control'.

Before we proceed, a few clarifications are in order. First, the purpose of this chapter is not to argue *against* MHC. My claim is not that we should abandon the notion. My contention, rather, is that the concept of MHC does not do nearly as much to mitigate concerns about compliance, dignity, and responsibility as is commonly assumed. Hence, the question of what precisely MHC can do, and my claim – not least to steer the debate in new directions – that MHC is not everything it might appear.

Second, since nobody, presumably, wants *meaningless* control, one might wonder why the mainstream notion is that of 'meaningful human control' rather than simply 'human control'. In my understanding of the debate, a standard assumption is that what is at stake goes beyond merely 'perfunctory' control and is grounded in certain moral and legal principles. The label 'meaningful', I take it, is supposed to help make this clear

(Amoroso & Tamburrini, 2019a; Santoni et al., 2018). But we need not attach too much weight to this particular label. Little would change if we substituted 'meaningful' with 'significant' or 'effective' or something of that sort. I will refer to MHC in line with mainstream practice.

Third, emerging technologies are a moving target. Not only may today's technological limits be gone tomorrow, but terms of art also rapidly evolve. Terms like 'in the loop' or 'on the loop', for example, may soon be outdated. What is becoming more frequent is appeals to MHC and, relatedly, 'appropriate levels of human judgment'.[3] Though terms of art in this field are difficult to avoid, this chapter attempts to eschew them as far as possible, to focus on foundational issues that will withstand technological change and the field's terminological fluctuations.

More substantively, one might question whether there is ever a lack of human control at all. Isn't there, one might ask, some human involvement at every stage of development and deployment? After all, human programmers carry out the initial mission programming and human operators activate AWS. Indeed, as Huq puts it, 'all machine-learning tools are at their origin the fruit of specific human design and engineering choices'. Thus, 'There is simply no such thing as a wholly endogenous algorithm' (Huq, 2020, p. 646).

Indeed, the deployment of weapons and operators follows a multi-stage decision-making process in which various human persons exercise different forms of control (Ekelhof, 2019). Human agents formulate objectives; they specify how targets are to be selected and how collateral damage is to be estimated; human agents make proportionality assessments, and they impose operational constraints (Ekelhof, 2019, p. 347; Roorda, 2015). Hence, Ekelhof (2019) claims that

> Understanding the distributed nature of control is an important factor when developing a concept such as meaningful human control, because it illustrates that human control does not need to have a direct link with the weapon system. In other words, humans typically exercise different forms of control over important decisions (such as target selection and engagement), even before weapons are activated
>
> *(Ekelhof, 2019, p. 347).*

But the nature of the relationship between human control and the weapon system is precisely what is at issue. The fact that humans are in control *before* a weapon system is activated hardly shows that no human control is required *after* a weapon system is activated.

AWS can produce a potentially lethal kinetic effect without any human involvement *post activation* (Leveringhaus, 2022). Programmed with its mission parameters, once activated, an AWS can deliver a payload and cause lethal harm without direct, real-time human involvement. Put another way, the systems of concern are those that have conventionally been described as 'out-of-the-loop' systems, which stand in contrast to in-the-loop systems and on-the-loop systems.[4] What are conventionally described as out-of-the-loop systems, once deployed, function without a human operator. MHC, for the purposes of this discussion and in line with mainstream appeals to the notion, implies direct, real-time human control over life-and-death decisions.

A final preliminary note. This chapter primarily confronts *moral* rather than legal or policy questions about AWS. While moral and political philosophers have a critical role to play in clarifying and defending the values at stake, the discussion offered here is but a part, though an important one, of a considerably bigger picture.

Section 11.2 addresses the Compliance Problem, Section 11.3 the Dignity Problem, and Section 11.4 the Responsibility Problem. These sections explain why popular appeals to MHC do less to address each of these problems than typically assumed. Section 11.5 offers a first step toward rethinking the relationship between human control and autonomy in weapon systems. Section 11.6 concludes.

COMPLIANCE

One common concern is that existing robots lack the cognitive capacities, including vital contextual understanding, necessary to comply with key principles of IHL, notably those of distinction and proportionality. AI systems might not be able to distinguish between combatants and non-combatants, and military and civilian targets; and they might not be able to make value judgments concerning what harms are excessive in relation to some military advantage, as required by the proportionality constraint. Hence, the prominent worry that AWS would likely not be able to comply with the laws of war. HRW put the issue as follows:

> The ability to distinguish combatants from civilians or from wounded or surrendering soldiers as well as the ability to weigh civilian harm against military advantage require human qualities that would be difficult to replicate in machines, including fully autonomous weapons
>
> *(Human Rights Watch, 2014; 2016).*

By maintaining MHC over targeting decisions, one might think, human agents are able to ensure that any harms caused are in accordance with key principles of distinction and proportionality. Call this the *Compliance Argument* for MHC. The rationale behind the Compliance Argument seems straightforward. Human agents who exercise control may, for example, prevent a malfunctioning weapon from attacking civilians; and they may also prevent the engaging of targets that would involve a degree of collateral harm that human agents would judge to be excessive to the military advantage gained (Scharre, 2018; Amoroso & Tamburrini, 2020).

However, even if AWS could be made to comply perfectly with IHL – whether by themselves or through MHC – not all would be well. Even if MHC may help ensure that AWS do not violate IHL, focusing just on compliance with IHL may be ill-considered. Many acts of killing in war are morally impermissible, despite being lawful – for example, the 'collateral' harming of innocent civilians in pursuit of an unjust military objective. The law, this is to say, is no failsafe moral authority. Certain acts of killing, which are morally impermissible but not legally prohibited are defensible only for human combatants but not for AWS (Eggert, 2022). If certain acts are defensible only for human combatants, some lawful acts of killing are not morally defensible if they are performed by AWS.

Another issue concerns the relevance of intentions and the so-called doctrine of double-effect. One question is whether we can distinguish between intended harms and unintended side effects in the context of harms inflicted by AWS. Not only is it unclear to what extent non-human, artificial agents can be said to have intentions, though they might have 'objectives'; but their decisions might also not be sufficiently transparent to be explainable, let alone to provide assessments and justifications of what harms count as 'excessive' (Eggert, 2023).

That aside, if many acts of killing in war are morally impermissible despite being lawful, even full compliance with IHL is compatible with morally wrongful killing. For example, the law permits the 'collateral' harming of civilians in the absence of a just cause; and it also effectively permits the targeting of combatants who possess moral rights against being harmed.[5] It follows that, even if MHC could ensure compliance with IHL, the potentially lethal acts in question might remain morally impermissible. In other words, compliance with IHL does not entail that the harms AWS could *lawfully* inflict are also *morally* permissible. My aim, to be clear, is not to criticize IHL but to caution against unthinkingly treating it as the be-all and end-all standard. To the extent that our concern is with the *moral* permissibility of using AWS, ensuring compliance with IHL through MHC will not get us far, because compliance with IHL is not always the same as complying with morality.

One response is to broaden our understanding of the Compliance Problem. Given how much harm to morally innocent people IHL permits, it is not clear that IHL's are the principles with which AWS should comply. Our worry should not just be that AWS might not be able to abide by the principles enshrined in IHL, but rather that they might not be able to abide by whatever principles ultimately apply to them. Call this the *Broad Compliance Problem.*

The Broad Compliance Argument for MHC, accordingly, says that MHC is necessary for ensuring that AWS comply with the moral principles that apply to them. This requires us to rethink two things: first, as a general matter, whether IHL is adequate to the task of governing just conduct by AWS; and, second, what the applicable principles are with which MHC is meant to ensure compliance. Our task, then, is twofold: (i) to look beyond compliance with IHL so as to distinguish between moral and legal permissions to harm, and (ii) to work out what the standards are with which AWS should comply.

The Broad Compliance Argument thus requires us to work out what the appropriate principles are with which human control should ensure compliance in the first place. This will likely raise questions about how the military use of AI might challenge traditional understandings of the appropriate relationship between IHL and IHRL, which goes beyond the scope of this chapter. Suppose, for the sake of argument, that the international human rights framework offers an alternative to IHL, at least insofar as IHRL better tracks individual, rights-based moral principles than IHL. The purpose of MHC, then, would include observing the difference between moral and legal permissions to kill; as well as ensuring that AWS act in accordance with more restrictive principles of

IHRL rather than IHL, and comply with moral prohibitions rather than legal permissions, such as not to harm people who have moral rights not to be harmed.

One difficulty with this is that the relationship between IHL and IHRL is notoriously contested. Another difficulty is that, if human agents are to exercise the kind of control necessary to ensure compliance with any principles that are ultimately applicable, AWS cannot really be autonomous. MHC would ensure precisely that weapon systems do *not* identify and engage targets independently from – that is, without the involvement of – human agents. We will return to this in Section 11.5.

DIGNITY

Consider the Dignity Problem. Deploying lethal AWS – 'killing by algorithm' – many critics worry, presents an affront to human dignity. Hence, life-and-death decisions must be limited to human agents. HRW describes the concern as follows:

> Ceding human control over decisions about who lives and who dies would deprive people of their inherent dignity. Inanimate machines, such as fully autonomous weapons, could truly comprehend neither the value of individual life nor the significance of its loss. Permitting them to make determinations to take life away would thus conflict with the principle of dignity and could "denigrate the value of life itself." . . . Because meaningful human control over weapons allows for ethical and unquantifiable factors to play a role in targeting decisions, it protects the dignity of civilians and soldiers alike
>
> *(Human Rights Watch, 2016).*

So, the claim is that respecting human dignity requires limiting life-and-death decisions to humans because humans understand the special value of life and the gravity of depriving another person of it. As Amoroso and Tamburrini put it,

> from the principle of human dignity respect, it follows that human control should operate as a moral agency enactor, by ensuring that decisions affecting the life, physical integrity, and property of people (including combatants) involved in armed conflicts are not taken by non-moral artificial agents
>
> *(Amoroso & Tamburrini, 2020, p. 189).*

Call this the *Dignity Argument* for MHC. The Dignity Argument says that outsourcing decisions about life and death to machines would violate human dignity and that such decisions must therefore be limited to humans (Asaro, 2012; Docherty, 2014; Sharkey, 2019; Sparrow, 2016).

Leveringhaus (2022) has voiced skepticism about the Dignity Problem. He distinguishes between a 'macro-level' of programming and a 'micro-level' of specific operations (Leveringhaus, 2022, p. 481). At the macro-level, Leveringhaus argues, it is human programmers who decide to program AWS to identify and eliminate enemy combatants.

So, on this level, it is human agents rather than machines that make life-and-death decisions. At the micro-level, by contrast, as Leveringhaus puts it, 'the machine has some leeway in translating [the programmer's] instructions into actions' (Leveringhaus, 2022). Within the category of enemy combatants, for example, the AWS might – for reasons that might be impossible to ascertain – target one combatant rather than another. So, an AWS will carry out its programming; but it might be unpredictable 'when, where, and whom it will kill', as Leveringhaus puts it (Leveringhaus, 2022). So, in this sense, on the micro-level, it is the machine making life-and-death decisions, but within the human-imposed constraints of its programming.

The question, then, Leveringhaus says, is whether 'the machine's micro-choice, rather than [the programmer's] macro-choice' violates that combatant's dignity. Since, in war, combatants are typically targeted for the reason that they are enemy combatants, the answer, Leveringhaus says, is no (Leveringhaus, 2022). There is no relevant difference between a human combatant targeting an enemy combatant because they are an enemy combatant and a machine targeting an enemy combatant because they are an enemy combatant. So, AWS would only pose a threat to human dignity if they were used deliberately to target non-combatants or if they caused excessive, disproportionate harm (Leveringhaus, 2022, p. 483). The threat to human dignity, in these cases, however, would not be inherent in AWS; it would be no different from other ways of killing that also violate human dignity.

But it is not obvious that the same justifications of self-defense and other defense that are available to human combatants also apply to AWS. Besides, one might say, human combatants at least have the capacity of *comprehending* the gravity of killing another human being. Only humans have the capacity to truly appreciate the value of human life and, hence, the gravity of taking one. Therein, one might think, lies the violation of dignity of outsourcing life-and-death decisions to machines. Leveringhaus is sceptical. Recognizing the value of life is so demanding under in conditions of armed conflict, he suggests, that it is not clear that even human combatants could meet it (Leveringhaus, 2022). To be sure, battlefields are hardly conducive to careful moral reflection. But this is no challenge to the claim that the basic capacity to appreciate the value of life is limited to human agents, whether or not they exercise it, and that this constitutes one relevant difference between human and artificial agents that might militate against outsourcing life-and-death decisions to non-human, artificial agents.

The twofold question remains whether (i) AWS pose a distinct threat to human dignity (ii) which MHC would successfully avert. Leveringhaus denies (i) that AWS pose a distinct threat to human dignity, on the grounds that many other forms of killing in war also violate human dignity (Leveringhaus, 2022). But it does not follow from the fact that many human acts of killing in war violate human dignity that deploying AWS would not violate human dignity. Replacing human combatants with AWS might just replace human violations of human dignity with AI-powered violations of human dignity. The fact that human combatants commit violations of human dignity all the time just means that we have more than one type of dignity violation to worry about.

Now, the question is not merely (i) whether AWS pose a special threat to human dignity; but also (ii) whether MHC helps to protect it. One reason for skepticism is indeed that humans are capable of extreme cruelty. Given the horrendous things people do in war, it will not do simply to assert that human dignity is better protected by limiting life-and-death decisions to human persons. While we may worry about the fact that robots lack emotions such as empathy and mercy, we should not underestimate the fact that they also lack emotions such as anger, cruelty, vengefulness, and panic. So, really, the fact that humans violate human dignity all the time does not tell us anything about (i) whether AWS might violate human dignity. Rather, it is a reason to doubt (ii) whether MHC would help solve the problem.

Proponents of AWS claim that AWS might significantly reduce the number of innocent deaths. Not only are robots free from emotions like vengefulness that have led human soldiers to commit terrible atrocities. AWS also save militaries from having to send soldiers into harm's way. If one's mission is just, presumably, it is preferable if those trying to accomplish it are not exposed to risks of lethal harm. AI may also enable a more careful selection of targets and more precisely timed attacks. In situations in which the human mind is simply too slow to process all relevant information, AI's superhuman speed and information-processing power will allow AWS to make split-second decisions that could save lives (Heyns, 2016). Couldn't such capacities, proponents of AWS might say, serve to protect rather than threaten human dignity, even if it comes at the price of MHC?

Some dignity-based arguments for MHC, according to which only humans should make life-death-decisions because only humans understand such decisions' gravity, seem to rest on an assumption about agents' attitude, their intentions, and whether they experience appropriate reactions. A sadistic person might fully comprehend the value of another person's life and the significance of its loss, and she might nonetheless – or precisely therefore – derive pleasure from taking it. But clearly that is not the idea. What we seem to care about is people's capacity to understand the *moral* gravity of ending another person's life and, importantly, the capacity to experience appropriate reactive attitudes, such as guilt and regret. Perhaps it is a certain kind of empathy that really matters to proponents of MHC – not simply the fact that it is a human rather than a machine performing a harmful, potentially lethal, act.

The Dignity Argument for MHC, then, might reconceive MHC as a necessary condition for a certain kind of causal connection: one needs to have been in control, or stand in a certain kind of causal relationship to a harmful outcome, for certain reactive attitudes such as regret to be apt. The idea would then be that those involved in the causal architecture leading to certain harms appropriately recognize victims and experience relevant reactive attitudes as a result and that this expresses or otherwise reflects some measure of respect for victims' dignity.

The difficulty then lies in weighing the value of (a) having life-and-death decisions be made by humans who understand the gravity of such decisions and have the capacity to experience appropriate reactive attitudes against the value of (b) reducing the number of overall deaths by using AI to render targeting more efficient and precise. Both considerations might appeal to the value of human dignity but, while the former militates in favor of MHC, the latter may not be compatible with it.

There is thus more to the relationship between AWS, dignity, and human control than meets the eye. The assumption that MHC matters because there is something inherently wrong with delegating life-and-death decisions to algorithms, while intuitive, is not the whole story. My claim, to be clear, is not that the assumption that there is something inherently wrong with delegating life-and-death decisions to algorithms is unsatisfying. Given the cruelty of which we know people are capable, MHC is no guarantee that human dignity will be respected.[6] Rather than undermining the widespread intuition that AI-generated life-and-death decisions may violate human dignity, this calls into question whether MHC would fix the problem.

Perhaps, instead of the supposedly intrinsic dignity-violating features of AWS, we should appeal to the sheer scale of harm AWS could cause, which far exceeds what individual human combatants would be able to do. Perhaps the straightforward moral imperative, familiar from debates about nuclear weapons, not to unleash such destructive power is ultimately what matters. In this case, however, the conclusion is that AWS should be banned; not that MHC will avert the issue.

RESPONSIBILITY

Removing human agents from the decision-making process, a prominent claim goes, would make it extremely difficult, perhaps altogether impossible, to ascribe responsibility for any harms wrongfully caused (Sparrow, 2007; Amoroso & Giordano, 2019; McDougall, 2019). This is the widespread worry that delegating certain acts and decisions to AI will result in responsibility 'gaps'. Call this the *Responsibility Problem*.

The worry at the heart of the Responsibility Problem is that neither operators, programmers nor manufacturers may be sufficiently directly morally or causally linked to AWS-inflicted harm to hold those human agents meaningfully accountable.

The solution that MHC is typically taken to provide takes the form of a basis for accountability. The reasoning is straightforward. If human agents are in control, they can be held accountable if something goes wrong. As HRW says in its report: 'Mandating meaningful human control would close the accountability gap and ensure that someone could be punished for an unlawful act caused by the weapon used.' (Human Rights Watch, 2016). Call this the *Responsibility Argument* for MHC.

The Responsibility Argument for MHC assumes that the possibility of holding people responsible for the use of force is a necessary condition for killing in war to be morally permissible. But what discussions about the Responsibility Problem typically disregard is that we might care about responsibility for at least two very different reasons. The first is that responsibility matters as a necessary condition for liability to punishment. Liability to punishment presupposes culpability, and culpability presupposes moral responsibility.

The second reason why we might care about responsibility is that we want to fairly allocate compensatory obligations, to mitigate harms that victims suffered. But it is possible to ground such compensatory duties without appealing to individual moral responsibility. While it would go beyond the scope of this chapter to provide a detailed defense of this claim, it suffices, for our purposes, to consider how we justify compensatory obligations on the grounds of collective responsibility, state responsibility, and even corporate responsibility – contexts in which we impose certain burdens on collective entities in the absence of individual moral responsibility. Examples include states acquiring duties to pay reparations for unjust wars and responses to historical injustice, in cases in which original perpetrators are no longer alive, but claims to reparation persist. In such cases, victims may still possess claims to compensation even if corresponding duties cannot be discharged by those morally responsible for the harms in question (Butt, 2013; Goodin & Barry, 2014; Miller, 2001; Stilz, 2011).

This is to say that there are routes to remedial obligations that do not presuppose the kind of individual moral responsibility possessed only by responsible moral agents. Insofar as remedial obligations may arise from, for example, mere causal responsibility, communal membership, or having benefited from injustice, the absence of MHC need not preclude the possibility of rectificatory obligations.

It matters what precisely we are after in enquiring about responsibility. While liability to punishment presupposes culpability, which in turn presupposes responsible moral agency, moral responsibility is not a necessary condition for liability to compensation. Compensatory obligations for harms caused by AWS need not presuppose that either machines or persons occupying some role in the causal chain leading up to some harmful event are *morally* responsible for the harm caused. This means that, so long as we are not concerned with punishment, responsibility gaps need not trouble us. If what matters is broader, compensatory obligations – such as duties to mitigate harms victims suffered – we need not rely on individual moral responsibility as a basis for rectificatory duties.

As a basis for responsibility, then, the appeal to MHC may not be what it seems. If the need for MHC arises from the need for a basis for accountability, and if there are other bases for accountability, such as those mentioned above, the need for MHC weakens. This means that we have at least one fewer reason to care about MHC than we might have assumed.

Leveringhaus argues that a main argument against AWS is that they make it impossible for human agents to revise a decision not to kill, as a result of 'removing human agents from the point of payload delivery'.[7] AWS cannot refuse to carry out an order to engage an enemy. An AWS, as Leveringhaus puts it, 'will kill once it has 'micro-chosen' a human target' (Leveringhaus, 2022). Humans, by contrast, have the capacity to refuse to carry out orders – sometimes with momentous consequences: Stanislav Petrov was the lieutenant colonel of the Soviet Air Defense Forces on duty in September 1983, when the Soviet nuclear early-warning system reported that the US had launched a missile. Petrov correctly judged the report to be a false alarm and, disobeying orders, against Soviet military protocol, decided not to launch a retaliatory nuclear attack on the US and its NATO allies. His decision prevented what would likely have resulted in large-scale nuclear war.

Before we proceed, then, I want to propose a different version of a Responsibility Argument for MHC. The most promising sense in which MHC addresses concerns about responsibility, I submit, is by allowing us to categorize harms wrongfully inflicted by AWS as *human omissions to prevent such harms*. The assumption here is that there are cases in which humans are responsible for failing to prevent harm: cases in which someone should have intervened but failed to do so. These are cases in which humans omit to act; instances of allowing harm to be caused by an AWS which human agents should prevent. This, I propose, is the critical sense in which MHC allows us to categorize harms wrongfully inflicted by AWS as human omissions to prevent such harms.

This proposal, however, not only requires us to rethink the normative purpose of MHC; it also requires us to specify human obligations that come with exercising judgement or control over AWS. This presupposes an account of when those in control ought to intervene to prevent harms that might otherwise be caused by AWS. This calls into question nothing less than whether weapon systems should have the capacity to identify and engage targets *without human intervention*.[8] If human agents must remain capable of exercising the kind of control that would allow them to intervene to prevent AWS from causing certain harms, this might altogether rule out autonomy in weapon systems.[9]

AUTONOMY AND CONTROL

One response to worries about AWS simply claims: humans are better than machines at complying with the law, humans are better at protecting human dignity, and humans are easier to hold responsible. So, if we care about compliance, dignity, and responsibility, the answer is straightforward. Machines should not be made to be autonomous, and 'select and engage targets without human intervention' in the first place. Indeed, this is what the Compliance, Dignity, and Responsibility Arguments for MHC ultimately seem to suggest.

While proponents of AWS see significant benefits in the potential of deployment, such as speed and efficiency that AWS may bring to the protection of innocent lives, the very attributes that make the use of AI desirable may also be at odds with the possibility of MHC. To be clear, it is not the lack of controllability that renders AWS morally desirable. The worry, rather, is that what makes AWS desirable, such as their superhuman speed, may have as its necessary flipside a lack of controllability.

Some scholars have proposed a differentiated approach to MHC, according to which different degrees of human supervision are appropriate in different kinds of contexts and for different kinds of weapon systems, depending on their degree of autonomy from human agency (Amoroso & Tamburrini, 2019b, c; Amoroso & Tamburrini, 2020; United States, 2018). To the extent that concerns about the military use of AI are bound up with AWS' independence from human agency, and insofar as human control effectively counteracts machine autonomy, MHC will present itself as a natural solution. The result is that demands for MHC then collapse into the claim that weapons simply should not be autonomous, in the sense of being decoupled from human agency to produce a potentially lethal kinetic effect without human involvement post activation. This means that, instead of relying on MHC as a universal solution, policy-makers should confront the question of whether autonomy in weapon systems is ultimately compatible with whatever measures are necessary to address the risks they pose. If autonomy in AWS is incompatible with the kind of control necessary to protect the values we ultimately think are at stake, those values will prohibit the use of AWS.

There is an uncomfortable tension between autonomy in weapon systems and human control. Amoroso and Tamburrini claim that 'meaningful human control (MHC) over weapon systems should be retained exactly in the way of their critical target selection and engagement functions' (Amoroso & Tamburrini, 2020). But this just seems to say that weapon systems should not be autonomous. Now, Taddeo and Blanchard (2022) claim that human control is compatible with autonomy in weapon systems. On their view, 'An artificial system can be, in principle, fully autonomous, insofar as it can operate independently from a human or of another artificial agent, and yet be deployed under some form of meaningful human control' (Taddeo & Blanchard, 2022, p. 11).[10] What precisely it means to 'be deployed under some form of meaningful human control' – though independently from human agents – remains unclear.[11] Nonetheless, they insist, 'human control is not antithetical to the autonomy of AWS and can be exerted over AWS at different levels, from the political and strategic decisions to deploy AWS to the kind of tasks delegated to them'. The question, they conclude, 'is which form of control is ethically desirable and should, ideally, be considered by decision- and policy-makers in designing the governance of AWS' (Taddeo & Blanchard, 2022, p. 18). Indeed, 'Many of the problems posed by AWS', according to Taddeo and Blanchard, 'do not concern the desirable level of autonomy of these systems, but the desirable level of control over these systems' (Taddeo & Blanchard, 2022, p. 11).

Concerns about control arise precisely from the possibility of autonomy in weapon systems. Retaining control over kitchen knives is not a worry, because kitchen knives are not autonomous. By contrast, retaining control over AWS may well be a worry precisely to the extent that they may have the capacity to operate with a certain degree of independence of human agency. To show that autonomy in weapon systems is compatible with human control, it will not do to define autonomy as compatible with human control. Otherwise, the premise is equivalent to the conclusion.

We consequently find ourselves confronted with an awkward question. Is there such a thing as human control over systems that are autonomous, as Taddeo and Blanchard's definition suggests? Or is *control* over *autonomous* weapons not merely an apparent but a real contradiction, meaning that we must ultimately choose between autonomy in machines and human control? What if we cannot have one without giving up the other?

If MHC is ultimately incompatible with autonomy in weapon systems, understood as the capacity to identify and attack targets *without human intervention* post activation, then the question is – is there any difference between arguing for MHC and arguing for a ban on AWS? If it makes no sense to think of MHC as a constraint on how certain weapons are to be used, perhaps we should just insist that, in the absence of MHC, a weapon simply must not be used. If the latter, the aim should be to ban AWS; not to work out under what conditions their use is permissible.[12]

We need to consider solutions other than MHC to each of the concerns AWS have raised. In response to the Compliance Problem, our task is to look beyond compliance with IHL, and to take seriously the difference between moral and legal permissions to harm. We first need to work out what the appropriate principles are with which human control should ensure compliance in the first place. Questions that this line of inquiry will likely raise include how the military use of AI might challenge traditional understandings of the appropriate relationship between IHL and IHRL.

In response to the Dignity Problem, we saw that the relation between respecting human dignity and MHC may be one of correlation rather than causation. The question that remains is whether a lack of human control necessarily violates dignity, even if MHC is no guarantee that human dignity will be respected.

In response to the Responsibility Problem, I argued, we should separate concerns about compensation from concerns about punishment. While moral responsibility is a prerequisite for liability to punishment, moral responsibility is not a prerequisite for compensatory obligations. Since the moral basis for punishment is not coextensive with the moral basis for compensatory obligations, addressing worries about a lack of a basis on which to ascribe responsibility requires clarifying what is at stake. And MHC may only be relevant to the extent that individual moral responsibility is.

In addition, I proposed that the moral purpose of MHC is to transform cases of wrongful harms caused by AWS into cases of wrongful human omissions. If this is right, and 'ought implies can', it must remain possible for human agents to intervene upon autonomized agency. This, however, means that we may need to forgo some of the advantages, such as superhuman efficiency, that proponents of AWS see as militating in favor of using AWS to begin with (Heyns, 2013).[13] This brings us back to the question of whether human control – or, indeed, the preservation of human judgement – as a basis for human responsibility necessarily requires decreasing autonomy in weapon systems.

CONCLUSION

The aim of this chapter was to examine how much MHC can, morally speaking, do to address key concerns about AWS. One possible answer, this chapter has begun to suggest, is – less than we might think. For one, MHC only partially addresses the problems it is typically assumed to help solve. First, MHC may help address concerns about compliance; but it is an open question what the relevant principles are that AWS would need to comply with. And the degree of control necessary to ensure compliance with whatever principles are actually applicable may ultimately be incompatible with autonomy in weapon systems. Second, MHC is no guarantee that human dignity will be protected. As a measure against violating dignity, appeals to MHC risk being unhelpfully vague. Since human persons are capable of extreme cruelty, it will not do simply to claim that human dignity is better protected by limiting life-and-death decisions and acts of killing to human persons. Finally, MHC may help provide a basis for accountability. It may do so largely by rendering wrongful harms caused by AWS cases of wrongful omissions by human agents who failed to intervene and prevent those harms. What, in the end, remains is an awkward tension between human control and autonomy in machines.

The GGE has so far adopted a non-binding set of eleven Guiding Principles on lethal AWS, which include broad statements on the importance of international law (Principles (a), (c), (e); human responsibility (Principles (b) and (d); and human control (Principle (d)).[14] This chapter cautioned against relying on MHC as the ultimate solution to the range of issues arising from increasing military applications of AI. MHC cannot solve all the problems its proponents sometimes take it to address. Other considerations we need to take seriously include the difference between lawfulness and moral permissibility, the various purposes of ascribing responsibility, and the degree to which control as a basis for responsibility is ultimately compatible with autonomy in weapon systems.

In the spirit of raising questions to advance the debate, one possibility now on the table is that we should stop appealing to MHC – convenient though it may be – as an apparently universal solution to disparate ethical concerns about compliance, dignity, and responsibility, because MHC does less to address each of these concerns than common appeals to it suggest. And to the extent that it does address them, it does so differently from what the mainstream seems to suppose. This does not mean that MHC is not important; indeed, this chapter went some way toward clarifying some of the moral reasons why it might matter. It just means that we should not treat MHC as the ultimate solution to more problems than it can solve.

NOTES

1 The list of concerns on which this chapter focuses is not comprehensive. Other worries include increased risks to peace and international stability.

2 See Section 11.4.

3 Both the 2012 and 2023 US DoD directives highlight the importance of designing AWS so as 'to allow commanders and operators to exercise appropriate levels of human judgment over the use of force'. See 'Directive 3000.09' (2023), p. 5, section. 1.2(a). HRW notes that, although human judgment is essential for ensuring compliance with IHL, neither directive clearly specifies what constitutes an 'appropriate level' of human judgment. See HRW, 'Review of the 2023 US Policy on Autonomy in Weapons Systems' (14 February 2023); available at: https://www.hrw.org/news/2023/02/14/review-2023-us-policy-autonomy-weapons-systems#_ftn14.

4 In the case of in-the-loop-systems, the decision to apply force to a target is made by the weapon's operator in real time. In the case of on-the-loop systems, operators are active supervisors on stand-by, who can override the weapon if something goes wrong. Leveringhaus, 'Morally Repugnant Weaponry?', 477.

5 In the language of contemporary debates about the ethics of war, I am taking what is known as a 'revisionist' standpoint.

6 One might object that this is true in individual, exceptional cases and that MHC is a distributed process, which may significantly lessen individual cruelty. Ekelhof, for example, describes targeting as a distinctly distributed

process that extends over time and in which numerous actors are involved. MHC, one might point out, is similarly distributed both in time and in organizational space. Thanks to Jan Maarten Schraagen for this point. The question then is what this 'distributed' sense of MHC is meant to achieve. If it is to avoid the concentration of power in any one individual or a small number of individuals, it looks like we're in the unlikely business of democratic control.

7 For his defence of the significance of the ability to revise one's decision not to kill, see Leveringhaus, A. (2016). *Ethics and autonomous weapons.* Palgrave Macmillan.

8 One might question this standard definition of AWS, given the distributed nature of human involvement in the causal chain leading up to targeting. The standard definition, one might point out, ignores that there has been human intervention, perhaps not at the moment of engaging a target, but in the days, weeks, or months prior to that moment. Since targets are programmed into the AWS by humans, the AWS are under MHC. Thanks to Jan Maarten Schraagen for this point. The issue is precisely the lack of human involvement at the time of targeting. Humans might program an AWS to target enemy combatants at t1, but it is still up to the machine, at t2, to select individual targets, and it may not always be clear why it targeted one person rather than another. The fact that a human person, at t1, programmed an AWS to behave a certain way does not mean that the AWS is under human control at t2.

9 This, indeed, explains much of the opposition to autonomy in weapon systems. The main exception is in the use of AI in anti-ballistic missile defence, where speed is of the essence, or in defence of naval platforms against incoming high-velocity missiles. Thanks to Jan Maarten Schraagen for this point.

10 Thanks to Jan Maarten Schraagen for pressing me to discuss Taddeo and Blanchard's account. For another account that proposes to reconcile autonomy in weapon systems with human control, see Santoni de Sio, F., & Van den Hoven, J. (2018). Meaningful human control over autonomous systems. *Frontiers in Robotics and AI 5*, 15.

11 Intriguingly, Taddeo and Blanchard seem to suggest that we should distinguish between human intervention and human control, as well as between machine autonomy and lack of human control (Taddeo and Blanchard, 'A Comparative Analysis,' 11). What remains obscure is precisely what these distinctions ultimately come down to.

12 One challenge is that, to ban something, it must be defined. To the extent that it is not in powerful states' interest to ban AWS, it is also not in their interest to clearly define AWS, hence the remarkable degree of vagueness that characterizes the international policy debate concerning AWS.

13 See §41.

14 Final Report of the 2019 Meeting of the high contracting parties to the CCW. UN Doc CCW/MSP/2019/CRP2/Rev1, Annex III (15 November 2019).

REFERENCES

Amoroso, D., & Giordano, B. (2019). Who is to blame for autonomous weapons systems' misdoings? In E. Carpanelli, & N. Lazzerini, *Use and misuse of new technologies* (pp. 211–232). Springer.

Amoroso, D., & Tamburrini, G. (2019a). Filling the empty box: A principled approach to meaningful human control over weapons systems. *ESIL Reflections, 8*(5), 1–9.

Amoroso, D., & Tamburrini, G. (2019b). *What makes human control over weapon systems "meaningful"?* International Committee for Robot Arms Control, Report to the CCW GGE.

Amoroso, D., & Tamburrini, G. (2020). Autonomous weapons systems and meaningful human control: Ethical and legal issues. *Current Robotics Reports, 1*(4), 187–194.

Amoroso, D., & Tamburrini, G. (2021). Toward a normative model of meaningful human control over weapons systems. *Ethics & International Affairs, 35*, 245–272.

Anderson, K., & Waxman, M. C. (2017). Debating autonomous weapon systems, their ethics, and their regulation under interntional law. In R. Brownsword, E. Scotford, & K. Yeung, *The Oxford Handbook of Law, Regulation, and Technology* (pp. 1097–1117). Oxford University Press.

Article 36. (2017). *Autonomous weapon systems: Evaluating the capacity of 'meaningful human control' in weapon review processes*. Discussion paper for the Convention on Certain Conventional Weapons (CCW) Group of Governmental Experts Meeting on Lethal Autonomous Weapons Systems.

Asaro, P. (2012). On banning autonomous weapon systems: Human rights, automation, and the dehumanisation of lethal decision-making. *International Review of the Red Cross, 94*, 687–709.

Asaro, P. (2020). Autonomous weapons and the ethics of artificial intelligence. In S. M. Liao, *Ethics of artificial intelligence* (pp. 212–236). Oxford University Press.

Birnbacher, D. (2016). Are autonomous weapon systems a threat to human dignity? In N. Bhuta, S. Beck, R. Geis, H.-Y. Liu, & C. Kreis, *Autonomous weapons systems: Law, ethics, policy* (pp. 105–121). Cambridge University Press.

Butt, D. (2013). Inheriting rights to reparation: Compensatory justice and the passage of time. *Ethical Perspectives, 20*, 245–269.

Department of National Defense. (2018). *Autonomous systems for defence and security: Trust and barriers to adoption*. Innovation Network Opportunities.

Docherty, B. (2014). *Shaking the foundations: The human rights implications of killer robots*. Human Rights Watch.

Eggert, L. (2022). *Handle with care: Autonomous weapons and why the laws of war are not enough*. Technology & Democracy Discussion Paper, Harvard University.

Eggert, L. (2023). Autonomised harming. *Philosophical Studies*. https://link.springer.com/article/10.1007/s11098-023-01990-y

Ekelhof, M. (2019). Moving beyond semantics on autonomous weapons: Meaningful human control in operation. *Global Policy, 10*, 343–348.

France. (2016). *Non-Paper on Characterisation of a LAWS*. Presented during the CCW Meeting of Experts on LAWS in Geneva, 11–15 April 2016.

Geiss, R., & Lahmann, H. (2017). Autonomous weapons systems: A paradigm shift for the law of armed conflict? In J. D. Ohlin (Ed.), *Research handbook on remote warfare* (pp. 371–404). Elgar.

Germany. (n.d.). *German commentary on operationalising all eleven guiding principles at a national level as requested by the chair of the 2020 GGE on Emerging Technologies in the Area of Lethal Autonomous Weapons Systems (LAWS) within the CCW*.

Goodin, R. E., & Barry, C. (2014). Benefiting from the wrongdoing of others. *Journal of Applied Philosophy, 31*, 363–376.

Heyns, C. (2013). *Report of the Special Rapporteur on extrajudicial, summary or arbitrary executions* (UN Doc. A/HRC/23/47).

Heyns, C. (2016). Human rights and the use of autonomous weapons systems (AWS) during domestic law enforcement. *Human Rights Quarterly, 38*, 350–378.

Heyns, C. (2017). Autonomous weapons in armed conflict and the right to a dignified life: an African perspective. *South African Journal on Human Rights, 33*, 46–71.

HLS International Human Rights Clinic. (2020). *New weapons, proven precedent: Elements of and models for a treaty on killer robots.* Human Rights Watch and Harvard Law School.

Horowitz, M. C., & Scharre, P. (2015). Meaningful human control in weapon systems: A primer. *CNAS Working Paper.*

Human Rights Watch. (2016). *Killer robots and the concept of meaningful human control, memorandum to convention on conventional weapons (CCW) delegates.*

Human Rights Watch, & ICRC. (2014). *Advancing the debate on killer robots: 12 key arguments for a preemptive ban on fully autonomous weapons.*

Huq, A. Z. (2020). A right to a human decision. *Virginia Law Review, 106*, 611–688.

ICRC. (2018). *Ethics and autonomous weapon systems: An ethical basis for human control?* CCW/GGE.1/2018/WP.5. 29.

ICRC. (2020). International Committee of the Red Cross (ICRC) position on autonomous weapon systems: ICRC position and background paper. *International Review of the Red Cross, 102*, 1335–1349.

Krupiy, T. (2015). Of souls, spirits and ghosts: Transposing the application of the rules of targeting to lethal autonomous robots. *Melbourne Journal of International Law, 16*(1), 145–202.

Leveringhaus, A. (2016). *Ethics and autonomous weapons.* Palgrave Macmillan.

Leveringhaus, A. (2022). Morally repugnant weaponry? Ethical responses to the prospect of autonomous weapons. In *The Cambridge Handbook of Responsible Artificial Intelligence: Interdisciplinary Perspectives* (pp. 475–487). Cambridge University Press.

McDougall, C. (2019). Autonomous weapon systems and accountability: Putting the cart before the horse. *Melbourne Journal of International Law, 20*, 58–87.

Miller, D. (2001). Distributing responsibility. *Journal of Political Philosophy, 9*, 453–471.

Moyes, R. (2016). *Key elements of meaningful human control.* Article 36 Background Paper to comments prepared for the CCW Informal Meeting of Experts on Lethal Autonomous Weapons Systems.

Roff, H. M., & Moyes, R. (2016). *Meaningful human control, artificial intelligence and autonomous weapons.* Briefing paper prepared for the Informal Meeting of Experts on Lethal Autonomous Weapons Systems, UN Convention on Certain Conventional Weapons.

Roorda, M. (2015). NATO's targeting process: Ensuring human control over (and lawful use of) autonomous weapons. In A. P. Williams, & P. Scharre, *Autonomous systems: Issues for defence policymakers* (pp. 152–168). Norfolk: NATO HQ SACT.

Sassóli, M. (2014). Autonomous weapons and international humanitarian law: Advantages, open technical questions and legal issues to be clarified. *International Law Studies, 90*(1), 1.

Scharre, P. (2018). *Army of none: Autonomous weapons and the future of war.* W. W. Norton.

Schmitt, M. N., & Thurnher, J. S. (2013), "Out of the loop": Autonomous weapon systems and the law of armed conflict. *Harvard National Security Journal, 4*, 231–281.

Sharkey, A. (2019). Autonomous weapons systems, killer robots and human dignity. *Ethics and Information Technology, 21*, 75–87.

Sparrow, R. (2007). Killer robots. *Journal of Applied Philosophy, 24*(1), 62–77.

Sparrow, R. (2016). Robots and respect: Assessing the case against autonomous weapon systems. *Ethics & International Affairs, 30*(1), 93–116.

Stilz, A. (2011). Collective responsibility and the state. *Journal of Political Philosophy, 19*, 190–208.

Switzerland. (2016). *Informal working paper submitted by Switzerland: Towards a 'Compliance-Based' Approach to LAWS.* Informal meeting of experts on lethal autonomous weapons systems (LAWS), Geneva, 11–15 April 2016.

Taddeo, M., & Blanchard, A. (2022). A comparative analysis of the definitions of autonomous weapons systems. *Science and Engineering Ethics, 28*(5), 37.

United States. (2018). *Human-machine interaction in the development, deployment, and use of emerging technologies in the area of lethal autonomous weapons systems, working paper submitted to the group of governmental experts on lethal autonomous weapons of the CCW, Geneva.* UN Doc. CCW/GGE.2/2018/WP.4.

AlphaGo's Move 37 and Its Implications for AI-Supported Military Decision-Making

12

Thomas W. Simpson

INTRODUCTION

The most dramatic use-case for Artificial Intelligence (AI) in military contexts is weapons systems that have the capacity to identify targets, including human combatants, and engage them autonomously. Imagine a drone with image recognition technology, able to identify enemy combatants by their uniform and authorized to target them with a rifle without any human involvement in either target identification or engagement. Such a system meets the US Department of Defense's Directive 3000.09 definition of an 'autonomous weapon system' in that, 'once activated, [it] can select and engage targets without further intervention by an operator' (2023, p. 21), and colloquially may be described as having humans 'off the loop'. Given the severe consequences of such a system malfunctioning and killing people whom the laws of war protect from lethal force—including civilians, enemy combatants who are surrendering or are *hors de combat*—and even posing a risk to friendly forces, it is unsurprising that normative debate on the use of AI

 DOI: 10.1201/9781003410379-15

in military contexts has been focused on whether lethal Autonomous Weapon Systems (AWS) could be, or are, morally permissible and legally compliant.[1]

While these debates are right and proper, it is noteworthy that the policy debate has largely moved from whether there should be a ban on AWS, to how they should be regulated, both in terms of the principles applying to their regulation and the relevant institutional structures (e.g., Trager & Luca, 2022). Debates around the moral permissibility and legal compliance of lethal AWS suffer from a further deficit, however. Technologically developed militaries will find important uses for AI in far more varied contexts, which may pose equally severe ethical challenges, than solely for lethal AWS. While 'killer robots' are attention-grabbing, that attention has occluded proper scrutiny of other use-cases, and a serious concern that militaries should be responsible in their use of AI must address the full gamut of likely use-cases. An outline goal of this chapter is, therefore, to contribute to remedying this neglect, by considering a vital use-case— one that exemplifies the far-reaching significance of AI and which poses, I will argue, a unique challenge to the moral principles we should adopt for evaluating the proper use of AI in the military.

More specifically, the use-case I focus on is the use of AI in enabling military decision-making. In military headquarters, 'command and control' is exercised over subordinate units, directing their activities. AI is likely to play a role throughout the decision-making processes by which command and control are exercised, generating recommended courses of action, which commanders will determine whether to reject, modify, or adopt. At first glance, the use of AI in military decision-making is not as controversial as its use in 'killer robots'. The AI would not be hunting anyone, giving rise to Terminator-style fears. Moreover, humans would be not just 'on' the loop, but 'in' it, and to all intents and purposes essentially so. (As I shall use the terms, humans are 'in' the loop when humans are actively engaged in both the target acquisition and engagement processes of a combined socio-technical system, with the system requiring explicit human authorization before engaging a target, and humans are 'on' the loop when a human monitors the operation of a system, which may itself be autonomous or semi-autonomous, with the human able to intervene to stop or change the action that the system would otherwise take. The US Department of Defense terms the latter 'operator-supervised' systems (2023, pp. 22–23). I use the short-hand designations of 'off', 'on', and 'in' to refer to systems with the properties as stated here.) ChatGPT may help staff write the voluminous operational orders, but no commander will abdicate to a machine their prerogative to issue orders, not least on pain of redundancy.

While this lack of controversy is indubitable, I will argue that this belies the moral reality. A large part of the advantage that AI is likely to realize for militaries on the battlefield depends on its capacity for what I call 'unpredictable brilliance', and it is the central problem of this chapter. To illustrate this problem, I start with a lesson from DeepMind's AlphaGo programme and apply it to the military. Then I turn to evaluate its significance. I identify the risks that AI-enabled decision-support systems may pose to militaries, preparatory to showing how these systems will redistribute responsibility within militaries. Much discussion of lethal AWS has already focused on the challenge of allocating responsibility; I show how the problem of unpredictable brilliance deepens this challenge, so that human-in-the-loop systems may require the allocation of responsibility to diverse responsibility-holders, just as much as human-off-the-loop systems do, and that the phenomenon of 'blameworthiness gaps' arises for them too. I close by considering an objection to my account, deriving from bioethics discussions of the role of AI-enabled diagnosis and treatment.

ALPHAGO'S MOVE 37

In 1997, IBM's Deep Blue became the first computer programme to beat a chess grandmaster, Gary Kasparov, and with subsequent improvements in computing power, computers now standardly beat human opponents (Kasparov, 2010). A harder challenge was therefore sought, as part of the long development of AI, and it was found in the game Go. While both are strategy games, the rules of Go are simpler than those of chess. One player has a bowl of black stones, and another of white, and they take it in turns to place them on a board. Stones of one color which are surrounded by those of the other are removed from the board, with the goal being for a player to surround a larger area, in total, than their opponent.

Although simpler than chess in respect of its rules, Go is nonetheless considerably more challenging in terms of what is required to analyze a position computationally and recommend a move. The complexity of a game like Go and Chess, where all the possible moves can be identified, can be defined in two terms, namely those of 'breadth' and 'depth'. From any single game position, a player must choose one move from a limited set of options, with the set of options its breadth. And, from a given position, a game will on average last a certain number of moves; this is its depth. A game's breadth and depth from a given position then expresses the number of branches down which the game could develop. On average, a typical chess game has a breadth of 35 moves and a depth of 80, meaning the number of possible moves that a computer might calculate, when evaluating which move will maximize the likelihood of victory, is vast, at 35^{80} (or 10^{123}). This is a huge number, exceeding the number of atoms in the universe (about 10^{80}), but still much, much smaller than Go. With Go played on a 19 by 19 board, and having a breadth of 250 and depth of 150, there are about 250^{150}, or 10^{360} moves. 'This is a number beyond imagination and renders any thought of exhaustively evaluating all possible moves utterly and completely unrealistic' (Koch, 2016; this is also the source for this quantitative comparison).

The problem of analyzing Go was, nonetheless, solved by the programme AlphaGo, the product of the company DeepMind. Using a combination of neural network algorithms and Monte Carlo tree search techniques (Silver et al., 2016), and with AlphaGo playing against itself repeatedly, the programme was able by 2015 to beat a professional Go player who was also the reigning European Go champion, Fan Hui. A series of five, highly publicized matches was subsequently set up in 2016, with a $1 million USD prize at stake, in which AlphaGo would compete against Lee Sedol, widely recognized as one of the best Go players in the world at the time. With AlphaGo having won the opening match, arguably the pivotal moment in the series occurred in the second game, in move 37, in which AlphaGo plays a 'fifth line shoulder hit'. The commentators' immediate reactions at the time were variously of shock and surprise. (The following reactions are cited from Kohs, 2017.) "Oh, totally unthinkable move." "That's a… That's a very surprising move. Coming on top of a fourth line stone is really unusual. I wasn't expecting that." Fan Hui, who was acting as a judge for the Sedol-AlphaGo matches, commented afterward, "When I see this move, for me it's just a big shock. What? Normally humans, we never play this—because it's bad. It's just bad; we don't know why; it's just bad." Another commentator, explaining the move, remarked, "It's the fifth line. Normally you don't make a shoulder hit on the fifth line." In short, among those watching at the time, who had the expertise required for interpreting the game to a general audience and responsibility for doing so, there was considerable confusion and indeed shock at the move.

AlphaGo's analysis of the game situation suggested that the commentators' surprise at move 37 was well-founded. According to AlphaGo's own assessment, there was a 1-in-10,000 chance that a human player would have made this move. Lee Sedol's own assessment of the move agreed that it was surprising, but added something further:

> I thought AlphaGo was based on probability calculation and that it was merely a machine. But when I saw this move, I changed my mind. Surely AlphaGo is creative. This move was really creative and beautiful. This move made me think about Go in a new light. What does creativity mean in Go? It was a really meaningful move.

Although the game was perceived by some as remaining balanced between the two players for a significant period after, subsequent commentary suggested that move 37 was the crucial move that ultimately won the second game for AlphaGo. Fan Hui, again, remarked, "This move was very special, because with this move, all the stones played before worked together. It was connected. It looked like a network, linked everywhere. It was very special." AlphaGo went on to win the series by four games to one. (The prize money was donated to charity by the DeepMind team). AlphaGo Master, a subsequent iteration of AlphaGo, went on to beat the acknowledged world No. 1, Ke Jie, in 2017. Also in 2017, DeepMind reported the development of AlphaGo Zero, a further iteration of the programme, which beat the original AlphaGo 100-0 (Silver et al., 2017).

How is AlphaGo able to make such surprising decisions that its moves are described as 'creative' and 'beautiful'? Two points are, I think, noteworthy. First, I noted above that AlphaGo was able to play itself repeatedly, and thereby learn what moves are likely to increase or decrease the chance of winning. But this understates the power of machine learning techniques. For AlphaGo, this consisted of playing against itself 'thousands of times', using reinforcement learning to build on the heuristics identified in a training data-set drawn from 30 million human games (Scharre, 2018, p. 125; DeepMind, 2023). AlphaGo Zero eschewed the training data and simply played itself, 4.9 million times (Silver et al., 2017, p. 355). In both cases, AlphaGo was able to experiment, at a speed and therefore on a scale vastly beyond that accessible to the individual human player, with different permutations of moves, discarding those which proved not to conduce to victory, but preserving the lessons from those which were—with some of these being types of move that humans had discounted as unwise.

Second, while AlphaGo tries to maximize its probability of winning, it does not care about the margin by which it wins. Go permits degrees of victory, in which one player may win by more or less, according to how much territory each player's stones control at the end of the game. Faced with the choice between an 89% chance of a substantial victory, or a 90% chance of marginal victory, AlphaGo will choose the higher-likelihood-but-marginal one. This 'willingness' to win by a small margin on AlphaGo's part is likely to contribute to its capacity to make seemingly surprising moves, with humans' preference to win by a substantial margin partly based on psychological preferences (the appeal of crushing one's opponent), and partly because the predicted degree of victory is a useful heuristic for estimating the likelihood of victory. But heuristics often fail, and AlphaGo seems able to identify those occasions when it is not reliable, and so make surprising recommendations. Of course, optimizing AlphaGo to maximize the odds of victory, rather than some weighted goal that takes account of both the odds of victory and the degree of victory, is a design decision. In the context of the game Go, the design decision is not an especially significant one, while in an unbounded, real-life context, the decision about what a machine learning algorithm should optimize on is likely to be highly consequential.

In an initial summary, then, we can describe AlphaGo's move 37 as possessing the property of 'unpredictable brilliance'. It was brilliant because it proved to be decisive, in the context of the game in which it was played, unbalancing the programme's opponent, and setting AlphaGo on a path to victory. It was unpredictable not only because the odds of a human making that same move were astonishingly slim—although they were—but also because even those watching at the time did not realize that the move would have the decisive effects on the game that it proved to and which, from AlphaGo's perspective, seeking to optimize the chances of victory, were welcome. Quite simply, it was a move that those human observers would not have chosen to have played, likely even if it had been recommended to them by an AI.

The phenomenon of unpredictable brilliance is observed in other forms of AI. While I have drawn the above portrayal of AlphaGo's move 37 from a single source, the *AlphaGo* documentary (Kohs, 2017), Paul Scharre documents other instances of the same phenomenon, to substantiate what he describes as AI's 'alien' and 'inhuman' form of cognition (2023a). The AI programme Libratus routinely beats humans in poker, and employs a different strategy, using betting tactics 'like limping and donk betting that are generally considered poor tactics, but it is able to execute them effectively because of a more fine-grained understanding of the game's probabilities' (Scharre, 2023a, p. 266). The same is true of AlphaZero's chess style, where it will even sacrifice a queen for positional advantage that pays off over the game.

I have suggested that AI's 'alien' cognition is likely to stem, in part, from the vastly greater dataset of possible gameplays than it can access which an individual human cannot, and from the fact that it may be programmed to optimize on a single outcome. But this is unlikely to be the major reason for the 'alien' nature of its cognitions, with other sources likely to be contributory as well. The efficiency gains that an AI can realize may be of such a degree that it makes decisions that feel qualitatively distinct. For instance, the company OpenAI has a programme, OpenAI Five, able to control agents in the *Dota 2* multiplayer online battle arena (MOBA). MOBA games are more complex again than games like Go, not only with players having vastly greater numbers of options open to them at any point, but also with vastly more players, playing for longer, so yielding much larger breadth and depth. In addition, information is asymmetric, with some information hidden from other players (unlike Go or chess). The teams of virtual agents controlled by OpenAI Five are—by now, perhaps unsurprisingly—able to beat teams of professional humans, being able better to coordinate their attacks, and to do so with more speed and precision, so overwhelming opposing teams. The bots are reported to play 'with unusual aggressiveness' relative to human players (Scharre, 2023a, p. 269). In an explicitly defense-related context, a series of simulated aerial dogfights between AI-controlled and human-controlled fighter jets resulted in a 5-0 win for the AI. The AI demonstrated a strong preference for 'forward-quarter gunshots', in which two aircraft fly directly toward each other. Human pilots invariably seek a 'rear-quarter' shot—i.e., from behind their opponent—largely because the alternative is incredibly dangerous to the pilot (and therefore forbidden in training), and also incredibly difficult to pull off, because in the split-second window of opportunity that exists to hit the target, the pilot's priority is invariably to avoid a crash (Scharre, 2023a, pp. 2–3). While the AI could pull off the forward-quarter shot, in the simulation at least, the human could not. (It is a further question how well performance in the simulation predicts performance in actual air-to-air combat.) The efficiency gains that an AI can realize can come from multiple sources: with no cognitive processing limitations, an AI is likely to possess greater situational awareness, a greater ability to manage its own resources, and an ability to carry out multiple tasks in parallel. It will also be risk neutral, exhibiting neither the natural risk aversion of military personnel who wish not to die, nor the risk-seeking behavior that is an indubitable feature of war, especially in elite units, which is normally functionally valuable in overcoming risk aversion, but which can lead personnel to make foolhardy decisions.

While efficiency gains, and access to vastly greater data-sets, are two sources of the counter-intuitive nature of AI cognition, it becomes truly 'alien' when it is the result of algorithms that are themselves developed by machine-learning techniques, to the point when they become, in effect, black boxes which are not inspectable by humans.

AI's 'alien' cognition is likely to be exhibited wherever it is found, and this includes one of the most significant use-cases for AI in the military, namely as an aid to decision-making. Part of the battlefield advantage that AI can offer will consist not just in individual platforms that can outcompete the equivalent human-controlled adversaries, as in a contest between an autonomous air-superiority fighter and one piloted by a human, but also in decisions about when and how to deploy which forces in such a way that one side can compel another to submit. In military headquarters, at levels spanning the tactical, operational, and strategic (i.e., from the battle group, in the land domain, or the ship, in the maritime, up to national or multi-national theatre commands), large staffs evaluate information which may reveal what the enemy is doing, in order to generate intelligence; maintain an awareness of the location and state of friendly forces; generate recommendations to commanders on possible courses of action; and once decisions are taken, then issue orders to subordinate units and track their implementation. All of this is summarized as the exercise of command and control. AI is likely to play a role throughout all elements of this decision-making process, and a transformative role in two parts in particular: namely, in generating intelligence, and in generating recommendations to commanders on possible courses of action, both for deliberate, pre-planned actions and in time-sensitive contexts. For present purposes, it is the latter which is of most interest.

Possessing far greater situational awareness, with an AI-enabled intelligence picture that is less susceptible to confirmation bias, AI-enabled decision-support systems are likely to give recommendations that, in some instances, confirm what a human commander would be likely to choose or modestly improve it, but in some instances, will offer highly surprising and counter-intuitive recommended courses of action (COAs). Just as much as individual platforms, like autonomous fighter jets, AI-enabled decision-support systems are likely to exhibit the property of unpredictable brilliance, with COAs that sometimes seem to be simply a waste of resources, and at other times also dangerous. And yet, this is likely to be precisely where the most significant advantage lies, for those forces that can field powerful AI-enabled decision-support systems, and effectively integrate their insights into their command and control. Decision-making which is both counter-intuitive—indeed, creative—and highly rational, being based on a sophisticated understanding of the probable outcomes from a range of possible actions, is likely to be able to exploit any errors an enemy has made in its own force disposition, and to unbalance the adversary. As Scharre writes, in summary, 'The militaries that will be most successful in harnessing AI's advantages will be those that effectively understand and employ its unique and often alien forms of cognition' (Scharre, 2023b; also Scharre, 2023a, p. 273).

RISKS POSED BY UNPREDICTABLE BRILLIANCE

Unpredictable brilliance is not just a property of AI cognition, but also a problem. The next section considers the problem that it poses in moral terms, but I start here with the problem that it poses for commanders. For commanders, the problem can be easily stated, at least in outline terms: to what extent should the AI-generated recommendation be trusted? In the context of AlphaGo's games against Lee Sedol, this problem did not arise, in part because the real-world stakes could effectively be discounted (the prize money was a trivial amount for a company like Alpha, Google's parent company and the owner of DeepMind, which had probably put up the money for publicity-related reasons), but largely because the game was set up so that the AI's recommendations determined what move was played. While AlphaGo was purely a piece of software and relied on a human player to place a black or white stone on the physical board, that person nonetheless had no more role other than enacting AlphaGo's decision as it was represented on a display screen. The series was an experiment, and trust played no part in it. But in the military context, the stakes would be substantial, and the human commander would not be required simply to follow an AI-enabled decision-support system's recommendation. Instead, the AI would generate recommended COAs, not instructions, and the commander would ultimately be accountable for the consequences of her decisions. She therefore faces a tricky decision: given an AI-recommended COA that has obvious risks, and for which the benefits are not easily discernible, should she follow the recommendation, or her own instincts and judgment? In effect, whom should she trust—the AI, or herself? Doing the latter means that she would be able to defend her decision, by the lights of accepted canons of military decision-making wisdom. But the consequence is that she would forego the prospect of a decisive battlefield advantage held out by the AI-enabled support.

Trust is necessary in part because, in the military context, there is significant risk. There are two forms of risk that are especially noteworthy, both of which arise because an AI-recommended COA could be unpredictable, or surprising, due to how it would allocate military resources on the battlefield. In more concrete terms, the COA may be surprising because it recommends deploying a type of unit that is unusual in a given context—for instance, it might recommend sending a main battle tank where normally dismounted infantry would go. (What would its reason for doing so be? Perhaps the AI-generated intelligence picture indicates an extremely low likelihood of dismounted 'red' infantry in the immediate urban environment, but the combined speed and firepower of the tank will be decisive in dislodging the enemy from the fixed positions on the other side of the town, which the enemy is in the process of preparing now, and which positions seek to deny a crucial bridge to 'blue'. But the assessments and probabilities taken account of in the AI's decision-making process may be hidden from the commander.) Or, its recommendation may be surprising because it would result in a significant amount of combat power being invested in a given location, which the commander and her staff assess as being unimportant, but which the AI assesses as being key terrain. Given that military commanders' decisions, ultimately, address the allocation of military resources, which equates to combat power, across the battlefield, the property of unpredictable brilliance would be realized through surprising, counter-intuitive recommendations about where that resource should be placed. One form of risk that a commander would have to accept, in following the AI's recommendation against her own judgment, is that which the decision would pose to the overall likelihood of achieving her campaign objectives. Whether this is at the tactical, operational, or strategic levels, in a world of scarce resources, if military assets are misallocated, the relevant objectives are less likely to be achieved. This is a risk that she should be highly sensitive to.

Another form of risk is that which the decision poses to the lives and safety of blue force personnel. This is especially obvious when an AI-generated recommendation would see personnel or crewed platforms deployed in ways where they would be significantly more vulnerable than would be the case for personnel deployed in the range of decisions that would be likely for a human commander. As implicitly suggested above, main battle tanks and other armored vehicles are particularly vulnerable to dismounted infantry in 'close' country—forests, mountains, and urban environments in particular. Conversely, dismounted infantry are particularly vulnerable in 'open' countries, such as plains and deserts. Deploying one form of combat power in terrain that is widely considered more suitable for the other is likely to lead to significantly greater casualties, at least according to widely accepted military heuristics. (A significant proportion of Russia's casualties during the first year of its invasion of Ukraine in 2022 is explained by poor decisions about how to deploy armored vehicles, which failed to follow the heuristic noted above.) Similar points apply to aerial and maritime platforms, which have differing combinations of firepower, mobility, and resilience, and therefore differ in how they are usually used effectively.

Although it is plain that an AI-generated recommendation would pose high levels of risk if personnel were ordered to go into situations that standard tactics, techniques, and procedures (TTPs) indicate would be unduly dangerous for them—being evident, not least, to those personnel themselves—the point is actually a more general one. Any situation in which an AI-generated recommendation is significantly discrepant from the range of likely decisions for a human commander involves a redistribution of risk, in which some people have a higher level of risk imposed on them. Where the baseline for comparison is the likely range of decisions that a human commander would make, and the AI's recommended COA is outside of this, the AI's COA will result in a set of some blue force personnel carrying less risk—perhaps because they are now part of a larger formation, or there is a larger reserve held back, and so on—but almost inevitably at the cost of an increase of risk for a set of some other blue force personnel. Even though the overall amount of risk is not fixed (because some COAs pose a greater level of risk overall, while others are less risky), nonetheless, one person's gain, in terms of risk reduction, is likely to be due to some other person's loss, in terms of increased risk. While that risk redistribution may be justified, perhaps in terms of its impact on the overall risk to blue force personnel, or its increase in the odds of mission success, nonetheless it will have occurred, even though that risk redistribution may also be invisible to those personnel who are affected. By hypothesis, when an AI's recommended COA is unpredictable or surprising, that COA is assessed by humans as riskier, in terms of its probabilities of either or both of mission success or blue force casualties, than the likely range of human-recommended COAs. (If it was not so assessed, it would be within the likely range of human COAs). What is unknown *ex ante*, and indeed, is unknowable, is whether the AI's recommended COA imposes, objectively, less risk to blue force personnel overall, and has a higher probability of mission success, than the likely range of COAs that a human would adopt, or whether the contrary is the case.[2]

I have claimed that trust would be necessary, when a military commander considers whether to adopt a COA recommended by an AI-enabled decision-support system, in part because there is significant risk. It may be objected, however, that this overstates the case. Although risk is inevitably involved in any COA adopted on the battlefield, the commander's trust of an AI-enabled decision-support system would be greatly minimized because she would not be interacting with the system from the beginning, but rather would have trained with it on multiple occasions. She would have a track record of its performance, enabling her to evaluate how much it should be trusted, under what conditions, and so on. This is of course correct; it would be highly unprofessional for a military to deploy an untested piece of equipment on an operation. Nonetheless, while this may mitigate the need for trust, it does not eliminate it, and the fundamental choice for a commander of trusting her own instincts, versus the recommendation(s) of an AI, is likely to remain. However much training has been conducted, a decision-support system is different from more conventional kinds of military equipment, such as platforms or missiles, in that its utility essentially turns on how accurately it represents and predicts an adversary's actions. Operations are qualitatively distinct from training because, in training, you actually fight against your actual enemy. Only a 'two-way range' permits that. As it is sometimes said, on operations, the enemy has a vote. Nor is this need for trust restricted to the start of operations on a given campaign. Rather, the need for trust will be recurring. War is one of the fastest drivers of human innovation, so as the red force adapts its TTPs and equipment to respond to blue force's TTPs and equipment, the COAs that will successfully unbalance and exploit red force weaknesses will be correspondingly new and, in some instances, counter-intuitive.

UNPREDICTABLE BRILLIANCE ERODES THE MORAL SIGNIFICANCE OF THE DISTINCTION BETWEEN HUMANS *OFF* AND *IN* THE LOOP

As noted earlier, much and perhaps most of the ethical debate about the use of AI in military contexts has revolved around lethal AWS, or 'killer robots', in which humans are out of the loop. The intuitive thought is that fully autonomous systems pose unique challenges, as contrasted with systems in which humans are on or in the loop. A central reason why one may be concerned about lethal AWS is that they seem to raise the prospect of killings in war, especially of those not liable to be killed, such as non-combatants or surrendering soldiers, for which no-one is clearly responsible. If humans are in the loop or on it, they can be held responsible; if humans are off the loop, it is less clear, and some have argued that in principle no-one is (Matthias, 2004; Sparrow, 2007).

While the distinction between autonomous systems and those which lack full autonomy is clear enough, its moral import is less significant than it seems. One of the tactical advantages of AI-enabled weapons systems is the speed at which they can operate. It is very plausible that systems might be developed in which humans remain in or on the loop, thus retaining the seeming moral advantages of having an identifiable person who can be held responsible for each targeting decision, while putting that person in a situation in which, given that the tactical situation is evolving at such speed, she has in effect no choice but to authorize the decision recommended made by the AI. A system that might satisfy this description is one which is designed to provide point defense against swarms of kamikaze drones, which the Phalanx Close-In Weapon System (CIWS) provides a model for. Mounted on ships to defend against incoming missiles by putting up a 'wall of lead', a human operator either approves fire recommended by the Phalanx, or the system can operate 'weapons free', engaging targets from 1.5 to 5.5 km away, with the operator monitoring the relevant conditions. Suppose the range was much tighter and the number of threats much greater, so that a future version of such a system was capable of and needed to engage 1,000 targets within five seconds from a range of 0 to 500 m, while a human operator was monitoring how the system operated. We may even suppose, in this counterfactual scenario, that the operator's express approval was continuously required—she must hold down a button for the system to strike incoming targets. The operator would be in the loop. But it is highly dubious that she would be responsible for those strikes, in the way required to hold her individually liable for any mistakes. *De facto* and *de jure* responsibility would diverge.

Putting soldiers, sailors, or aircrew in this kind of situation could be useful for militaries, in terms of how legal liabilities are distributed, by allowing the military to hold an individual liable for illegal or reputationally damaging targeting decisions, and so avoiding corporate responsibility. But it would be exploitative, as the soldier would lack the time or situational awareness to make an informed decision. Decision-making which is necessarily done at extremely high speed, then, and in reliance on system-generated recommendations, erodes the moral import of the distinction between humans off and in the loop. While the distinction provides a useful heuristic to distinguish between situations in which a human is individually responsible for specific targeting decisions and those in which someone is not, it is not determinative. There are cases where humans in the loop can nonetheless lack individual responsibility.

The phenomenon of AI's unpredictable brilliance, I contend, has the same erosive effect. The speed at which a decision may necessarily have to be made is one factor that can have the effect of removing responsibility from a human 'decision-maker', the scare quotes indicating that it is questionable to what extent the human is, indeed, the decision-maker. She retains the freedom, overall, to substitute her own judgment for that of the AI, but the consequences of doing so are likely to be severe. In the case of AI support to commanders in tactical, operational, and strategic headquarters, the same erosion of responsibility occurs. Faced with an unpredictable or surprising COA recommended by an AI, the military commander retains a decision as to whether to follow the AI's recommendation or to follow her own. All she has to guide her in that decision is whatever information she possesses about the reliability, in general, of this AI's recommended COAs—there is no more granular level at which she can assess this proposal because, by hypothesis, the proposal runs counter to the accepted principles of military decision-making. Further, determining the overall reliability of an AI is not, primarily, her responsibility, but the responsibility of the force development and procurement programmes that brought the system into service and approved it for operational deployment. So long as she uses it in a way that is compatible with the directions on appropriate use, it is by no means clear to what extent she can be held individually responsible for a decision, the consequences of which turn out to be bad.

What follows from this? It does not follow, at least not straightforwardly, that there is an in-principle, moral objection to the use of such AI-enabled decision-support systems. In other work, I have argued that, subject to some constraints being satisfied, lethal AWS could be permissibly deployed (Simpson & Müller, 2016). The crux of the issue is not whether someone identifiable is responsible for each killing by a lethal AWS. There are multiple roles the occupants of which may be responsible for such killings, including those who design and maintain lethal AWS, those commanders who deploy them on a given occasion, and most especially, those who authorize the use of a given system and regulate its use. And it is always the case that someone—either an individual or a corporate body—is responsible. But responsibility does not imply blameworthiness. It is wholly possible that there could be some killings by lethal AWS, which in technological terms would be classified as malfunctions and in human terms as tragedies, in which someone is killed who is not liable to be killed, like a non-combatant, and yet for which no-one is blameworthy. This is because the use of lethal AWS may satisfy the demands of 'wide proportionality', in which harms to those who are innocent are weighed against an act's expected good effects (McMahan, 2020, p. 15). Whether the use of a system satisfies the demands of wide proportionality may occur on occasions in which a non-combatant's predicted death is a proportional cost for a given strike on a valuable military target. Although that point is not controversial, it is also noteworthy that a system may satisfy those demands and yet impose the risk of harm to the innocent in a wider range of situations, such as a non-combatant being mistakenly identified by an AWS as a combatant. For lethal AWS to be used proportionately, however, it is not sufficient simply that the absolute level of risk of humans should be reduced. Rather, we should also be sensitive to how the use of lethal AWS would redistribute risk. In particular, it is not morally acceptable for AWS to reduce risk to blue force personnel but at the cost of increasing it to people who are not liable to be killed. It is a fundamental constraint on the just use of lethal AWS that the risk they pose to non-combatants should be lower than would be posed by a military not equipped with lethal AWS and, further, that the risk should be as low as is technologically feasible (Simpson & Müller, 2016).

This account of how responsibility and blame are allocated applies not only to lethal AWS, but also the use-case of concern here, of decision-support systems. Suppose that a surprising, counter-intuitive COA is recommended by an AI. Even though a human would be in the loop, it does not follow that that commander carries individual responsibility and blame for adopting that recommended COA. The surprising and counter-intuitive features of AI cognition erode her responsibility. So long as she is integrating the system into her command and control in accordance with the principles and guidance she has received, the commander is absolved from blame for the consequences. Those on whom responsibility principally rests, and on whom blame is most likely to fall, if it does, are those individuals or that corporate body which has authorized the use of the AI-enabled system and regulates its use. And it is possible that no-one should be blameworthy, even if the system results in harmful risks eventuating, so long as the demands of wide proportionality are met. Positively, then, what follows from AI's property of unpredictable brilliance is that there will be a more widespread distribution of responsibility, and liability to blame, within the military, away from commanders, and toward the institutions (both individual office-holders and as corporate bodies) that authorize and regulate the systems by which wars are likely to be fought in future, included in which are AI-enabled decision-support systems. The widespread distribution of responsibility is already a feature of the military, with established procedures and institutions responsible for testing, evaluating, validating, and verifying new equipment and TTPs. My contention is that, with the introduction of AI-enabled decision-support systems, the widespread nature of this distribution will become more accentuated still. At present, commanders retain a core responsibility for 'J3' decisions when deployed—those decisions that address directly how operations should be conducted. Yet even this is likely to be restricted. As a parallel, and in a use-case I consider further below, note the likely implications if AI-enabled diagnoses of illness and recommendations for treatment achieve better health outcomes than medical doctors. Doctors' responsibility then would be to assure themselves that the AI-enabled decision-support system is being used in a context in which it is appropriate, but once done, the individual's responsibility would be simply to adopt the AI-generated recommendation. Similarly, once a military commander has confirmed that the context in which she is operating is suitable for the decision-support system, little will remain for her to do other than accept the AI's recommendation. Her command responsibility will have been redistributed, to the procedures and institutions responsible for developing and maintaining the AI.

To make the point plain that, even with a human in the loop, a commander can nonetheless be absolved of blame, I have started with scenarios in which an AI makes a surprising, unpredictable recommendation. This is an expository device, however, to support a more general and perhaps less intuitive conclusion. Operationally deploying an AI-enabled decision-support system would have the effect of absolving the human commander from blame not only in situations in which the AI's recommended COA was outside the range recommended by humans, but also in situations in which it was within. The effect of authorizing a system for operational deployment, with responsibility for its reliability resting with those performing the authorization, is to remove blameworthiness as such for its decisions from commanders. Not only would the commander be absolved of blame, but as a corollary, she would also be absolved from praise.

The likely effects of this on how militaries function, in sociological and organizational terms, should not be underestimated. For centuries militaries have valorized and praised commanders who show an instinctive understanding of the battlefield, often disdaining personal risk, and have led their soldiers to great victories—in Europe, think of Alexander the Great, Napoleon, Nelson, Rommel, and Patton. As and when AI becomes not just embedded in the decision-making process, but the central driver of it, we will enter a post-heroic era of warfare, where leadership may be exercised at the lowest tactical levels, at which soldiers must still face risk and overcome their fear, but it will not be exercised at the operational level. The social dislocation will be profound.

AI DECISION-SUPPORT IN HEALTHCARE: AN OBJECTION

I conclude with an objection. Despite my foregoing account of how responsibility and blame are allocated for AI-enabled systems, could there nonetheless be an in-principle, moral objection to their use? Consider the related context of healthcare, in which AI-enabled decision-support systems can advise a doctor on possible diagnoses and treatments for a patient—a domain where, similarly, lives depend on the quality of decision-making. Examples include IBM's 'Watson for Oncology' and Aidoc Medical's triage and notification systems. A number of writers in bioethics have pointed to other moral risks associated with AI-enabled medical decision-support systems, in addition to the redistribution of responsibility that they impose (Grote & Berens, 2020, p. 209), even though such systems may achieve better results than a human doctor. It is claimed that they undermine trust in the doctor–patient relationship (Hatherley, 2020) and may conflict with core ideals of patient-centered medicine, undermining patients' autonomy, resulting in treatment no longer being the outcome of shared deliberation between both parties (Bjerring & Busch, 2021; Lorenzini et al., forthcoming; McDougall, 2019). Although these authors have differing accounts of the significance of these moral risks, their shared point is that while effective treatment is one valued feature of healthcare, it is not the only one and that these valued goals may trade-off against each other.

These moral risks of AI-enabled medical decision-support undoubtedly exist. In the context of healthcare, it is also plausible that these risks could be the basis for principled limits on the use of AI-enabled decision-support systems. But this derives from a core feature of medical practice, namely the consent of those exposed to risk, that does not have the same significance in the military situation. In healthcare, patients face some harm, in the form of injury or illness, and they undergo treatment with its attendant risks to improve their welfare. If a patient ignores an AI's diagnosis and recommended treatment because they wish to know the basis of that assessment, so accepting a risk of worse outcomes for the sake of preserving their autonomy, that is their trade-off to make. But in the military context, while consent is exercised when one joins up (at least in professional militaries), it is not relevant when deployed. Subordinates are under the authority of their superiors, duty-bound to follow orders, and part of what they have consented to when joining up is a liability to assume high levels of personal risk in the course of carrying out operations. In turn, superiors have a duty of care to ensure that the personal risk their subordinates are exposed to is only that which is necessary for the military, collectively, to achieve its campaign objectives. In the military context, then, commanders are not free to trade-off better campaign outcomes against 'softer' values, such as ensuring that the basis of a decision can be explained. Commanders' core goal is to achieve their campaign objective, and in doing so they are to minimize the risk to their personnel, and civilians. Insofar as an AI's recommended COA will enable them to do so better, and this is known within their military, commanders have a duty to comply with that recommendation.[3]

NOTES

1 While it is widely believed that weapons systems designed to kill and which are capable of operating autonomously exist, it is not publicly known whether AWS have in fact been used to kill a human autonomously. Robert Trager and Laura Luca state that "at least Israel, Russia, South Korea, and Turkey have reportedly deployed weapons with autonomous capabilities" (Trager & Luca, 2022). A UN report stated, in passing, that the Turkish STM Kargu-2 was used in Libya while 'programmed to attack targets without data connectivity between the operator and the munition: in effect, a true "fire, forget and find" capability', but this capability was denied by the manufacturer (United Nations Security Council, 2021, §63; Bajak & Arhirova, 2023). It is likely that the first uses of lethal AWS will not be publicly known about at the time, and likely for some time after, as there is no easy way to determine after a strike whether, for instance, a loitering munition, such as the Switchblade 'kamikaze' drone operated by the US or the Zala Lancet operated by Russia, was used with a human in the loop, or off it.

2 To make things even more complex, not only is the objective risk posed by an AI's COA unknowable *ex ante*, compared to the range of likely human

COAs, it is also unknowable *ex post*. That COA 1 was adopted means that COA 2 was not, and the alternative reality in which COA 2 was adopted is unavailable.

3 I am grateful to Jan Maarten Schraagen for comments and criticism.

REFERENCES

Bajak, F., & Arhirova, H. (2023, January 3). Drone advances in Ukraine could bring dawn of killer robots. *Los Angeles Times*. https://www.latimes.com/world-nation/story/2023-01-03/drone-advances-in-ukraine-dawn-of-killer-robots

Bjerring, J. C., & Busch, J. (2021). Artificial intelligence and patient-centred decision-making. *Philosophy & Technology, 34*, 349–371. https://link.springer.com/article/10.1007/s13347-019-00391-6

DeepMind. (2023, June 26). *AlphaGo*. deepmind. https://www.deepmind.com/research/highlighted-research/alphago

Grote, T., & Berens, P. (2020). On the ethics of algorithmic decision-making in healthcare. *Journal of Medical Ethics, 46*, 205–211. https://jme.bmj.com/content/46/3/205.abstract

Hatherley, J. J. (2020). Limits of trust in medical AI. *Journal of Medical Ethics, 46*, 478–481. https://jme.bmj.com/content/46/7/478

Kasparov, G. (2010, February 11). The chess master and the computer. *The New York Review*. https://www.nybooks.com/articles/2010/02/11/the-chess-master-and-the-computer/

Koch, C. (2016, March 19). *How the computer beat the Go master*. Scientific American. https://www.scientificamerican.com/article/how-the-computer-beat-the-go-master/

Kohs, G. (Director). (2017). AlphaGo [Film]. Moxie Pictures; Reel as Dirt.

Lorenzini, G., Ossa, L. A., Shaw, D. M., & Elger, B. S. (Forthcoming). Artificial intelligence and the doctor-patient relationship expanding the paradigm of shared decision-making. *Bioethics*. https://onlinelibrary.wiley.com/doi/full/10.1111/bioe.13158

Matthias, A. (2004). The responsibility gap: Ascribing responsibility for the actions of learning automata. *Ethics and Information Technology, 6*(3), 175–183. https://link.springer.com/article/10.1007/s10676-004-3422-1

McDougall, R. J. (2019). Computer knows best? The need for value-flexibility in medical AI. *Journal of Medical Ethics, 45*, 156–160. https://jme.bmj.com/content/45/3/156.abstract

McMahan, J. (2020). Necessity and proportionality in morality and law. In C. Kreß & R. Lawless (Eds.), *Necessity and proportionality in international peace and security law* (pp. 3–38). Oxford University Press. https://academic.oup.com/book/33456/chapter/287728709?login=true

Scharre, P. (2018). *Army of none: Autonomous weapons and the future of war*. New York: W. W. Norton

Scharre, P. (2023a). *Four battlegrounds: Power in the age of artificial intelligence*. New York: W. W. Norton

Scharre, P. (2023b, April 10). *AI's inhuman advantage*. War on the Rocks. https://warontherocks.com/2023/04/ais-inhuman-advantage/

Silver, D., Huang, A., Maddison, C. J., Guez, A., Sifre, L., Driessche, G. van den, Schrittwieser, J., Antonoglou, I., Panneershelvam, V., Lanctot, M., Dieleman, S., Grewe, D, Nham, J., Kalchbrenner, N., Sutskever, I., Lillicrap, T., Leach, M., Kavukcuoglu, K., Graepel, T., & Hassabis, D. (2016). Mastering the game of Go with deep neural networks and tree search. *Nature, 529*, 484–489. https://doi.org/10.1038/nature16961

Silver, D., Schrittwieser, J., Simonyan, K., Antonoglou, I., Huang, A., Guez, A., Hubert, T., Baker, L., Lai, M., Bolton, A., Chen, Y., Lillicrap, T., Hui, F., Sifre, L., Driessche, G. van den, Graepel, T., & Hassabis, D. (2017). Mastering the game of Go without human knowledge. *Nature, 550*, 354–359. https://doi.org/10.1038/nature24270

Simpson, T. W., & Müller, V. C. (2016). Just war and robots' killings. *Philosophical Quarterly, 66*(263), 302–322. https://doi.org/10.1093/pq/pqv075

Sparrow, R. (2007). Killer robots. *Journal of Applied Philosophy, 24*(1), 62–77. https://onlinelibrary.wiley.com/doi/abs/10.1111/j.1468-5930.2007.00346.x

Trager, R., & Luca, L. (2022, May 11). Killer robots are here - and we need to regulate them. *Foreign Policy.* https://foreignpolicy.com/2022/05/11/killer-robots-lethal-autonomous-weapons-systems-ukraine-libya-regulation/

United Nations Security Council. (2021). *Final report of the panel of experts on Libya established pursuant to Security Council resolution 1973 (2011).* https://documents-dds-ny.un.org/doc/UNDOC/GEN/N21/037/72/PDF/N2103772.pdf?OpenElement

US Department of Defense. 2023. *DoD Directive 3000.09: Autonomy in weapon systems.* https://www.esd.whs.mil/portals/54/documents/dd/issuances/dodd/300009p.pdf

Bad, Mad, and Cooked

13

Moral Responsibility for Civilian Harms in Human–AI Military Teams

S. Kate Devitt[1]

*Accountability can… be a powerful tool for motivating
better practices, and consequently more reliability and
trustworthy [computer] systems*
(Nissenbaum, 1996, p. 26)

INTRODUCTION

… as impressive and professional the Army's training and preparation processes are,
and no matter how tough and skilled our frontline soldiers are, there is a limit to their
mental resilience

(Fitzgibbon, 2020).

War puts human decision-makers in friction and fog; often without good choices, pick-
ing between the least bad options from a limited understanding of a complex and con-
gested battle space and strained cognitive capacities. AI has the potential to resolve some

 DOI: 10.1201/9781003410379-16

uncertainties faster, and to bring greater decision confidence. Human–AI teams, done well, could make compliance with international humanitarian law (IHL) easier, to help protect civilians and protected objects in conflicts (Lewis, 2023; Lewis & Ilachinski, 2022; Roberson et al., 2022).

AI will amplify at speed and scale whatever values are prioritized and programmed into it as well as those emerging from human–AI decision teams. For example, if AI software engineers spend a disproportionate amount of their time categorizing military targets and little time classifying protected objects, then civilians and objects of high humanitarian value will not be prioritized to be seen or to protect. If system engineers and decision scientists do not consider risk to civilians from the beginning of the development of new capabilities, then the risk of causing unintentional human harms is increased. AI has limited reasoning and ability to respond to surprise, as demonstrated by protesters placing traffic cones on the hood of Waymo driverless cars in San Francisco rendering them immobile (Kerr, 2023). Networked autonomous assets set to fire and the humans set to oversee the application of force, may see only the military objectives, and fail to see the proportionate effect on humanity and culture within the domain of attack.

This chapter argues that we risk putting humans into battlefields where their ability to act ethically, according to their own judgments of the correct course of action, is problematically reduced due to the integration of information and decision layers that nudge or coerce them into compliance with an algorithmic assessment. The chapter asserts that society is likely to punish human operators after incidents in its desire to ensure accountability. The risk is that there is little accountability for those who create, deploy, and maintain war-fighting decision systems. In the chapter these human operators are 'cooked apples' and they are at risk of a range of psychological effects from disengagement from their role to moral injury. We must first acknowledge the risk of 'cooked apples' and seek to ensure mechanisms for accountability of decision systems are employed by militaries.

The chapter will first focus on the sort of AI likely to affect human decision-making and attributions of responsibility: AI that takes over typically cognitive tasks of humans in a warfighting context. The chapter will provide definitions of responsibility and AI, some background regarding responsibility in war, and will introduce the concepts of 'bad', 'mad' apples and 'cooked apples'. Before delving into how AI can contribute to cooked apples, I introduce ways in which AI can reduce civilian harm – indicating that the chapter is not a whole-hearted critique of AI, only that its implementation and operation must be scrutinized. The chapter goes on to discuss the challenges of accountability in the history of computers and the specific risk of Large Language Model (LLM) AI models before suggesting ways to measure and anticipate the human experience of working with AI. Modifications to the critical decision method in cognitive task analysis to include Decision Responsibility Probes is a means to explore moral responsibility attributions. The chapter finishes with a tool to assess AI workplace risks relevant to military contexts of use to indicate how a risk-based approach could contribute to measures of an individual's engagement and thus responsibility for targeting decisions as well as when they feel overly responsible for technologies that are part of a complex system. Combining tools from human factors and workplace health and safety (WHS) will enable militaries to track, measure, and take responsibility for ensuring personnel remain 'good' and not 'cooked'.

DEFINITIONS

Responsibility

The *Edinburgh Declaration on Responsibility for Responsible AI* (Vallor, 2023, 14 July) differentiates between five types of responsibility for AI and autonomous systems:

- Causal responsibility: What event made this other event happen?
- Moral responsibility: Who is accountable for answerable for this?
- Legal responsibility: Who is, or will be, liable for this?
- Role responsibility: Whose duty was it (or is it) to do something about this?
- Virtue responsibility How trustworthy is this person or organization?

The Declaration argues that responsibility is best understood as relational rather than an agent or system property. So, rather than saying an isolated agent is responsible, a better question is to ask about the context of responsibility and to consider who an agent is responsible to and how that responsibility manifests within the context of decision-making.

Responsibility in this way means "articulating the ongoing, evolving duties of care that publics rightly expect and that people, organizations and institutions must fulfil to protect the specific relationships in which they and AI/AS are embedded" (Vallor, 2023).

Moral Responsibility

This chapter grounds the concepts of moral responsibility within the philosophical and legal tradition that encompasses liability, blame, and accountability. A person is morally blameworthy if their actions cause the harm or constituted a significant causal factor in bringing about the harm and their actions were faulty (Feinberg, 1985). Faulty actions can be intentionally or unintentionally faulty. A person can be held liable for unintentionally faulty actions if an incident occurs due to recklessness or negligence. A reckless action is one where a person can foresee harm but does nothing to prevent it. A negligent action is one where a person does not consider probable harmful consequences.

Morally blameworthy individuals must have situational awareness (they must know the situation within which they are making decisions), they must have the capability to intervene to make the right moral action and they must have agency and autonomy over their decision. Thus, an individual who is coerced, forced, restrained, or unable to act morally, cannot be held morally accountable for unethical acts. An individual who has been poorly informed, misinformed, partially informed, or uninformed cannot be morally blameworthy. Unless, by their actions, they prevented themselves from being relevantly informed. Finally, considerate of the findings of the Edinburgh Declaration (Vallor, 2023), any attributions of moral blameworthiness must be situated within the duties and expectations of organizations and institutions. This accords with the central premise of the chapter that the moral blameworthiness of decisions of individuals must be considered within the broader decision environments within which they act.

Artificial Intelligence

This chapter draws on the *OECD taxonomy of AI systems* (OECD, 2022) reconsidered for military AI as described in the *Responsible AI in Defense (RAID) Toolkit* (Trusted Autonomous Systems, 2023) to define an AI system. In this way a military AI system can be described by moving through the questions in the *RAID Toolkit D. Checklist Complete*, pp. 9–22:

- AI: What is the AI capability and how does the AI component function? Explain what the AI capability comprises, and how the specific AI element will function?
- Development inputs: What is the composition of the AI functionality? Explain the complexity and structure of the algorithms, and interlinked hardware that constitute the AI functionality
- Human–Machine Interaction: How does the AI capability (and more specifically the AI functionality) interact with the human operators across the spectrum of human involvement?
- AI Use Inputs: Identify the foreseeable inputs for the AI capability to operate when in use. Input is the data that is required for the AI functionality to operate
- AI Use Outputs: Identify the foreseeable outputs from the AI capability. Output is the data resulting from the execution of the AI functionality, or the ways that people or things receive the data resulting from the execution of the AI functionality (in whatever form the data is represented)
- Object of AI Action: Ascertain on what or whom the impact of the AI action will be. It requires consideration of the external actors that will be influenced or affected by the AI's actions within the environment of its anticipated use case
- Use case environment: Describe the military context in which the AI will be employed
- System of control: control measures, system integration, and AI frameworks. Explain how the AI capability fits within the broader system of control applied to military operations and activities

The AI I consider in this chapter is not just software engineering, algorithms, models, data, or use of specific training methods such as machine learning, but the systemic replacement of cognitive functions that humans would normally be tasked with in targeting decisions to be morally and legally responsible for those decisions. Because of this, the types of AI not considered in this paper include AI unrelated to targeting, such as that used in the functions of autonomous systems to plan, navigate, self-monitor, or self-repair. It is envisioned that AI in a targeting role, by potentially fusing multiple information sources and types of information, some beyond human perceptual faculties (e.g. hyperspectral imagery, IR), may be able to perform targeting assessments with greater fidelity and reliability than humans can achieve which motivates their use.

Perceiving. AI that brings situational awareness and clarity. Examples include the classification of simple objects such as a road, chair, building, bridge, traffic light, or person. A human being might need to verify the classified object or need to trust that the AI has the classification correct. This is particularly true for military objects because databases of military objects have significantly less data to draw on for model training than databases for regular objects in the real world such as dogs and cats. If the AI misperceives an object, then the human may have additional cognitive load and responsibilities in determining when a misperception by the AI has occurred.

Remembering. AI that has historical information on prior cases upon which to constrain hypotheses for decision-making. This includes the rules, processes, and procedures including rules of engagement and IHL. For instance, an LLM AI trained on the corpus of documents within a military is asked questions about precedents. A human expert (or multiple experts) would need to verify that the AI has interpreted the relevant documents correctly or reasonably and then vouch for them.

Understanding. AI that integrates information from multiple sources into a probable scenario including causal hypotheses. For instance, interpreting thermal imagery, location, historical movements, and purpose into a theory of who is traversing a particular location and for what purpose. Examples of this include the integration of data sources into the classification of a complex scenario or situation, such as suggesting the identity and purpose of a group of persons in a meeting at a particular location. A human would need to know the edge cases where information integration is likely to be unreliable and the most reliable circumstances when information integration is mostly likely to provide a suitable explanation of the situation.

Imagining. AI that creates new concepts from prompts and parameters. For instance, an LLM AI trained on the corpus of documents within a military is asked to create innovative tactics to achieve a military objective given constraints of personnel, equipment, timeframe, conditions, and rules of engagement. A human would need to seize upon the creative output of the AI and consider its value against their own knowledge and experience, military objectives, their own goals, values, and ethical assessments.

Reasoning. AI that is able to apply higher order considerations in the interpretation of multiple hypotheses as well as request more information to fill in data gaps, to recommend caution, and is aware of the requirements on human beings in the manner in which information is used to make decisions. For instance, an LLM AI built in the style of AutoGPT that can be tasked with a goal and create its own sub goals and projects to break the task into distinct components including a specific set of actions to question its own assumptions

(Ortiz, 2023; Wiggers, 2023). A human would need to audit the decision steps and reasoning of the AI to vouch for it.

Systematic influences are solidified, codified, embedded, and amplified through the introduction of AI and autonomous systems into military decision-making and action, and are unlikely to be scrutinized or engaged with critically by operators in the 'fog of war'. Thus, militaries must be clear on their objectives and design frameworks for AI and clear on where the AI is helpful to help with human analytical skills to ensure decision processes and outcomes lead to military and political success. Lessons from the medical field on where AI assists and where it makes outcomes worse may be abstracted and applied to the miliary domain. For example, during COVID-19 an AI model was trained using data that included patients who were scanned standing up and lying down. The lying down patients were more likely to be seriously ill, so the AI learned wrongly to predict serious COVID risk from a person's position (Heaven, 2021). The point is if the human can't critically engage (which is difficult in a battlefield situation) in AI outputs, then human-in, human-on, or human-constraining-the-loop are not sufficiently informed to be responsible. Even knowing how to define or evaluate 'the loop' is challenging (Leins & Kaspersen, 2021). A similar point has been made about the dangers of people using LLM such as ChatGPT to answer queries in domains for which they lack sufficient expertise to evaluate outputs (Oviedo-Trespalacios et al., 2023).

RESPONSIBILITY IN WAR

[Commanders need the] courage to accept responsibility either before the tribunal of some outside power or before the court of one's own conscience"

(Clausewitz et al., 1976)

Nations are responsible for decisions to go to war (*jus ad bellum*) in a way that individual soldiers cannot be held accountable for, whereas soldiers are responsible for their conduct within war (*jus in bello*) (Walzer, 2015). This chapter takes the importance of taking responsibility for one's own actions as a given, as a premise. The question then becomes: how is responsibility for decisions determined when the actions of war and decision-making become dominated by cognitive AI and complex systems, adding to human experience and awareness already mediated by information layers? The worry is if humans in warfare begin to eschew responsibility, then the connective tissue of war as a political act begins to fray. Thus, responsibility must be retained.

Miller (2020) argues that commanders must retain meaningful human control (MHC) of weapons if AI systems begin to select and engage targets autonomously, a task previously done by humans, in order to maintain commitments to IHL. He says, "commanders will remain obligated to take necessary and reasonable measures to prevent and suppress violations of IHL by their forces. Therefore, MHC should be defined as the control necessary for commanders to satisfy this obligation" (p. 545). A responsible commander must use weapons that they understand including their purpose, capabilities, and limitations of the system. The level of direct control required will depend on the context of operations within which they are used. Commanders must apply geographic and temporal constraints to uphold distinction and proportionality. In more complex and civilian-saturated environments, Miller argues MHC may require commanders to apply additional control measures or human supervision. There is academic debate regarding what 'meaningful human control' must consist of, but from a legal perspective, command responsibility has a set of clear requirements that human–AI teams must comply with (Liivoja et al., 2022).

Moral Responsibility for Civilian Harms

In this section the role of the war-making institution with regard to civilian harm is unpacked because AI for targeting will instantiate and amplify system organizational targeting norms, values, and processes. The principle of civilian immunity applies to the deliberate targeting of civilians and civilian objects. However, civilians may legally be harmed if the harm is unintentional and incidental to the military's objectives. Collateral damage is often seen as a necessary and ordinary consequence of war. However, Neta Crawford (2013) pushes back against the fatalism of this premise. She distinguishes three kinds of collateral damage (CD):

- Genuine accidents (CD1)
- Systemic collateral damage (CD2)
- The foreseeable if unintended consequence of rules of engagement, weapons choices, and tactics; and double effect/proportionality killing accepted as military necessity (CD3)

She claims that systemic collateral damage has become a moral blind spot, miscategorized as accidental, or treated as natural and unaffected by policy choices. She also argues that proportionality/double effect (Quinn, 1989) collateral damage also occurs in a moral blind spot, because of the wide legal latitude to harm civilians afforded by prioritizing 'military necessity'. Her argument is that CD2 and CD3 are produced both by the expansive and permissive conceptions of military necessity and by the organization of war making. Systemic collateral damage is decided at the organizational and command level and stems from institutionalized rules, procedures, training, and stresses of war.

Her point is that there are great gains to be made by altering these systematic influences on collateral damage. Recent changes to the Department of Defense Handbook requiring a 'presumption of civilian nature until proven' military (Leins & Durham, 2023) also create a dependence on data sources and automation that make this complexity incredibly real – from the data sets and sources used to their security, as well as their analysis and any AI or automation, all of these raise questions regarding reliance and 'governance by data' or by assumption.

Crawford notes that typically militaries explain incidents where civilians are harmed disproportionately as being the result of individuals, not processes. So, blame-worthiness is laid upon a 'few bad apples' or in some cases, even, 'mad apples'. The bad apples are those who intentionally and deliberately kill civilians. The mad apples have lost their ability to navigate decisions in war. Her point however is that "to blame indi-vidual soldiers for snapping is to be, in a sense, blind to how the moral agency of soldiers is shaped and compromised by the institutions of war making" (p. 466). Unfortunately, because collateral damage is legal under international criminal law, with its focus on individual moral and legal responsibility or intentional acts, the bar is set very high for proving deliberate intention.

However, Crawford points out that it is "not only the 'reality of war' but also the structure of the law with regards to non-combatant immunity and military necessity that creates the potential for large-scale, regular, systematic collateral damage in war as the 'ordinary' consequences of military operations" (p. 466). She locates CD2 in

- Rules of engagement
- Standard operating procedures
- International humanitarian law

Crawford's ambition is to extend moral responsibility for decisions beyond individuals and into military organizations. She notes that the way militaries are organized both enables and constrains individual moral agency. She also claims that military organiza-tions should themselves be moral agents. Such a claim has arguments for and against. Some in the 'against' camp say that an organization cannot be a moral agent because it cannot feel guilt, shame, and so forth, unlike the individuals who comprise it (Orts & Smith, 2017). On the other hand, individuals do seem to blame corporations and organi-zations for decisions enabled by the entity that cause moral harms. For example, in this chapter I take Crawford's supposition as a premise. I treat both individuals and military organizations as bearing moral responsibility. In doing so I acknowledge that there are philosophical concerns with such a premise.

Good apples follow orders but may systemically cause avoidable harms to civilians due to the systems within which they are making decisions:

- AI/CD1 AI-enabled systems can increase or decrease accidental collateral damage by autonomously minimizing harms and offering greater situational awareness (SA) for human oversight and intervention
- AI/CD2 AI can systematically increase or decrease collateral damage if it is hosted within decision systems that reduce the activation of higher-order reasoning regarding targeting decisions or reduce other psychological barri-ers to targeting
- AI/CD3 Parameters coded within AI can increase or decrease the acceptabil-ity of double effect/proportionality killing as military necessity

In this chapter I will consider how AI systems can exacerbate CD1, CD2, & CD3 by otherwise good apples. I will briefly touch on how AI systems can abet, reveal, or hide bad apples. But, also, note how easily incidents judged to be accidents are really the result of systemic failures, thus miscategorized.

The consequences I analyze are both whether individuals are unfairly blamed for actions for which they should not be held morally accountable for; and whether individuals themselves blame themselves and suffer moral injury for their role within systems that afforded them little recourse to avoid the scale of civilian harms that may occur. Individuals suffering misattributed accountability and/or moral injury, I refer to as 'cooked apples'. Military personnel ought to be on the lookout for environments that 'cook' them and to blow the whistle on systems where harms are magnified by these technologies.

AI to Reduce Civilian Harm

A premise of this chapter is that good apples intend to prevent civilian harm and conduct their professional duties in accordance with rules of engagement, the laws of armed combat, IHL and their own moral compass. Therefore, good apples would wish to reduce harm to civilians and be interested in improving decision-making so that this was achieved – including considering using AI for that purpose. The Centre for Naval Analysis (Lewis, 2023) analyzed over 2,000 real-world incidents of civilian harm and classified them into twelve pathways divided into misidentification and collateral damage. From their analysis they identified four applications of AI in the role of *perceiving, remembering,* and *understanding* that could reduce civilian harms and potentially moral injury, empowering personnel to make more informed decisions without adding to their cognitive load:

1. *Alerting the presence of transient civilians.* Many civilians are harmed because intelligence is unable to keep up with the movements of civilians. Better monitoring of persons around the target area would bring them to the attention of operating forces that can fixate on a target rather than protect persons in the area.
2. *Detecting a change from collateral damage estimate.* Collateral damage estimates can be wrong. AI can be used to identify the difference between earlier estimates and new images closer to the time of an attack that might indicate a change in who is within the area.
3. *Alerting a potential miscorrelation.* Militaries can have correct information about a threat, but then misidentify the location of that threat (e.g. a vehicle with legitimate targets within may be swapped out with civilians). Better AI-enabled surveillance could help identify that a miscorrelation has taken place.
4. *Recognition of protected symbols.* Modern battlefield sensors are attuned to the infrared spectrum because it allows the confident identification of machine signatures such as engines as well as humans in varying light conditions and the apparatus does not itself (thermographic camera) have a signature like radar. AI methods could identify protected symbols for designating protected objects alerting the operator or chain of command.

Additionally investing in a trusted communications network for human–AI teams would enhance the potential to protect protected persons and objects (Devitt et al., 2023). The opportunity of AI to reduce civilian harm must be kept in mind as the chapter considers the way that teaming with AI might reduce human responsibility and accountability as well as causing harms to personnel. To begin this investigation, I will consider that computers have long presented a challenge to accountability mechanisms.

COMPUTERS, AUTONOMY, AND ACCOUNTABILITY

Concerns about the effect of computers on accountability have been identified for decades as computational devices have become ubiquitous. In 1966, Josef Weizenbaum referred to the 'magical thinking' around basic automated systems, and the challenge of humans often being uncritical of their outputs (Weizenbaum, 1966). In 1996, Nissenbaum argued that computers contribute to obscuring lines of accountability stemming from both the facts about computing itself and the situations in which computers are used. She identified four barriers to accountability: (i) the problem of many hands, (ii) the problem of bugs, (iii) blaming the computer, and (iv) software ownership without liability (Nissenbaum, 1996). Computational technologies create systemic accountability gaps. Note the similarities of concern expressed by Crawford (2013), Parsons & Wilson (2020) and Nissenbaum. All three identify the *systems* of decision-making as problematic and requiring a need to be addressed. The Uber self-driving car example for its recency and the lessons we might take for the future of soldiers increasingly serving with autonomous systems.

The Uber Self-Driving Car Incident

In the Uber case, a pedestrian, Elaine Herzberg, walking her bicycle across a highway in Arizona, was killed because, first, the experimental autonomous driving system failed to recognize her as a pedestrian (confused, the AI kept reclassifying the moving object as separate cars exiting the highway rather than a continuous entity); second, the safety parameters were turned down because they were deemed as overly sensitive and distracting to the test drivers (causing a 'boy who cried wolf' problem where test drivers stopped believing that there was a genuine risk when the alerts sounded) and finally, the test driver was on her mobile phone when the incident occurred and therefore did not see the pedestrian herself in order to try and allay the accident.

The AI in the car finally identified the object as a bicycle 2.6 seconds from hitting the object, but then switched it back to classifying it as 'other' at 1.5 seconds before impact. At this time the system generated a plan to steer around the unknown object but decided that it couldn't. At 0.2 seconds to impact the car let out a sound to alert the human operator. At two-hundredths of a second before impact the operator grabbed the steering wheel which took the car out of autonomy and into manual mode (Smiley, 2022). When the legal case was finalized, there was no liability against Uber or any of those responsible for decisions regarding the perceptive or safety systems of the autonomous vehicle. The judge determined that the solo human 'operator' was solely liable for the incident and she was entirely blameworthy (Ormsby, 2019). Was this fair attribution of responsibility? Let us consider the systems in play:

In the years before the accident, federal regulators were standing back allowing companies to voluntarily report their safety practices and recommended that states, such as Arizona, did the same (Smiley, 2022). So, Uber was able to make their own safety policies.

A year before the incident, in 2017, Uber had changed their safety practices from requiring two humans in each test car (one driver and one to look out for and discuss hazards and to write up issues for the company to review), to only having one human in each self-driving car. This policy change had several impacts. Now, a single operator had to manage the complexities of each test drive. Additionally, there was no longer a conversational partner to share SA with and to prevent automation complacency. The Uber AI's self-driving competence had increased sufficiently by this time that the cars made mistakes much more rarely. Being alone for hundreds of miles increased the allure of the single operator's mobile phone and increased the likelihood of attentional safety transgressions.

In this instance, the test driver became the locus of blame for the incident even though the technical factors contributing to the incident still required investigation and intervention. Still, from a legal perspective, and consistent with the definition provided regarding being informed, if the test driver had been watching the road rather than being on their phone, then they could have seen the pedestrian and taken control of the vehicle and averted the accident. The legal maneuver reveals that the driver was 'cooked', the human driver was expected to behave in a way that was difficult and unrealistic for a reasonable human to do – and over 70 years of human factors research on automation bias would support this (Endsley, 2017; Hoff & Bashir, 2015; Lee & See, 2004; Merritt et al., 2013). Human attention tends to wander when they are not actively engaged in a task. When systems operate as expected, humans become complacent. The difficulty of sustaining attention for monitoring purposes over longer periods of time is also known as vigilance decrement (Martínez-Pérez et al., 2023).

What is egregious about allowing accountability to fall on only an inattentive human operator is that organizations should know better. Attention wandering was first noted in British naval radar operators who increasingly missed critical radar signals (enemy combatants) as their watch periods progressed. This phenomenon has been studied scientifically since the end of World War II (Cummings et al., 2016; Mackworth, 1948; Mackworth, 1950; Thomson et al., 2015) should be anticipated and its effects mitigated.

Risks of Large Language Models

LLMs trained on vast data sets that predict sentences based on natural language inputs. While in existence for a number of years, have burst into public consciousness in 2022 and current iterations such as 2023's GPT-4 with one trillion parameters exhibit a remarkable degree of higher-order reasoning and complex thinking compared with other forms of AI to date (Bastian, 2023; Hardy et al., 2023; Ott et al., 2023). LLMs will increasingly be used to improve strategic and operational effectiveness with better data, models, and scenarios and are likely to affect responsibility attributions.

To gain the trust of users, the interface of LLMs is designed around the cognitive architecture of human users using natural language explanations rather than statistical or didactic responses. Such approaches are promising because the information is clear and the logic more transparent for human users. LLM integration with robots may also enable better ability for robots to choose actions that comply with human intent and are relevant to solving the task (Rana et al., 2023). The complexity and seamless presentation of information to operators and the ease of LLM query response suggest a step-change in human-machine integration.

However, the ease of legibility belies the limits of the underlying models for recommending courses of action including partial causal variables, inductive error, limits in data, calibration errors, context insensitivity, simplification, and extrinsic uncertainties. Also worryingly, the better the cognitive fit of a tool, the more influential outputs can be on user acceptance (Giboney et al., 2015) regardless of system limitations. Cognitive fit refers to where information is presented in a way most easily processed by a cognitive agent. For example intelligence briefs prepared with more pictures can be easier for some to process than wordy text (Barnes et al., 2022). Paradoxically, the ease of interaction between chat LLMs and humans may reduce the engagement of higher-order functions that enable humans to reflect and question the information provided. Thus, LLMs appear to be interacting with end users in a rational dialectic, but in fact are using rational persuasion as a form of paternalism (Tsai, 2014) creating a particularly sophisticated and nuanced form of automation bias. LLM outputs may intrude on the users' deliberative activities in ways that devalue their reflective decision-making processes and keep decision-making concordant with system-level values.

So, a danger of designing AI tools that mimic higher-order human reasoning and produce outputs aligned to the cognitive preferences of decision-makers is that users may be manipulated to assent. When an LLM/AI agent rationally persuades a user, it offers reasons, evidence, or arguments. It is possible to construct an AI to rationally persuade a human operator to choose a right action, yet the information represented is paternalistic or disrespectful by being incomplete, simplified, or obfuscating. Therefore, Systems that appear very confident, authoritative, and omniscient need to self-identify their own limits to human operators and modulate their confidence in their answers, and humans need to be trained sufficiently to know when and how to be skeptical of AI reports. LLMs that use rational persuasion are likely to be attractive to militaries to achieve cohesion and conformity of response and to offset human decision-maker's limited capacity to gather, weigh, or evaluate evidence, as well as for efficiency reasons – where human hesitation, cogitation, unreliable reactions will slow decision-making down in high tempo conflicts.

AIs deployed with humans need to gain and maintain trust and avoid paternalism. How should this be achieved? AI systems use vastly more data and operations on that data including data integration than a single decision-maker can understand, rendering individual reflective cognition problematic. When action recommendations align with expert human intuition, there may be little cause for concern. However, when recommendations diverge from expert human intuition, then humans must either trust the system – and follow its dictates – without necessarily knowing why they are agreeing; or reject the system risking a suboptimal alternative. AI system makers may try to improve trust with honest articulation of how decisions are generated, but it is likely that information will necessarily be simplified and manipulated to facilitate consent. And, while users benefit from gaining more decision control, explanations can sometimes increase cognitive load without assisting their decision-making (Westphal et al., 2023).

The fast integration of LLM as chat agents that appear to exhibit cognitive functions of thinking, understanding, correcting themselves, etc… as well as robotic systems able to autonomously adjust their actions to changing circumstances, increases the likelihood that humans working with machines will attribute agency to the AIs driving these systems or even considering them moral agents; potentially avoiding responsibility for decisions made by the AI when things go wrong and taking more credit for outcomes when things go right.

Johnson (2023) argues that human-like human-robotic interfaces make people feel less responsible for the success or failures of tasks and use AI agents as scapegoats when bad outcomes occur. He points out that the use of AI will require more (rather than less) contributions and oversight from the human operator to mitigate the contingencies that fall outside of an algorithm's training parameters or fail in some way. Given the risks of LLMs for responsibility attributions, it is worth reviewing how a military LLM are being marketed.

Palantir's Artificial Intelligence Platform

Palantir's Artificial Intelligence Platform [AIP] for Defense (Palantir, 2023) has been marketed in the context of future military decision-making. Palantir emphasizes that "LLMs and algorithms must be controlled in this highly regulated and sensitive context to ensure that they are used in a legal and ethical way" [0:18-0:25] and claim that their AIP has:

> industry leading guardrails to control, govern and trust in the AI. As operators and AI take action in platform, AIP generates a secure digital record of operations. These capabilities are crucial for mitigating significant legal, regulatory and ethical risks posed by LLMs and AI in sensitive and classified settings.

[0:57-1:17]

In the scenario demonstration, Palantir notes that using their system effectively requires a deep understanding of "military doctrine, logistics and battle dynamics" [5:51-5:55]. Implicit is the notion that doctrine contains the legal and ethical values required to ensure human responsibility over decisions. Human agency is recorded through the process of human interrogation of the system described by Palantir as "reasoning through different scenarios and courses of action safely and at scale" [6:28-6:32]. However, the video and associated documentation do not provide transparent guidance for users or decision-makers on how operators can verify or validate the data they are receiving from the LLM. The 'content protection' tab offers obscure acronyms and a 'validation' option [7:23], but without any explanation – at least from the publicly available materials. To trust these systems, acquirers of these technologies would ensure appropriate test and evaluation, verification, and validation of these systems, to be assured that they are fit for purpose under the anticipated range of uses within a defined context of operations. They would also need a tailored curriculum and training program for operators to ensure that operators knew the parameters of the system and the bounds of their own role in decisions being made in the battlespace. Instead, the Palantir marketing material focuses on their claims of trustworthiness, auditability of human reasoning, LLM outputs and decision-making, and ultimately putting responsibility for decisions back on humans using the technology. Are militaries prepared to manage LLM integration ethically, so that individuals are capable of holding moral responsibility for decisions made with LLMs?

So far the risks of LLMs discussed include soldiers making decisions without adequate ability to engage their higher-order critical functions and possibly also making soldiers feel less responsible for their decisions. I will now consider how AI can make operators feel even more responsible – particularly when things go wrong.

MORAL INJURY

Moral injury refers to the psychological effect on soldiers who feel ethically compromised through their professional conduct in warfare. Human–AI teams change the decision-making environment and will change the risk calculus of moral injury. This chapter takes as a premise that militaries ought to reduce the conditions of moral injury. This premise is controversial. Some in society, such as absolute pacifists would expect soldiers to always suffer moral injury in a conflict because they believe that the act of killing is unethical under any circumstances. This chapter assumes a just war perspective that acts of harm by a state are, in some cases, justified. In those cases, the military ought to ensure soldiers are fully cognisant of why war is being waged and why it believes that the conduct expected of soldiers is justified. A soldier's individual sense of ethical conduct is constructed within these institutional conditions.

Moral injury can be caused by ethical discordance between the individual regarding whether entering a conflict itself is just (*jus ad bellum*), how soldiers judge their own conduct during a conflict, but also how soldiers perceive the ethical conduct of the military they serve during a conflict (*jus in bello*). This chapter will focus on moral injury *jus in bello* considered broadly to not just include the individual responsibilities of soldiers, but also the strategic and organizational war-waging responsibilities of the military organization. Parsons and Wilson (2020) divide these into responsibility for aligning war aims with means of war, achieving the organizational capacity to achieve war aims at the least cost in lives and resources, and the maintenance of legitimacy (see Box 13.1).

**BOX 13.1   THREE WAR-WAGING RESPONSIBILITIES
IN *JUS IN BELLO'S* STRATEGIC DIMENSION**

1. Achieve and sustain coherence: war aims must be aligned with means as well as strategies, policies, and campaigns in order to increase the probability of achieving the aims set.
2. Generate and sustain organizational capacity: initial aims and decisions must be translated into actions that achieve the war aims at the least cost in lives and resources and the least risk to the innocent and one's political community. These decisions and actions must adapt to changing conditions as the war unfolds and bring the war to a successful end.
3. Maintain legitimacy: war must not only be initiated for the right reasons and observe the laws of war; additionally, public support must be sustained, and the proper integration of military and civilian leadership must be ensured. Executing these responsibilities sufficiently well is the second way political leaders exercise their responsibilities to their soldiers and their nation as well as the innocent put at risk by war (Parsons & Wilson, 2020).

Soldiers may be critical of the way their militaries are conducting war including how they are teamed with technologies. For example negative psychological effects may occur for remote pilots operating physically alone, rather than working side-by-side with a team. Militaries might have personnel working unnatural and jarring military shifts only to switch to a civilian one as they move in-between their family homes to secure remote military locations each day (Enemark, 2019). Moral injury may stem from the fact that operators are physically safe, whereas their targets are in situ within a conflict. The physical separation may amplify moral dissonance (French, 2010) due to perceptions of what it is to have courage and to be 'at war' from both operators themselves and their peers who may judge them (Holz, 2021). While these problems exist regardless of the use of AI, they point to how technologies do not merely change decision-making, they change the decision-making environment with often unintended negative consequences for the operators.

This chapter is especially interested in how individual decision-making by the soldier is affected within the larger *jus in bello* processes of war including how the military sustains coherence of effort, organizational capacity, and maintains legitimacy (*Box 13.1*). I argue that moral injury can occur from a perception of being let down by the systems of war fighting (see section 'civilian harms') as well as the individual conduct of soldiers in a conflict.

Reference to *jus in bello* principles may not be enough to manage moral risk. Contrary to some public commentary that supposes that remote pilots are less morally engaged (termed 'moral disengagement') because they are physically distant from their targets, many operators do struggle with taking human life regardless of what others might assure is morally permissible. Enemark (2019) points out that military personnel are able to judge themselves by reference to deeply held beliefs about right and wrong and how the betrayal of those beliefs causes moral injury and post-traumatic stress disorder (Fani et al., 2021; McEwen et al., 2021; Williamson et al., 2018). For example, an air force study in 2019 found that 6.15% of remotely piloted aircraft pilots suffer from post-traumatic stress disorder (Phelps & Grossman, 2021). Drone crews have a higher incidence of psychiatric symptoms than pilots of traditionally crewed aircraft (Saini et al., 2021). But as Phelps and Grossman (2021) point out, while remote pilots may have higher rates of psychological effects than regular pilots, it is important to acknowledge that most remote pilots do their job and do not suffer either moral injury or PTSD. So, the claims of harms ought to be considered proportionately.

The opportunity of this chapter is to consider what lessons to take from historical precedent to consider the future operating environment. Without care and consideration, some good apples operating AI and autonomous systems may become 'cooked apples', suffering from a range of negative psychosocial harms from frustration, moral disengagement, lost agency all the way to post-traumatic stress disorder and moral injury due to the way that teaming with technologies changes the environment of decision-making and the human role within in.

Considering the above, I argue that moral injury can come in two forms from working with AI, that of either being a moral witness or becoming a moral crumple zone.

Extreme Moral Witness

An extreme moral witness is an operator who experiences the humanity of their target intensely, feels emotionally connected to their target, and finds the intentional act of harming them challenges their moral beliefs. This may be because the oversight functions required for compliance with IHL and the laws of armed conflict expose operators to be more situationally aware of the humanity they are harming or putting at risk of harms (Phelps & Grossman, 2021).

A research question is to what degree should operators experience the humanity of their targets? What is the normative ideal toward which designers of human–AI teams ought to aspire? Considering an Aristotelian approach (Aristotle, 350 B.C.E., n.d.) – a virtuous human–AI team will create optimal cognitive and affective conditions to ensure human agency and accountability but not put humans through unnecessary suffering or trauma in the course of doing their job. The vices include moral disengagement on the one hand and experience of moral injury on the other hand (see Table 13.1). Moral disengagement generates a lack of a sense of responsibility whereas moral injury creates a disproportionate sense of moral responsibility, where individuals take on more responsibility for their actions than is perhaps reasonable or required given their role as a part of complex targeting systems, command structures, and national imperatives to act that

TABLE 13.1 Virtue and Vices of Moral Engagement

VICE	VIRTUE	VICE
Moral disengagement	Moral engagement	Moral injury
Lack sense of responsibility	Feels responsible	Overdeveloped sense of responsibility

they are obliged to follow. Note extrinsic and intrinsic influences can produce virtuous and vice behaviors. That is to say, virtuous behaviors can be amplified and made more likely within cognitively aligned decision-making environments and be diminished in misaligned cases.

Again, there will be skeptics, say, absolute pacifists, who wish that soldiers directly and intensely experience any harms they do to others. I argue that soldiers ought to be morally engaged in their decisions, but not morally injured (to the extent possible). That is, there is a difference between what is fair for a murderer to experience (as a consequence of their acts) versus a soldier compelled to harm in accordance with their professional duty. That a nation ought to avoid subjecting service personnel to elevated risk of trauma, depression, anxiety, and suicide (Jamieson et al., 2021; Kaldas et al., 2023).

Note, the main argument in this chapter, however, does not require a firm determination with regards to how much soldiers should experience the harms they cause others – I leave this normative framework to each nation and each military to decide. Instead, I argue that whatever the national ethical standards are for engagement, then militaries owe it to their personnel to align their human–AI interface to that standard. Not thinking about a standard and just building AI systems without sensitivity to their effect on human operators is negligent. Determining that operators ought to experience the humanity of their targets to a greater or lesser extent and then making technologies align with this, is responsible conduct.

In terms of how AI might help militaries achieve an ethical standard of engagement, I draw on the social media content moderation literature (Gillespie, 2020). Some theorists support using AI to classify and remove inappropriate content on social media before human moderators experience it. AI could also be used in a similar way in the battlespace, such as evaluating the extent of damage in a battle damage assessment and limiting how much disturbing content human operators must observe. Another technique, called visibility moderation (rather than content moderation), algorithmically alters the content prioritized to viewers of social media (Zeng & Kaye, 2022). A similar approach using AI could limit the amount of disturbing information fed to operators by number of seconds or the severity of the content. AI could also dynamically distribute content across a team of human operators so that collectively, the entire battle space and actions within it, were appropriately (i.e. not disproportionately) witnessed, and that the experience was shared by the team. Sharing narratives between human teammates is another way of human experience that is known to help manage post-traumatic stress recovery. Thus, AI can be used to ensure moral engagement rather than moral injury.

Moral Crumple Zone

The term 'moral crumple zone' was created to explain cases where human operators are blamed for errors or accidents that are not entirely in their control (Elish, 2019). Moral injury from being a moral crumple zone occurs because an operator shoulders the blame for an incident that has occurred rather than society acknowledging the complex and systemic factors that can lead to adverse outcomes – where the human operator is a component part, and potentially blameworthy, but could not have caused the harm without the technological apparatus within which they made decisions. This concept is not new and has been explored since the 1980s (Dekker, 2013; Reason, 2016; Woods et al., 2017) but gains new nuance with the rise of AI taking over human decision-making at higher cognitive levels in ever more complex systems. Operators who fall into a moral crumple zone may have SA, but lack a sense of agency over their situation, that they were a cog in the machine. Or they may lack SA. Perhaps this is because they had to decide too quickly in order to achieve high priority mission objectives, they were slow to intervene on a system that they had previously trusted to operate with little need for oversight, or they had the wrong information upon which to base their decisions. A military example is the punishments to personnel in the Médecins Sans Frontières Kunduz hospital incident[2] with no responsibility taken at higher levels for the human-technical systems in place that lead to the tragedy (Donati, 2021).[3] In the future, two risk situations need to be considered: (i) military responsibility for errors may fail to be attributed due to operational complexity (Bouchet-Saulnier & Whittal, 2018) and/or (ii) responsibility is unfairly apportioned to manage political fallout. In each case operators might feel unfairly treated by society who may pick out the human as a poor decision-maker, negligent, or malfeasant after an incident of civilian harm.

So far in this chapter I have defined the key concepts, considered responsibility requirements, and the advantages and risks to civilian and military personnel from the introduction of AI into systems. I will now move to new methods to measure human–AI decision-making and conditions needed for the attribution of responsibility.

HUMAN FACTORS

New methods are needed to explore the degree to which both operators and those who may judge their behavior will consider them responsible for decisions made with AI and the factors that affect these attributions. In both the moral witness and moral crumple zone cases, modifications of human factors research methods engaged in advance of deployment of human–AI teams can help identify responsibility risks including moral disengagement and moral injury as well as identify optimal decision environments to encourage moral engagement and responsibility for decisions made. These new methods are recommended to build on existing best practice human factors methods both to ensure alignment as well as to increase the usefulness of new methods within existing test and evaluation paradigms in Defense contexts of use.

One way to extend is by taking existing human factors *Naturalistic Decision-Making* (Klein et al., 1993; Zsambok & Klein, 1997) methodologies of cognitive task analysis and adding additional probes to critical decision method (CDM) (Table 4.5 CDM probes, Stanton et al., 2017) relevant to evaluating moral responsibility.

Drawing on philosophical analysis (Talbert, 2016), moral responsibility requires:

- Free will (autonomy and agency over one's decisions)
- SA (knowledge of the circumstances)
- Capable action (capacity to act during events involving ethical risk in accordance with intent).

I will start with measuring SA because a long tradition already exists within human factors research to measure it such as SART, SAGAT, and SPAM (Salvendy & Karwowski, 2021). Humans working with AI need to be engaged to be effective (Endsley, 2023) and to be accountable for decisions made with AI. Yet, with increasingly complex human–AI systems, operators may not simply lose SA but lose understanding of the systems themselves including the functions and parameters of their operation. This is problematic because humans have the responsibility of overseeing the performance of the AI and the ultimate responsibility for decisions made with AI. Humans need to have SA over all aspects of the task including what aspects the AI is undertaking and how well they are working. Humans working with AI systems need to be prepared for its perceptual limitations; hidden biases; limits of causal models to predict evolving situations and the extent of the AI's brittleness operating in new situations (Endsley, 2023).

Humans need to be aware of the SA of their human and AI teammates to ensure that team responsibilities are upheld. Humans need to be situationally aware in order to achieve their work responsibilities individually and in teams including perceiving their situation, comprehending it, and projecting from the current situation to inform potential future situations. In order to achieve human–AI team SA, the functioning of the AI must be suitably transparent and its actions explainable to the human (Endsley, 2023).

A morally disengaged operator might achieve level 1 SA, perceiving their situation, but not level 2 SA where level 2 requires interpreting what the data means relevant to the goals and decision requirements of the individual, or level 3 where the ramifications for moral disengagement might affect future situations.

There are choices 'cooked' operators may make to try and relieve their experiences such as over-trusting automatic functions (possibly relinquishing a sense of responsibility) or deliberating circumventing AI functions to try and gain more control and SA.

Once SA is measured, this leaves measures of free will and capable action to be determined. Critical Decision Method is an established protocol that allows researchers to better understand human decision-making within sociotechnical systems (Table 4.5 CDM probes, Stanton et al., 2017).

Critical Decision Method Responsibility Probes (CDM-R)

Responsibility probes (CDM-R) are described in Table 13.2 under headings of: 'decision agency', 'decision capability' and 'decision responsibility' and are recommended to be used in conjunction with existing SA tools to measure conditions for moral responsibility.

Answers to CDM-R prompts will explicate how and when operators feel agency and autonomy over decisions, when they felt capable to act, and how informed they were of the situation and technology they were using. Through the use of CDM-R human factors researchers would identify points in the decision-making process where operators are more or less likely to experience a sense of responsibility for their actions and may predict how responsibility attributions would play out in post-incident reviews for both legal and ethical determinations. Prompted answers will provide information about

TABLE 13.2   Decision Responsibility Probes for Critical Decision Method (CDM-R)

Decision responsibility	Describe your sense of responsibility for the decision. At what decision points did you feel responsible? At what decision points did you not feel responsible? How did you determine who or what system was responsible at various decision points? If there was a decision point at which you felt particularly responsible, describe what aspects of the process affected this feeling? If there was a decision point at which you did not feel responsible? Describe precisely when you no longer felt responsible. What features affected your sense of decision responsibility?
Decision Agency & Autonomy	Describe your sense of agency and autonomy over the decision. At what decision points did you feel free to apply your own considerations? At what decision points did you not feel free to apply your own considerations? If there was a decision point at which you felt a strong sense of agency or autonomy, describe what aspects of the process affected this feeling? If there was a decision point at which you did not feel agency or autonomy? Describe precisely when you no longer felt free to apply your own considerations. What features affected your sense of agency and autonomy?
Decision Capability	Describe your sense of capability to make the decision. Include skills, knowledge, and abilities. At what decision points did you feel capable? At what decision points did you not feel capable? If there was a decision point at which you felt particularly capable, describe what aspects of this point affected this feeling? If there was a decision point at which you did not feel capable? Describe precisely when you no longer felt capable. What features affected your sense of capability to make the decision?

gaps in training, skills, or knowledge. CDM-R answers combined with SA results will provide system and interaction designers areas for improvement of the UX/UI.

More research needs to be done on the experience of responsibility for operators in human–AI teams as well as what factors affect how external adjudicators will judge the degree to which they have acted responsibly and bear responsibility for decisions. It may be that a human operator feels more responsibility for decisions when they under-trust automation and less responsibility when they defer to automated functions. However, a sense of responsibility may vary throughout different decision points as well as exist at different observation perspectives, transcending a specific decision.

Future research may identify clusters of psychological experiences working with AI systems to allow them to predict the likely effects on operators during decision-making. The next section provides information on an AI WHS framework to identify risks to operators.

AI WORKPLACE HEALTH AND SAFETY FRAMEWORK

Human safety issues need to be considered when AI is designed ahead of deployment. Otherwise, AI risks being introduced based primarily on its professional significance rather than its human impact. This section will describe a risk assessment tool toward human safety and AI systems developed by SafeWork NSW (an Australian government regulator) to create an AI WHS score card (Cebulla et al., 2022; Centre for Work Health and Safety, 2021). To create the tool, researchers used qualitative and quantitative methods including literature review and consultations with AI experts, WHS professionals, regulators and policymakers, representatives from organizations adopting or having adopted AI, and others with knowledge in the field.

Stakeholders agreed that AI influences workflows, both automating tedious and repetitive tasks and creating a new intensity of work, creating new hazards. AI may augment work tasks, offering personnel new methods to improve the quality of their work, which in turn may affect the way tasks are assigned by management, e.g. AI will be used by managers to optimize workflows. The report findings suggested that while a range of psychological, physical, and social risks are associated with the introduction of AI in the workplace, it was in the realm of the psychological that AI was perceived to likely have the greatest impact[4]. Cognitive hazards include information processing, complexity, and duration of tasks. Social factors reducing safety were noted including that if AI takes over traditional managerial tasks, it may reduce interactions between workers and managers that could have WHS implications. The effect on social interactions and communication between personnel and their chain of command is relevant for Defense forces looking to implement AI systems (King, 2006). The report noted that "little evidence was found of organizations taking strategic approaches to anticipate the impacts of AI on workplaces beyond the intended process or product change" (p. 2).

With regards to the method used relevant to achieving responsible use of AI in military systems, the initial draft of the score card drew on two frameworks, (i) *Australia's AI Ethics Framework* (Department of Industry Innovation and Science, 2019; Devitt & Copeland, 2023; Reid et al., 2023) and (ii) the *AI Canvas* (Agrawal et al., 2018). A simplified framework was developed with three broad categories (human condition, work safety, and oversight) and three higher-level steps (ideation, development, and application—see Table 13.3).

This new framework was then imbued with principles of good work design considering the physical, cognitive, biomechanical, and psychological characteristics of a

TABLE 13.3 Risk Domains Aggregate eight Australian AI Ethics Principles, Appendix G: AI WHS Protocol (Centre for Work Health and Safety, 2021)

HUMAN CONDITION	WORKER SAFETY	OVERSIGHT
Human, social, and environmental wellbeing Human-centered values Fairness	Privacy protection and security Reliability and safety	Transparency and explainability Contestability Accountability

task (Safe Work Australia, 2020, p. 9). The authors note that AI risks may not be visible, detectable physical risks and points of hazards (Cebulla et al., 2022, p. 923). AI is more likely to produce psycho-social risks resulting from AI's dehumanizing application. Psycho-social risks involve subjective assessments and are situation-specific, making them harder to measure (Jespersen, Hasle, et al., 2016; Jespersen, Hohnen, et al., 2016). Highly demanding jobs where employees have little control are likely to increase strain. Whereas giving employees more control of the work (such as the timing, sequencing, and speed) reduces strain. This is at odds with the anticipated accelerated decision-making environments envisioned in mosaic warfare (Clark et al., 2020; Devitt, 2023; Hall & Scielzo, 2022) – increasing the likelihood of psychosocial risks to personnel working alongside AI. To increase wellbeing, Human–AI teams ought to prioritize human autonomy, build competence and confidence to be effective with job tasks and to feel connected with people involved with on-the-job tasks (Calvo et al., 2020). Humans find their work meaningful when they have autonomy (Cebulla et al., 2022) and dignity (Bal, 2017) at work. Factors affecting a sense of dignity at work means equality, contribution, openness, and responsibility. Key to this chapter is the link between psycho-social health and a sense of responsibility. Dignity at work signifies work that is meaningful with a degree of responsible autonomy and recognized social esteem. Militaries introducing AI need to consider the wellbeing of their personnel offering purpose and avoiding demeaning, arbitrary authority, unhealthy or unsafe conditions, or physical or mental degradation (Autor et al., 2020). The researchers produced the score card with risk rating and an AI WHS Protocol as a guideline for its use.

The Scorecard is extensive (Centre for Work Health and Safety, 2021, pp. 59–71 Appendix F. AI WHS Scorecard) Examples of Risks from AI WHS Scorecard (version 2.0) relevant to military applications are in Table 13.4.

The introduction of AI into military contexts of use has the potential to reduce repetitive and mundane jobs. Through increasing AI, military personnel may experience greater surveillance and loss of privacy in their roles, which may have less of an impact than if the same conditions were applied in a civilian context. This is because militaries generally have greater leeway on how service persons are expected to behave and be treated including the degree to which their actions are scrutinized. Militaries ought to pay close attention to the effects on the war fighter through the introduction of AI such as declines in productivity, efficiency, morale, and cohesion. If personnel become disengaged or even hostile within AI-enabled operational circumstances they may threaten the achievement of military and socio-political objectives in the first instance and, if risks are not managed, may threaten the reputation of the military organization. Personnel may even sue the military for the working conditions they experience while being deployed (Fairgrieve, 2014). Up to and including dying in preventable training exercises and routine non-combat incidents (Mohamed, 2021). Thus, while WHS considerations may be different in military contexts than civilian ones from a legal perspective, the future of AI within the military workspace has a myriad of potential harmful effects on personnel in both combat and non-combat circumstances.

And while WHS has limited application in conflicts, there is a developing line of British case law that refers to decisions made in peacetime that result in adverse safety consequences in conflict.

TABLE 13.4 Examples of AI Ethics Risks Relevant to Military Applications from NSW AI WHS Safety Scorecard (Centre for Work Health and Safety, 2021)

ETHICS RISKS TO WHS	EXAMPLES	POTENTIAL MILITARY CONTEXT
Risk of overconfidence in or overreliance on AI system resulting in loss of diminished due diligence	After a six-month trial of a new AI product without incident preventative safety measures are no longer prioritized	An autonomous logistics robot is introduced in a warehouse without consistent safety training and skilling
Risk of AI being used out of scope	A productivity assessment tool designed to improve workflow efficiency is used for penalizing or firing people	An AI targeting tool offers confident and reliable determinations of lawful combatants within a specific context of operations. AI tools to measure human performance are unfairly used to recommend promotions, postings, and awards including medals An AI surveillance tool designed to identify the presence of transient civilians is used to target all humans moving within a target area.
Risk of AI system undermining human capabilities	AI system automates processes, assigning workers to undertake remaining tasks resulting in progressive de-skilling	Pilots no longer accrue sufficient manual flying hours to respond quickly and competently when an incident occurs. An integrated targeting AI using data fusion from multiple sources degrades human reasoning with regard to the likelihood that objects are lawful combatants
Risk of (in)sufficient consideration given to interconnectivity/ interoperability of AI systems	Multiple data sources need integrating, each quality assessed and assured	Network-centric or mosaic warfare fails to coordinate across multiple AI Assets and value systems across allied forces.
Risk of no offline systems or processes in place to test and review the veracity of AI predictions/decisions	An AI tool is used to triage incoming calls to an organization but the tool provides incomplete answers unable to resolve the query; dissatisfied client complaints.	An AI targeting system takes incoming intelligence and produces recommended orders through an LLM. Personnel have no mechanism to review or verify the quality of the recommendations before sending them through their chain of command.

It will be easier to find that the duty of care has been breached where the failure can be attributed to decisions about training or equipment that were taken before deployment, when there was time to assess the risks to life that had to be planned for, than it will be where they are attributable to what was taking place in theatre. The more constrained he is by decisions that have already been taken for reasons of policy at a high level of command beforehand or by the effects of contact with the enemy, the more difficult it will be to find that the decision-taker in theatre was at fault

(Smith and others (FC) v The Ministry of Defense, Supreme Court UK, 2013).

If so, then best practise WHS frameworks might be drawn on in development, acquisition, and training ahead of deployment of systems in conflicts to predict and mitigate risks of human–AI military teams. A WHS Scorecard for AI systems (modified for military use) could be used by militaries to estimate effects on operators from working side-by-side with AI such as feeling disempowered, demotivated, exhausted, a sense of a reduction in status or value, anxiety, boredom etc., as well as a sense of reduced or enhanced responsibility.

DISCUSSION

The future risks a more distributed and fluid-like attack by militaries seeking to confuse their enemies and adapt in theatre. This type of command and control, sometimes called 'Mosaic warfare' involves switching control across multiple platforms in an algorithmically optimized operation (Clark et al., 2020). Rather than a war of attrition, the 'decision-centric' mosaic warfare achieves two advantages, it imposes multiple dilemmas on an enemy to prevent it from achieving its objectives and speeds up decision processes. The risk to human connection to these decisions is obvious, if decision-making across AI-enabled autonomous platforms is sped up beyond human cognitive capacities, then loss of SA is inevitable. From this there is a loss of knowledge and competence required for moral responsibility for decisions made. Militaries need to consider the human factors of AI-enabled mosaic warfare. Extensive simulation, training, and exercises with assets ahead of a conflict must occur so that commanders understand the left and right of arc of decisions made in theatre (Devitt, 2023). Human operators need to feel connected to the technology stack they are responsible for. They need to feel their intention expressed through the behaviors of these systems, even if the way the systems achieve this intent is too complex for individual humans to grasp.

CONCLUSION

Militaries are responsible for the decision to go to war and the conduct of war itself. The moral blameworthiness of decisions of individuals in conflicts must be considered within the broader decision environments within which they act. AI has the potential to reduce civilian harm and protect soldiers from undue levels of suffering leading to moral injury. However, processes must be instituted to measure changes to decision-making in human–AI teams. This is particularly acute with the systemic replacement of cognitive functions by AI that humans would normally be tasked with in targeting decisions to be morally and legally responsible for those decisions. The thesis of this chapter is that good apples can become 'cooked' if they are poorly paired with AI systems including detaching from responsibility for decisions, becoming moral crumple zones, suffering moral injury, or becoming extreme moral witnesses. Therefore, militaries must carefully scrutinize AI systems that take away human agency to ensure that safe, meaningful, satisfying, and engaging work is found for humans responsible for decision-making and working alongside AI. Combining tools from human factors and WHS will enable militaries to track, measure, and take responsibility for ensuring personnel remain 'good' and not 'cooked'.

NOTES

1 Thanks to Jan Maarten Schraagen, Alec Tattersall, and Kobi Leins for their review comments that improved this chapter immeasurably.
2 "Twelve of the sixteen personnel involved in the bombing of the hospital had been punished with removal from command, letters of reprimand, formal counselling, and extensive retraining. The list included a general officer, the AC-130 gunship aircrew, and the US Special Forces team on the ground" (Donati, 2021, p. 234).
3 Medicines San Frontières "asked to know who in the chain of command was ultimately responsible for the forty-two people killed in the hospital that night. That question, along with all the others, was never answered" (Donati, 2021, p. 235).
4 However, a physical risk was noted if workers felt obliged to work faster due to increased surveillance and monitoring.

REFERENCES

Agrawal, A., Gans, J., & Goldfarb, A. (2018, 17 April). A simple tool to start making decisions with the help of AI. *Harvard Business Review*. https://hbr.org/2018/04/a-simple-tool-to-start-making-decisions-with-the-help-of-ai

Aristotle. (350 B.C.E.) (n.d.). *Nicomachean ethics* (W. D. Ross, Trans.). https://classics.mit.edu/Aristotle/nicomachaen.html

Autor, D., Mindell, D., & Reynolds, E. (2020). The work of the future: Building better jobs in an age of intelligent machines. In *MIT task force on the work of the future*. https://workofthefuture.mit.edu/wp-content/uploads/2021/01/2020-Final-Report4.pdf

Bal, M. (2017). *Dignity in the workplace: New theoretical perspectives.* Springer.

Barnes, J. E., Bender, M. C., & Haberman, M. (2022, 1 September). Trump's tastes in intelligence: Power and leverage. *New York Times.* https://www.nytimes.com/2022/09/01/us/politics/trump-intelligence-briefings.html

Bastian, M. (2023, 25 March). GPT-4 has more than a trillion parameters - Report. *The Decoder.* https://the-decoder.com/gpt-4-has-a-trillion-parameters/#:~:text=Further%20details%20on%20GPT%2D4's,Mixture%20of%20Experts%20(MoE).

Bouchet-Saulnier, F., & Whittal, J. (2018). An environment conducive to mistakes? Lessons learnt from the attack on the Medecins Sans Frontieres hospital in Kunduz, Afghanistan. *International Review of the Red Cross, 100*(907–909), 337–372. https://doi.org/10.1017/S1816383118000619

Calvo, R. A., Peters, D., Vold, K., & Ryan, R. M. (2020). Supporting human autonomy in AI systems: A framework for ethical enquiry. In C. Burr & L. Floridi (Eds.), *Ethics of digital well-being: A multidisciplinary approach* (Vol. 140, pp. 31–54). Springer.

Cebulla, A., Szpak, Z., Howell, C., Knight, G., & Hussain, S. (2022). Applying ethics to AI in the workplace: The design of a scorecard for Australian workplace health and safety. *AI & Society, 38,* 919–935. https://doi.org/10.1007/s00146-022-01460-9

Centre for Work Health and Safety. (2021). *Ethical use of artificial intelligence in the workplace - AI WHS Scorecard.* NSW Government. https://www.centreforwhs.nsw.gov.au/knowledge-hub/ethical-use-of-artificial-intelligence-in-the-workplace-final-report

Clark, B., Patt, D., & Schramm, H. (2020). *Mosaic warfare: Exploiting artificial intelligence and autonomous systems to implement decision-centric operations.* https://csbaonline.org/research/publications/mosaic-warfare-exploiting-artificial-intelligence-and-autonomous-systems-to-implement-decision-centric-operations

Clausewitz, C. V., Howard, M., & Paret, P. (1976). *On war.* Princeton University Press.

Crawford, N. (2013). *Accountability for killing: Moral responsibility for collateral damage in America's post-9/11 wars.* Oxford University Press.

Cummings, M. L., Gao, F., & Thornburg, K. M. (2016). Boredom in the workplace: A new look at an old problem. *Human Factors, 58*(2), 279–300. https://doi.org/10.1177/0018720815609503

Dekker, S. (2013). *Second victim: Error, guilt, trauma, and resilience.* CRC Press.

Department of Industry Innovation and Science. (2019). *Australia's Artificial Intelligence Ethics Framework.* Retrieved 25 September, from https://www.industry.gov.au/data-and-publications/australias-artificial-intelligence-ethics-framework/australias-ai-ethics-principles

Devitt, S. K. (2023). Meaningful human command: Advance control directives as a method to enable moral and legal responsibility for autonomous weapons systems [pre-print]. In D. Amoroso, J. van den Hoven, D. Abbink, F. Santoni de Sio, G. Mecacci, & L. C. Siebert (Eds.), *Meaningful human control of artificial intelligence systems.* Edward Elgar Publishing. https://arxiv.org/abs/2303.06813

Devitt, S. K., & Copeland, D. (2023). Australia's approach to AI governance in security and defense. In M. Raska & R. Bitzinger (Eds.), *The AI wave in defense innovation: Assessing military artificial intelligence strategies, capabilities, and trajectories.* Routledge. https://arxiv.org/abs/2112.01252

Devitt, S. K., Scholz, J., Schless, T., & Lewis, L. (2023). Developing a trusted human-AI network for humanitarian benefit. *The Journal of Digital War, 4,* 1–17. https://doi.org/10.1057/s42984-023-00063-y

Donati, J. (2021). *Eagle down: The last Special Forces fighting the forever war.* Hachette UK.

Elish, M. C. (2019). Moral crumple zones: Cautionary tales in human-robot interaction. *Engaging Science, Technology, and Society, 5,* 40–60.

Endsley, M. R. (2017). From here to autonomy: Lessons learned from human-automation research. *Human Factors, 59*(1), 5–27. https://doi.org/10.1177/0018720816681350

Endsley, M. R. (2023). Supporting Human-AI Teams: Transparency, explainability, and situation awareness. *Computers in Human Behavior, 140*, 107574.

Enemark, C. (2019). Drones, risk, and moral injury. *Critical Military Studies, 5*(2), 150–167.

Fairgrieve, D. (2014). Suing the military: The justiciability of damages claims against the armed forces. *Cambridge Law Journal, 73*(1), 18–21. https://doi.org/10.1017/S0008197314000117

Fani, N., Currier, J. M., Turner, M. D., Guelfo, A., Kloess, M., Jain, J., Mekawi, Y., Kuzyk, E., Hinrichs, R., & Bradley, B. (2021). Moral injury in civilians: Associations with trauma exposure, PTSD, and suicide behavior. *European Journal of Psychotraumatology, 12*(1), 1965464.

Feinberg, J. (1985). Sua culpa. In D. G. Johnson & J. W. Snapper (Eds.), *Ethical issues in the use of computers* (pp. 102–120). Wadsworth Publ. Co.

Fitzgibbon, J. (2020, 27 November). Diggers in Afghanistan: Where did it go wrong? *Newcastle Weekly.* https://newcastleweekly.com.au/joel-fitzgibbon-where-did-it-go-wrong/

French, P. A. (2010). *War and moral dissonance.* Cambridge University Press.

Giboney, J. S., Brown, S. A., Lowry, P. B., & Nunamaker Jr, J. F. (2015). User acceptance of knowledge-based system recommendations: Explanations, arguments, and fit. *Decision Support Systems, 72*, 1–10.

Gillespie, T. (2020). Content moderation, AI, and the question of scale. *Big Data & Society, 7*(2). ttps://journals.sagepub.com/doi/pdf/10.1177/2053951720943234

Hall, B., & Scielzo, S. (2022). Red rover, red rover, send an F-35 right over: Assessing synthetic agent trust in humans to optimize mission outcomes in mosaic warfare. In *MODSIM World 2022,* Norfolk, VA, USA. https://www.modsimworld.org/papers/2022/MODSIM_2022_paper_13.pdf

Hardy, M., Sucholutsky, I., Thompson, B., & Griffiths, T. (2023). Large language models meet cognitive science. In *Proceedings of the annual meeting of the cognitive science society,* Sydney.

Heaven, W. D. (2021). Hundreds of AI tools have been built to catch covid. None of them helped. *MIT Technology Review.* https://www.technologyreview.com/2021/07/30/1030329/machine-learning-ai-failed-covid-hospital-diagnosis-pandemic/

Hoff, K. A., & Bashir, M. (2015). Trust in automation: Integrating empirical evidence on factors that influence trust. *Human Factors, 57*(3), 407–434.

Holz, J. (2021). Victimhood and trauma within drone warfare. *Critical Military Studies, 9*(2), 175–190.

Jamieson, N., Usher, K., Ratnarajah, D., & Maple, M. (2021). Walking forwards with moral injury: Narratives from ex-serving Australian Defense Force members. *Journal of Vetrans Studies, 7*(1), 174–185. https://doi.org/10.21061/jvs.v7i1.214

Jespersen, A. H., Hasle, P., & Nielsen, K. T. (2016). The wicked character of psychosocial risks: implications for regulation. *Nordic Journal of Working Life Studies, 6*(3), 23–42. https://vbn.aau.dk/en/publications/the-wicked-character-of-psychosocial-risks-implications-for-regul

Jespersen, A. H., Hohnen, P., & Hasle, P. (2016). Internal audits of psychosocial risks at workplaces with certified OHS management systems. *Safety Science, 84*, 201–209. https://www.sciencedirect.com/science/article/abs/pii/S0925753515003422

Johnson, J. (2023). Finding AI faces in the moon and armies in the clouds: Anthropomorphising artificial intelligence in military human-machine interactions. *Global Society, 38*(1), 1–16. https://doi.org/10.1080/13600826.2023.2205444

Kaldas, N., Douglas, J., & Brown, P. (2023). Royal commission into defense and veteran suicide. https://defenseveteransuicide.royalcommission.gov.au/

Kerr, D. (2023, 26 August). *Armed with traffic cones, protesters are immobilizing driverless cars.* NPR. https://www.npr.org/2023/08/26/1195695051/driverless-cars-san-francisco-waymo-cruise

King, A. (2006). The word of command: Communication and cohesion in the military. *Armed Forces & Society, 32*(4), 493–512.

Klein, G. A., Orasana, J., Calderwood, R., & Zsambok, C. E. (Eds.). (1993). *Decision making in action: Models and methods.* Ablex Publishers.

Lee, J. D., & See, K. A. (2004). Trust in automation: Designing for appropriate reliance. *Human Factors, 46*(1), 50–80. https://doi.org/10.1518/hfes.46.1.50_30392

Leins, K., & Durham, H. (2023, 28 August). 2023 *DOD manual revision - to shoot, or not to shoot... automation and the presumption of civilian status.* Articles of War, Lieber Institute West Point. https://lieber.westpoint.edu/shoot-not-shoot-automation-presumption-civilian-status/

Leins, K., & Kaspersen, A. (2021, 30 August). *Seven myths of using the term "human on the loop": "Just what do you think you are doing, Dave?"* Artificial Intelligence & Equality Initiative (AIEI), Carnegie Council for Ethics in International Affairs. https://www.carnegiecouncil.org/media/article/7-myths-of-using-the-term-human-on-the-loop

Lewis, L. (2023). *Emerging technologies and civilian harm mitigation.* Center for Naval Analysis. https://www.cna.org/quick-looks/2023/EMERGING-TECHNOLOGIES-AND-CIVILIAN-HARM-MITIGATION.pdf

Lewis, L., & Ilachinski, A. (2022). *Leveraging AI to mitigate civilian harm.* CNA. https://www.cna.org/reports/2022/02/leveraging-ai-to-mitigate-civilian-harm

Liivoja, R., Massingham, E., & McKenzie, S. (2022). The legal requirement for command and the future of autonomous military platforms. *International Law Studies, 99*(1), 27.

Mackworth, N. H. (1948). The breakdown of vigilance during prolonged visual search. *Quarterly Journal of Experimental Psychology, 1*(1), 6–21.

Mackworth, N. H. (1950). *Researches on the measurement of human performance* (Med. Res. Council, Special Rep. Ser. No. 268.). His Majesty's Stationery Office.

Martínez-Pérez, V., Andreu, A., Sandoval-Lentisco, A., Tortajada, M., Palmero, L. B., Castillo, A., Campoy, G., & Fuentes, L. J. (2023). Vigilance decrement and mind-wandering in sustained attention tasks: Two sides of the same coin? *Frontiers in Neuroscience, 17*, 122406.

McEwen, C., Alisic, E., & Jobson, L. (2021). Moral injury and mental health: A systematic review and meta-analysis. *Traumatology, 27*(3), 303.

Merritt, S. M., Heimbaugh, H., LaChapell, J., & Lee, D. (2013). I trust it, but I don't know why: Effects of implicit attitudes toward automation on trust in an automated system. *Human Factors, 55*(3), 520–534. https://doi.org/10.1177/0018720812465081

Miller, M. T. (2020). Command responsibility: A model for defining meaningful human control. *Journal of National Security, Law and Policy, 11*, 533–546. https://jnslp.com/2021/02/02/command-responsibility-a-model-for-defining-meaningful-human-control/

Mohamed, S. (2021). Cannon fodder, or a soldier's right to life. *Southern California Law Review, 95*(5), 1037. https://southerncalifornialawreview.com/2023/02/23/cannon-fodder-or-a-soldiers-right-to-life/

Nissenbaum, H. (1996). Accountability in a computerized society. *Science and Engineering Ethics, 2*, 25–42. https://link.springer.com/article/10.1007/BF02639315

OECD. (2022). OECD framework for the classification of AI systems: A tool for effective AI policies (OECD Digital Economy Papers, February 2022 No 323, Issue). https://oecd.ai/en/classification

Ormsby, G. (2019, 11 March). Uber fatality unveils AI accountability issues. *Lawyers Weekly.* https://www.lawyersweekly.com.au/biglaw/25213-uber-fatality-unveils-ai-accountability-issues

Ortiz, S. (2023, 14 April). What is Auto-GPT? Everything to know about the next powerful AI tool. *ZDNET.* https://www.zdnet.com/article/what-is-auto-gpt-everything-to-know-about-the-next-powerful-ai-tool/

Orts, E. W., & Smith, N. C. (2017). *The moral responsibility of firms.* Oxford University Press.

Ott, S., Hebenstreit, K., Liévin, V., Hother, C. E., Moradi, M., Mayrhauser, M., Praas, R., Winther, O., & Samwald, M. (2023). ThoughtSource: A central hub for large language model reasoning data. *Scientific Data, 10*(1), 528–528. https://doi.org/10.1038/s41597-023-02433-3

Oviedo-Trespalacios, O., Peden, A. E., Cole-Hunter, T., Costantini, A., Haghani, M., Rod, J. E., Kelly, S., Torkamaan, H., Tariq, A., David Albert Newton, J., Gallagher, T., Steinert, S., Filtness, A. J., & Reniers, G. (2023). The risks of using ChatGPT to obtain common safety-related information and advice. *Safety Science, 167*, 106244. https://doi.org/https://doi.org/10.1016/j.ssci.2023.106244

Palantir. (2023). *Artificial intelligence platform for defense*. Palantir. Retrieved April 4, 2023, from https://www.palantir.com/platforms/aip/

Parsons, G., & Wilson, M. A. (2020). Fighting versus waging war: Rethinking *jus in bello* after Afghanistan and Iraq In G. Parsons & M. A. Wilson (Eds.), *Walzer and war: Reading just and unjust wars today* (pp. 125–155). Palgrave Macmillan. https://doi.org/10.1007/978-3-030-41657-7_7

Phelps, W., & Grossman, D. (2021). *On killing remotely: The psychology of killing with drones*. Little, Brown and Company.

Quinn, W. S. (1989). Actions, intentions, and consequences: The doctrine of double effect. *Philosophy & Public Affairs, 18*(4), 334–351. https://www.jstor.org/stable/2265475

Rana, K., Haviland, J., Garg, S., Abou-Chakra, J., Reid, I., & Suenderhauf, N. (2023). SayPlan: Grounding large language models using 3D scene graphs for scalable task planning. *arXiv preprint arXiv:2307.06135*.

Reason, J. (2016). *Managing the risks of organizational accidents*. Routledge.

Reid, A., O'Callaghan, S., & Lu, Y. (2023). Implementing Australia's AI ethics principles: A selection of responsible AI practices and resources. https://www.csiro.au/en/work-with-us/industries/technology/National-AI-Centre/Implementing-Australias-AI-Ethics-Principles-report

Roberson, T., Bornstein, S., Liivoja, R., Ng, S., Scholz, J., & Devitt, K. (2022). A method for ethical AI in defense: A case study on developing trustworthy autonomous systems. *Journal of Responsible Technology, 11*, 100036. https://doi.org/10.1016/j.jrt.2022.100036

Safe Work Australia. (2020). Principles of good work design: A work health and safety handbook. Retrieved from https://www.safeworkaustralia.gov.au/resources-and-publications/guidance-materials/principles-good-work-design

Saini, R. K., Raju, M., & Chail, A. (2021). Cry in the sky: Psychological impact on drone operators. *Industrial Psychiatry Journal, 30*(Suppl 1), S15.

Salvendy, G., & Karwowski, W. (2021). 17.2 Situation awareness defined. In G. Salvendy & W. Karwowski (Eds.), *Handbook of human factors and ergonomics* (5th ed.). John Wiley & Sons. https://app.knovel.com/hotlink/pdf/id:kt012XFF03/handbook-human-factors/situation-awareness-defined

Smiley, L. (2022, 8 March). 'I'm the Operator': The aftermath of a self-driving tragedy. *Wired*. https://www.wired.com/story/uber-self-driving-car-fatal-crash/

Stanton, N. A., Salmon, P. M., Rafferty, L. A., Walker, G. H., Baber, C., & Jenkins, D. P. (2017). *Human factors methods: A practical guide for engineering and design*. CRC Press.

Smith and others (FC) v The Ministry of Defense, 37 (2013). https://www.supremecourt.uk/cases/uksc-2012-0249.html

Talbert, M. (2016). *Moral responsibility: An introduction*. John Wiley & Sons.

Thomson, D. R., Besner, D., & Smilek, D. (2015). A resource-control account of sustained attention: Evidence from mind-wandering and vigilance paradigms. *Perspectives on Psychological Science, 10*(1), 82–96.

Trusted Autonomous Systems. (2023). *Responsible AI in defense (RAID) toolkit*. Trusted Autonomous Systems Defense Cooperative Research Centre. Retrieved August 1, 2023, from https://tasdcrc.com.au/responsible-ai-for-defense-consultation/

Tsai, G. (2014). Rational persuasion as paternalism. *Philosophy & Public Affairs, 42*(1), 78–112. https://doi.org/10.1111/papa.12026

Vallor, S. (2023). Edinburgh declaration on responsibility for responsible AI. *Medium*. Retrieved 15 July, from https://medium.com/@svallor_10030/edinburgh-declaration-on-responsibility-for-responsible-ai-1a98ed2e328b

Walzer, M. (2015). *Just and unjust wars: A moral argument with historical illustrations*. Little Brown.

Weizenbaum, J. (1966). ELIZA - a computer program for the study of natural language communication between man and machine. *Communications of the ACM, 9*(1), 36–45.

Westphal, M., Vössing, M., Satzger, G., Yom-Tov, G. B., & Rafaeli, A. (2023). Decision control and explanations in human-AI collaboration: Improving user perceptions and compliance. *Computers in Human Behavior, 144*, 107714. https://doi.org/10.1016/j.chb.2023.107714

Wiggers, K. (2023, 22 April). What is Auto-GPT and why does it matter? *TechCrunch*. https://techcrunch.com/2023/04/22/what-is-auto-gpt-and-why-does-it-matter/

Williamson, V., Stevelink, S. A., & Greenberg, N. (2018). Occupational moral injury and mental health: Systematic review and meta-analysis. *The British Journal of Psychiatry, 212*(6), 339–346.

Woods, D., Dekker, S., Cook, R., Johannesen, L., & Sarter, N. (2017). *Behind human error* (2nd ed.). CRC Press.

Zeng, J., & Kaye, D. B. V. (2022). From content moderation to visibility moderation: A case study of platform governance on TikTok. *Policy & Internet, 14*(1), 79–95.

Zsambok, C. E., & Klein, G. (Eds.). (1997). *Naturalistic decision making*. Lawrence Erlbaum Associates.

Neglect Tolerance as a Measure for Responsible Human Delegation

14

Christopher A. Miller and Richard G. Freedman

INTRODUCTION AND BACKGROUND

Responsible use of an AI-based weapon system is similar to the responsible use of any weapon albeit with novel, but not completely unfamiliar, complexities and challenges. We have been developing a process and a metric for assessing the degree to which a decision to rely on another agent (whether human or machine) can be said to be "responsible" – that is, technically and ethically sound, informed, and made with reasonable expectations of avoiding physical and ethical hazards. This approach places the emphasis on deciding whether to place an agent in autonomous control of behavior in context and not (directly) upon whether the agent can or will make ethical decisions on its own. Our approach leverages prior work (Goodrich et al., 2001) on Neglect Tolerance (NT) – the degree to which one can rely on a robot to maintain performance above a

 DOI: 10.1201/9781003410379-17

threshold for a period of time under conditions of human "neglect" (i.e., without direct human attention and intervention).

This work derives from and extends a focus on *"Meaningful Human Control"* (MHC). In April 2016, the United Nation's Convention on Certain Conventional Weapons Meeting of Experts on Lethal Autonomous Weapons Systems was held in Geneva. The group of delegates to that convention converged on MHC as a possible standard or prerequisite for the ethical deployment of AI systems in military contexts (Memorandum to Convention on Conventional Weapons [CCW] Delegates, 2016). The MHC concept itself was not completely defined at the meeting, but it has generally "been used to describe a threshold of human control that is considered necessary" for a system to be considered ethically in warfare (Roff and Moyes, 2016) regardless of whether the machine itself reasons ethically.

The phrase "Meaningful Human Control" has been criticized for its lack of concrete or consistent definition (Cummings, 2019) and for whether it represents any difference beyond the legal and ethical requirements for all weapons (Horowitz & Scharre, 2015). It has been suggested that the phrase was advanced in the belief that it would effectively prohibit *any* autonomous weapon system deployment (Horowitz and Scharre, 2015). Suggested alternatives include *Autonomous System Certification* (Cummings, 2019) accentuating the idea that the usage pattern and division of roles is what needs to be certified (as in a pilot's certification to fly a specific aircraft), not the technology exclusively. *Meaningful Human Involvement* is another alternate concept stressing that humans must be ethically involved in "meaningful and context-appropriate ways" that might not entail "control" (personal communication from Mike Boardman, Principal Advisor, Human Sciences Group, CBR Division, Defence Science and Technology Laboratory on November 11, 2022). The U.S. Department of Defense has recently updated DoD Directive 3000.09 on Autonomy in Weapon Systems which uses the phrase *Appropriate Levels of Human Judgement*. It states that U.S. defense policy will be: "Autonomous and semi-autonomous weapon systems will be designed to allow commanders and operators to exercise appropriate levels of human judgment over the use of force" (Department of Defense, 2023). A recent workshop, forthcoming book, and a multi-stakeholder, international summit in The Hague in February 2023, used the term "Responsible AI" to refer to ethical human use of AI systems (cf. www.reaim2023.org). It is in this latter spirit that we have adopted the term *Responsible Human Delegation* (RHD) to highlight the decision (which may take place at many alternate points in an organization or incident) to use an agent in a role without direct oversight, though perhaps with a set of instructions or a bounded space of behavioral options. Delegation, as we have argued elsewhere, is the act that makes Supervisory Control operational (cf. Sheridan, 2002; Miller and Parasuraman, 2007). The decision of whether and how to delegate is an act of control, but it is also a choice to forego direct control for a period and/or a region of operational space. RHD must be based on the ability to make an informed decision about likely agent performance within that space given the conditions that are likely to hold. In the remainder of this chapter, we explore and propose approaches to ascertaining and measuring RHD.

A CORE DISTINCTION AND PROBLEM FRAMING

A practical distinction in RHD arises over the difference between *"real-time (RT) control"* vs. *"non-RT control"*. RHD implies the delegation of some control authority to another agent, yet the decision-making and behavioral "space" of that delegated control is usually very adjustable, at least in principle. A driving instructor has "delegated" manipulation of all automobile control devices (steering wheel, brakes, accelerator, etc.) to the student, but the instructor remains very actively involved in the real-time control of the vehicle through instructions and, if necessary, grabbing the wheel or employing an auxiliary braking pedal.

In an RT control situation involving automation, the human user is (or is supposed to be) aware of and able to exert control over the system *as it performs* its tasks – such as in a telepresence robot or a current autopilot. Note that this control does not have to be immediate, such as in fly-by-wire aviation systems, which can interpret, adjust, and provide safeguards that "translate" human inputs.

Non-RT control systems, however, are more readily apparent as instances of delegation. In a non-RT control situation, the supervisor's control or influence must be exerted *before* releasing the system to perform its tasks via policies, rules, permissions, and selection of the place and conditions of release. In the driving instructor example above, this might be equivalent to the instructor saying "Drive us home, but don't take the freeway; I'm going to take a nap".

Some might argue that there cannot be human "control" during intervals and conditions where the human cannot exert (immediate) control at all. This seems unrealistic, especially as applied to military systems (cf., Horowitz and Scharre, 2015). Ever since the first proto-human threw a rock at another, we have been using weapons over which we don't have immediate and proximal RT control. Instead, we exert control before releasing the weapon through the choice of whether, when, and how to release it. Judging whether the weapon's use was ethical and "responsible" has less to do with the behavior of the weapon and more to do with whether the human operator's choice to release that weapon at that target in those circumstances was justified and *appropriate*. It should be made with a reasonable expectation of accurately striking only its intended target. The situation is similar with modern AI-based weapons operating out of immediate RT-control either through lack of communications or in some "autonomous" modes. The chief difference between a modern autonomous weapon system and a rock, in this regard, is that autonomous systems typically permit some kind of programming or policy creation about when and how they will behave, while rocks follow a largely ballistic trajectory once released. This, however, is similar to the situation that has existed in human communication with human subordinates throughout human history. The creation of a policy and the decision of when to release the agent with it are, therefore, channels through which human control is afforded. The delegator remains responsible for the decision to delegate.

AI systems go one step further than traditional automation in that they can be capable of learning behaviors, decision-making policies and even setting their own goals. This makes them more complex and unpredictable to a human operator, overseer, or co-worker. Insofar as such systems remain subordinate to some human's will and intent, however, the situation remains similar: control is through the delegation of a "space" of possible actions, and the decision to release the system with instructions bounding that space is where human control remains.

These two conditions, RT and non-RT control, give rise to different considerations for whether and when RHD might exist. When control is exerted in real-time, a human who is thoroughly "in the loop" delegates – maintaining awareness of the world context, the state of work, and the behaviors of the automation if any. Delegation typically involves a smaller scope and shorter time horizons. The human operator may delegate/command that the landing gear be lowered instead of lowering it manually, but the decision and behavior are very "local". In such contexts, the presence or absence of RHD revolves around the ability of the human to achieve their intent via the capabilities of the system. Therefore, familiar Human–Machine interaction parameters such as handling qualities, requisite variety, workload, situation awareness, etc. determine whether or not the human exhibited RHD when issuing a command. Were these attributes sufficient to allow an informed choice to command that the landing gear be lowered? Was the human aware of potential implications? Could they override and revert the decision in time? Etc.

On the other hand, when non-RT control is concerned, RHD must be available and exerted through the creation of programs or policies to which the AI-based system adheres and through the decision whether to deploy the system in existing and foreseeable conditions at specific targets. This implies that humans should be capable of making reasonable predictions about system behaviors in the possibly dynamic operational conditions in which it will be deployed, thus, requiring a deep understanding of the system's decision-making and behavioral processes. If this is not possible, then the decision to deploy the system under conditions where the proximal "operator" cannot know what it will do must be made by the operator's superior – and responsibility will or should lie there.

So how can we judge whether or not a specific combination of system and usage context affords RHD? Put another and more specific way: how can we know when a proposed combination of human knowledge and capabilities, interface affordances, system behaviors (including the instructions/programming which guide and constrain those behaviors), and world state variations combine to provide adequate "space" for a human to make a responsible decision about delegating or not? We are not asking whether deciding to throw a specific rock in a specific context was a responsible decision, but rather whether we've chosen and trained the human rock-throwers, put them in a location and context where rock throwing at targets is sufficiently predictable so that *they* can make responsible decisions about whether and at whom to throw.

With RT-control systems, where the human retains immediate influence on the system, the situation is far from trivial but it is subject to traditional human certification and system verification and validation methods. We need just to ensure that the human is selected and trained to understand and operate the system and that the system affords adequate control authority and situational transparency for the human to exert their will in all likely circumstances and contexts of use. While we are far from perfect at achieving these ends, the problems are at least familiar and are addressed through existing military standards and procedures such as U.S. Military Standard MIL-STD 1472 (Department of Defense, 2020).

For human interaction with non-RT AI-based control systems, the problem is more challenging. Because there will be periods during which the humans will not exert influence over the system's behavior, humans must exercise control by making responsible decisions about the conditions in which to deploy the system and the behavioral policy with which it is released. As with throwing a rock, the question would seem to hinge on whether the human had reasonable knowledge of the target, their own accuracy in hitting it, and the degree of threat to other individuals and property, etc. As the duration and complexity of autonomous system behavior increases, this becomes more difficult. A rock has very little behavioral variability: it travels in a largely ballistic arc making predictability of its endpoint comparatively simple, though even then wind gusts and the movement of individuals in and out of the trajectory complicate the problem. By contrast, the interaction of behavioral policy and "programming" of autonomous agents with unexpected world states has already produced a wide range of unpredicted breakdown conditions with unexpected and sometimes fatal results. The advent of autonomous learning systems, which may result in constantly changing behavior with little or no transparency, makes this situation worse still.

EVALUATING RHD CONDITIONS

With the understanding of RHD as sketched above, we need the ability to answer the core question: does a given combination of human awareness and abilities, machine behaviors, and world state variables permit responsible delegation decisions – that is, *informed* decisions in which the probability of producing a desired outcome or avoiding an undesirable one are accurately understood by the decider?

Neglect Tolerance (NT) (Crandall & Goodrich, 2002) may be a useful concept for evaluating RHD. NT was introduced in 2001 (Goodrich et al., 2001; Olsen & Goodrich, 2003) to quantitatively characterize the degree of autonomy of a system. The core notion is that the longer a human operator can leave a machine unattended and behaving autonomously (i.e., its "neglect tolerance") in a given context and still receive acceptable performance against some standard, the more autonomous the machine is. While there have been various formulations of NT over time and various factors shown to affect it (Elara, 2011; Elara et al., 2009; Wang & Lewis, 2007), Figure 14.1 illustrates the basic formulation (derived from Crandall et al., 2005 and adding our own explanatory commentary). The x-axis shows some measure of overall system effectiveness. For us, this will be the likelihood of ethical behavior from the human-machine system, but the initial formulation (cf. Figure 14.1) used simple task performance. The y-axis shows the time elapsed. The system has some level of effectiveness against the performance metric

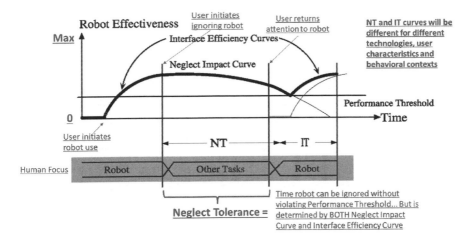

FIGURE 14.1 Neglect Tolerance as a Measure for Robot Autonomy. The base figure is adapted from Crandall et al., (2005). Underlined annotations are provided by the authors.

with human oversight – generally (but not necessarily) assumed to be higher than without human attention. When human oversight is suspended, that performance frequently degrades probabilistically according to a specific *Neglect Impact Curve* unique to the specific task, machine, and world context. Furthermore, there is a presumed minimal threshold, defined on the effectiveness dimension, below which system behavior is no longer acceptable. Human intervention is required to bring performance back above that threshold, but because this intervention will take some time and be only probabilistically required, the human will have to inspect the system with sufficient frequency to allow for detection and intervention to prevent falling below the threshold, further reducing the NT interval. Detection and intervention time requirements will be governed by the interfaces and controls available, human workload and training, cognitive "switching costs" between tasks, etc. which enable the operator to become sufficiently familiar with the system and world state to exert effective influence. This trajectory also adheres to a technology- and context-specific *Interface Efficiency Curve*. Olsen and Goodrich's (2003) measure of NT is the temporal interval during which the system can be ignored (i.e., "neglected" or allowed to behave in an autonomous fashion with non-RT human control) while maintaining a reasonable probability that acceptable behavior will.

We can usefully define a similar concept identifying the likelihood that a machine system will behave in an ethically responsible fashion when acting autonomously in a non-RT control context – an **Ethical Neglect Tolerance (ENT)** interval as illustrated in Figure 14.2. To compute an ENT score, we must begin by defining and agreeing on a set of "ethical hazard" states that the system, in its context(s) of use, might be prone to, such as killing a non-combatant, responding with disproportional force, etc. This **Ethical Hazard Analysis** step might well draw from techniques for traditional system analysis such as Failure Modes Effects Analysis (Ben-Daya, 2009) but applied instead to the probabilities of the identified ethical risks. Such risks are generally already known and quantified to some extent under military doctrine and rules of engagement, and there

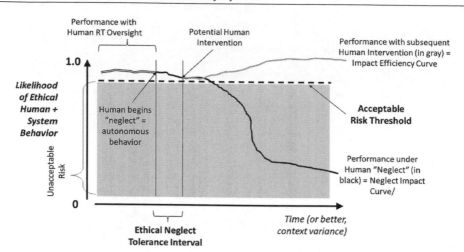

FIGURE 14.2 "Ethical" Neglect Tolerance – Adapting Neglect Tolerance to problems of determining whether conditions exist to afford responsible delegation decisions.

has already been substantial work on techniques for estimating especially collateral damage (Dillenburger, 2012; Humphrey et al., 2008).

After identifying ethical hazards, we must estimate the Neglect Impact Curve for each – that is, the likelihood that the system under non-RT control (i.e., behaving "autonomously") will transgress into each hazard state, both with and without human oversight. Similar Interface Efficiency Curves for each hazard must provide the degree of improvement (usually, but not inevitably, decline) in ethical risk that human oversight provides over time. Such estimates can be provided through analysis, simulation, computation, prediction, observation (of historical events or behavior in simulation), or other methods. These estimates may be incomplete or inaccurate, but as with traditional hazard analyses, identifying, tracking, and reasoning about them will likely produce better systems than overall not doing so. We can also improve estimates over time with better models, experience, more precise definitions, and more constrained contexts. The Acceptable Risk "Threshold" must be defined as well – that is, an acceptable probability the overall system will avoid performing any of the ethical breakdowns identified.

In this framing of ENT, we can define the conditions for "*Responsible Human Delegation*" as existing whenever the human (operator, organization, designer, etc.) is able to either:

1. Attend to (i.e., not neglect) the automated system for intervals in which the probability of an excursion below the threshold exceeds acceptable risk, or
2. Create policies and programming that keep the likelihood of ethical behavior above the threshold during the full periods of human neglect

This requires having adequate knowledge, awareness, control authority, and time to avoid giving an autonomous system license to operate in conditions or for intervals that exceed the risk tolerance threshold. In other words, ethical risks are distilled into operational boundaries within which the system may operate autonomously. Note that it is still possible that the autonomous system may transgress one of the ethical hazards. Even the most proficient and professional stone throwers may experience an "unanticipatable" event (e.g., another stone that knocks the first off course) that results in a civilian casualty. These are unfortunate and horrible incidents, but we generally agree that, if the cause for the use of violent weapons was justified initially, the user was trained and proficient, they understood the context, and had reasonable expectations of avoiding collateral damage risk, etc., then the weapon was used in a responsible way and no blame adheres to the user. That is, the weapon was released under conditions of "acceptable risk" and the decision to deploy the system was a "responsible human delegation" decision.

Given this conception, it seems possible to compute an ENT interval for a system in a context without that system having any internal ethical reasoning capability of its own. Such a system would behave according to its default and/or programmed behavior without any adjustment (either from the human or from an "onboard" ethical reasoning capability), and the likelihood of performing an unethical action would obey the trajectory of the associated Neglect Impact Curve. Alternatively, a system with a sophisticated (and accurate) ethical reasoning capability would greatly extend the interval of ENT by making all and only ethical decisions for an indefinite duration – its curve would never fall below the threshold. We will likely always fall between those two extremes.

A WORKED EXAMPLE

In this section, we will work through an initial, artificial, and comparatively simple example to illustrate how the ideas of ENT may be applied and operationalized. First, we describe the scenario and then apply the ENT concept. We have chosen a non-military scenario to support near-universal familiarity with the situation and open discussion of it. To handle the combinatorial scaling involved with the many factors affecting each other within even this simple analysis, we will discuss the benefits of modeling with Bayesian Networks for assessing the probabilities involved in ENT. We will also discuss other forms of modeling that complement situations that Bayesian Networks are unable to represent, opening the possibilities for capturing a variety of "what if" inquiries during the design process.

A Simple Scenario: Nibbles vs. the Robot Vacuum Cleaner

As a simple situation, let's consider an apartment owner who has a cherished hamster named Nibbles and is interested in acquiring a much-needed autonomous vacuum-cleaning robot. She is concerned, however, about whether she can responsibly decide to "delegate" vacuuming to the automated robot via non-RT control (e.g., while she is away from the apartment or sleeping). This scenario was chosen, in part, to illustrate the broad applicability of the techniques we will present, and in part to avoid any difficulties with protected military information, tactics, and equipment capabilities. It nevertheless illustrates (in a simplistic fashion) the general problem of deciding whether to deploy an automated system under various conditions and configurations when some ethical hazards are possible.

Nibbles is a very clever hamster who is curious about the world, and he escapes from the cage more often than the owner likes. The apartment owner has found him all over the apartment, and her most serious worry is that Nibbles might roam freely at the same time as the Robotic Vacuum Cleaner (RVC) is in operation... and an unfortunate encounter might ensue. This is the sole ethical hazard we will consider in this simple ENT analysis. If she is present and attentive, or can return home in time, she can scoop up Nibbles and place him back in the cage before it is too late. But how long is it until "too late" happens? More generally, since she has some control over the design of the apartment, the regimen for the RVC, the enclosure for Nibbles, and even her own behaviors, she can ask under what circumstances such a delegation decision *would be* responsible and perhaps make modifications to make this so. This analysis is intended to let her know whether, and in what circumstances, she can responsibly use the automated vacuum cleaner, or to let her explore modification considerations to permit responsible delegation decisions involving it.

A Baseline Static Model

As a starting point, we will consider a few basic events and properties, and a few alternate design solutions, beginning with the apartment owner buying a basic RVC and letting it run whenever it sees fit. This analysis is designed to answer the question "Can the owner responsibly leave Nibbles alone with the vacuum cleaner?" All of the parameters modeled have some correlation with the others, which will enable us to craft a Bayesian Network that can infer how likely Nibbles and the vacuum are to encounter each other in a harmful way. The first step to developing a Bayesian Network is to identify the *random variables* and each one's possible values. Each random variable will be a single node in a graph. A fairly minimal set of variables can be used to model this minimal scenario:

1. Time of Day: morning, afternoon/evening, or night
2. Nibbles's Active Status: awake or asleep
3. Nibbles In Cage: true or false
4. Owner's Awareness: Nibbles in cage or Nibbles escaped
5. Owner's Active Status: Awake, Asleep, Cooking, Working
6. RVC's Operational Status: Operational (Plugged In and functional), Not Operational (unplugged and/or disabled)
7. Nibbles Encounters RVC: true or false

In addition, because we know we will want to explore some additional options (and for the ease of model presentation in this chapter), we will include some additional variables. In practice, these could be added later to expand the model:

1. Cage Escape Sensor System Installed: *true* or *false*
2. RVC Upgraded with Heat-Sensor-Based Braking: *true* or *false*
3. RVC's Awareness: *something warm is ahead nearby* or *nothing warm is ahead nearby*
4. RVC's Active Status: *cleaning*, *paused*, or *charging*
5. Owner Location: *living room*, *kitchen*, *bedroom*, *apartment building*, or *outside*

Next, we define the correlations between these random variables as **conditional probability tables** that specify how likely it is that some queried variable will take on each of its possible values based on the values assigned to other random variables. The conditional probability table is allocated to the same node in the graph as its corresponding queried random variable. Random variables that do not have any conditional dependence on other random variables have a simpler probability table that specifies its **prior**, a probability distribution over how likely the variable's assignment is one of the possible values. Although we will use roughly estimated numbers for the conditional probability and prior tables in this example (based on our intuition), it is possible to use sensors and/ or data records (from the real world, or from live, virtual, or constructive simulation) to fill in the tables with more realistic numbers that reflect the actual environment.

For example, a query for the time of day would be governed by a simple distribution with each of the values (day, afternoon/evening, night) taking on 1/3 of the probability distribution. By contrast, Nibbles' Active Status is conditional on the time of day (since hamsters are nocturnal), and whether Nibbles is in his cage is, in turn, conditioned on his active status. These conditions are presented in Table 14.1. Similar probability tables were prepared for each of the 12 variables listed above but are not included here for space considerations.

Figure 14.3 provides a visual representation of the resulting Bayesian Network. To represent the relationship between the queried random variable and all its conditional random variables per conditional probability table, we include directed edges in the graph. There is one edge per conditional relation, serving as the outgoing direction from the corresponding conditional random variable and being the incoming direction to the

TABLE 14.1 Conditional probability distributions for some of the variables in Nibbles' scenario

#2 – QUERY FOR NIBBLES' ACTIVE STATUS (CONDITIONED ON #1 – TIME OF DAY)		
P(AWAKE)	*P(ASLEEP)*	*GIVEN TIME OF DAY VALUE =*
0.6	0.4	*morning*
0.2	0.8	*afternoon*
0.9	0.1	*Night*

#3 – QUERY FOR NIBBLES IN CAGE (CONDITIONED ON #2 – NIBBLES' ACTIVE STATUS)		
P(TRUE)	*P(FALSE)*	*GIVEN NIBBLES' ACTIVE STATUS =*
0.7	0.3	*Awake*
0.95	0.05	*Asleep*

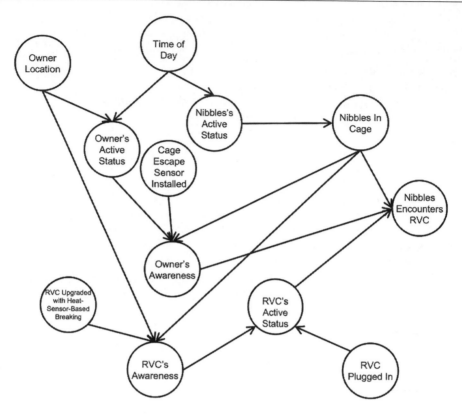

FIGURE 14.3   A static Bayesian Network representing the Nibbles vs. the vacuum scenario.

queried random variable. There should be no directed cycles in a Bayesian Network, which constrains how we define correlation relationships in the conditional probability tables.

The node and state in which we are most interested is the final output of the graph: #7 – Nibbles Encounters RVC. The value for a query of this node is directly dependent on three variables as shown in Figure 14.3. Table 14.2 shows the conditional probabilities for this node given its three inputs. Note that, per this table, Nibbles and the vacuum will not have any interactions unless Nibbles has escaped his cage and the RVC is not charging. Because they are both moving around the floor in their own patterns, it is also not a guarantee that they will interact even when they are both on the loose in the apartment. If the owner believes Nibbles has escaped, then she has a chance to stop their encounter (we will elaborate on more complex ways to represent this in a section below).

If we perform ***inference*** on this Bayesian Network, we will find some baseline probability distributions for every random variable. Because exact inference is a computationally complex problem (Cooper, 1990), we instead approximate the likelihood that a random variable will be assigned one of its values via ***sampling methods***. One effective, yet simple, method for Bayesian Networks is Gibbs Sampling (Geman & Geman, 1984); after an arbitrary initial assignment of values to all the random variables, we process the random variables in some predetermined order so that we visit every node in

TABLE 14.2 Conditional probability distribution for the variable Nibbles encounters RVC in Nibbles' scenario

#7 – QUERY FOR NIBBLES' ENCOUNTERS RVC (CONDITIONED ON #3 – NIBBLES IN CAGE, #4 – OWNER'S AWARENESS AND #11– RVC'S ACTIVE STATUS)				
P(TRUE)	P(FALSE)	GIVEN NIBBLES IN CAGE =	GIVEN RVC'S ACTIVE STATUS =	GIVEN OWNER'S AWARENESS = ...
0.0	1.0	true	cleaning	Nibbles in cage
0.8	0.2	false	cleaning	Nibbles in cage
0.0	1.0	true	paused	Nibbles in cage
0.3	0.7	false	paused	Nibbles in cage
0.0	1.0	true	charging	Nibbles in cage
0.0	1.0	false	charging	Nibbles in cage
0.0	1.0	true	cleaning	Nibbles escaped
0.5	0.5	false	cleaning	Nibbles escaped
0.0	1.0	true	paused	Nibbles escaped
0.1	0.9	false	paused	Nibbles escaped
0.0	1.0	true	charging	Nibbles escaped
0.0	1.0	false	charging	Nibbles escaped

the network. When visiting a node, we will compute the probability distribution over its random variable's RV possible values v_i with respect to the conditional probability table entries that align with the other random variables' currently assigned values as well as each possible value of the visited random variable. Due to the conditional independence assumptions inherent in a Bayesian Network's design as well as Bayes's Rule, we can compute the distribution as a product of conditional probability table entries – we only use the conditional probability tables that involve the visited node's random variable (i.e., the one querying that random variable and any whose conditions involve that random variable) using the formula:

$$P(RV = v_i \mid other \ RV_1 = v_{1,j}, ...) =$$

$$P(RV = v_i \mid conditioned \ other \ RV_n = v_{n,j})$$

$$\cdot \prod_{\{RV_k \mid RV_k \ is \ conditioned \ on \ RV\}} P(RV_k \mid RV = v_i, conditioned \ other \ RV_n = v_{n,j})$$

Sampling from the computed distribution is like rolling a dice that is custom-weighted, assigning the value that lands face-up to the visited random variable. After visiting all the nodes in order, we can save the assignments given to each random variable as counts toward how frequently that assignment occurred. We repeat this multiple times to get a ratio of counts-to-iterations.

This baseline probability distribution is not very useful because it does not have any situation-specific context. We account for context and explore "what if" questions through the application of ***evidence***, which forces a random variable assignment rather than sample based on the associated conditional probability tables and related random variables. For example, if the RVC is plugged in without the heat sensor upgrade, there is no cage escape sensor installed, and the owner is working outside her apartment, then we can instantiate the corresponding nodes in Figure 14.3 with these specified values. When we performed 10,000 iterations of Gibbs Sampling, this model inferred that the probability that Nibbles encounters the RVC is 0.0381. If the apartment owner is willing to tolerate an Acceptable Risk Threshold of 4%, then this would be a "responsible" machine configuration and NT behavior for neglecting the RVC and allowing it to perform autonomously.

Thus, Nibbles is relatively safe even without the owner around to watch the RVC, but 4% might not be an acceptable threshold given that this percentage allows Nibbles to run into the RVC at least one day per month. If the Acceptable Risk Threshold is set more strictly, some additional "what if" cases could be explored using different evidence instantiating different nodes in Figure 14.3. We can further infer that the encounter probability is 0.0416 if she installs the cage escape sensor versus 0.0202 if she upgrades the RVC to include heat sensor braking, and 0.00 if she simply (and reliably) unplugs the RVC before leaving (putting it in the RVC Plugged In: False state). The fact that the probability increased with the installation of the cage escape sensor may seem counter-intuitive, but this illustrates our choice to represent an issue with this owner's overreliance on an alert with a false positive rate of 0.05 and a false negative rate of 0.1. Inspecting other nodes' random variables, we see that the inference of Owner's Awareness increases the probability assigned to *Nibbles in cage* from 0.6983 (without sensor) to 0.8966 (with sensor) – this false sense of security slightly reduces how responsible she is to check on the situation at home while working. For a different owner who might be more skeptical of the cage escape sensor's accuracy or purchase a different sensor with greater accuracy, then these conditional probability tables might be different.

Accounting for Time in a Dynamic Model

A single, static probability distribution is informative, but insufficient to model whether the owner's neglect to monitor the vacuum cleaning robot over time is responsible. The longer she leaves Nibbles unattended, the greater the chance that he can escape and encounter the vacuum-cleaning robot. The longer the neglect interval, the less responsible the decision to neglect becomes since the probability of encounter is essentially cumulative. So, when does neglect reach the point of "too irresponsible", and how can her decision to watch (i.e., not neglect but rather supervise) the robot and Nibbles impact the likelihood of Nibbles' safety? We explore responses to these questions with extensions to the previous Bayesian Network that handle *dynamic transitions* over time (Dean & Kanazawa, 1989; Dagum et al., 1992).

Figure 14.4 illustrates this visual representation of the dynamic Bayesian Network based on the static version from Figure 14.3.

The first thing to notice is that the random variables now have subscripts denoting a sequence of instances or snapshots in time ($t-1$ for the previous instance, t for the current, $t+1$ for the next, etc.). We must establish the specific duration of time between consecutive instances in the sequence to use the second notable change: there are dashed

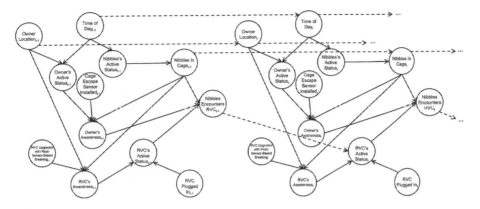

FIGURE 14.4 A dynamic Bayesian Network representing the Nibbles vs. the vacuum cleaning robot scenario.

edges between nodes representing correlations between random variables in consecutive time steps. For example, there is a correlation between the previous instance's Time of Day random variable and the current one. The common assumption to keep these models simple is that only the previous instance's random variables and the current instance's may express a correlation, but this is not a requirement. However, the amount of time that passes between consecutive instances matters for determining the entries in the conditional probability tables. If the time between instances is eight hours, then transitioning between the values for the Time of Day random variable is guaranteed. Otherwise, at a finer-grained scale such as every hour or minute, the likelihood of a transition will continue to be probabilistic based on that uncertainty of the final instance in the interval, as illustrated in Table 14.3. Because the initial instance's Time of Day random variable does not have previous information to use this conditional probability table, that random variable's node has no incoming edges and acts like a prior for the starting point.

TABLE 14.3 Conditional probability distributions for time of day given different sampling intervals

#1 – QUERY FOR TIME OF DAY 1 (CONDITIONED ON SAMPLING EVERY 8 HOURS)			
P(MORNING)	P(AFTERNOON)	P(NIGHT)	GIVEN THAT TIME OF DAY$_{T-1}=$
0.0	1.0	0.0	morning
0.0	0.0	1.0	afternoon
1.0	0.0	0.0	night
#1 – Query for Time of Day 1 (Conditioned on sampling every hour)			
P(MORNING)	P(AFTERNOON)	P(NIGHT)	GIVEN THAT TIME OF DAY$_{T-1} =$
0.875	0.125	0.0	morning
0.0	0.875	0.125	afternoon
0.125	0.0	0.875	night

Like before, marking observed random variables at any moment of time as evidence enables us to inspect what-if scenarios. Rather than determine a single probability, we can now infer into the future to see how the expectation changes following specific events. Performing inference on the query variable at each instance provides us with data to plot an ethical NT curve as hypothesized above. Each point will be of the form (k, p) where k is the number of instances since the start of the plot (denoting time) and p is the probability of the queried random variable being assigned the value in question (notionally, the risk probability of a hazard state under automation neglect). Based on the Acceptable Risk Threshold chosen for responsible neglect, we can determine how long the owner may neglect the RVC before putting Nibbles into irresponsible danger. Whenever the owner intervenes or has some change in mental state, we can apply new evidence in that instance of the Bayesian Network to have further impact on the inferred probability distributions. These changes will propagate into future time steps and, if advantageous, keep the risk acceptable for a while longer. The temporal delay between instances to which we applied such evidence signifies the tolerance in this system.

Figure 14.5 considers a duration of one-hour intervals to investigate "what if" cases to a deeper extent with the dynamic transitions.

While the owner is outside and working, suppose that Nibbles does escape from his cage. This clearly increases the risk of an encounter over time, but we can now infer the increase as time progresses. What is the tolerance beyond which it would be irresponsible to leave Nibbles unattended and not checked on? How would installing the cage escape sensor or upgrading the RVC with a heat sensor-based avoidance system

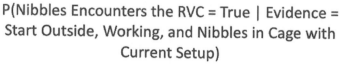

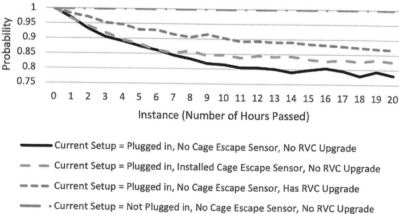

FIGURE 14.5 The Dynamic Bayesian Network's assessment of how likely Nibbles will encounter the RVC each hour under various sensor and RVC setups. The evidence for the owner is set in the first instance, but the evidence for the setup is set at every instance to ensure model consistency.

improve this tolerance? If the Acceptable Risk Threshold is a 90% chance of acceptable RVC+Nibbles behavior (i.e., the probability of an encounter is 0.1 or less), then the owner has no concerns during any time interval if the RVC is not plugged in. That threshold will be exceeded in about 11 hours with the heat sensor upgrade, and about four hours with the cage escape sensor or without any such sensors. These durations are an upper bound because the owner would need some time (which we have not modeled) to context switch and leave work to return home before intervening (as with the Interface Efficiency curves in Figure 14.2). With this analysis, we would say that a RHD decision is based on the equipment package installed, but that any delegation (neglect) of more than four hours with the baseline RVC is irresponsible at a 90% risk threshold.

In the current model, we have assumed that the only way the owner can intervene is if she ensures Nibbles is in his cage, which we assume takes less than one hour even going back home from her office outside the apartment. If we define her involvement through setting evidence at some time step's instance that Nibbles is in his cage (either confirming this is the case or putting him back in if found on the floor), then we can perform inference again and observe how this affects the probability that Nibbles encounters the RVC. However, we must carefully redesign the Bayesian Network to start at this instance to avoid tampering with the past – the correlation between random variables of different instances in time means that later evidence would imply facts about the previous instance's unobserved random variables. Intervention by means outside the system of random variables is not possible without a formalization for actions, such as a Markov Decision Process (Bellman, 1957). Figure 14.6 illustrates the curve with annotations for the owner's interventions in each environment setup and it becomes

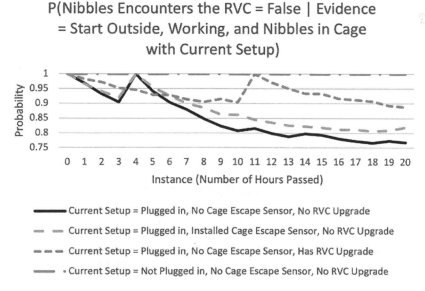

FIGURE 14.6 The change in how likely Nibbles will encounter the RVC over time when the owner checks on him before surpassing the acceptable risk threshold. We only include one intervention, but each setup needs at least one more intervention later.

apparent how various frequencies of intervention might serve to keep the ENT above a chosen threshold.

Increasing Model Complexity

Bayesian Networks, dynamic or static, are flexible enough to support modification depending on what else we need to include in the model. If there are additional factors to consider for the scenario design, then we can generate new random variables and insert their nodes into the network. Depending on the edges for updated correlations, we will also need to add new conditional probability tables (for each added node) and revise existing conditional probability tables (per existing node whose set of incoming edges now includes one whose outgoing edge coming from an added node). Suppose the cage is not very sturdy and Nibbles is hardy enough to survive a fall. Then besides trying to break free using traditional means, Nibbles might try to thrash once in a while and slide the cage off the table. Likewise, the RVC could bang into the table while cleaning and nudge the cage toward the edge. We could include random variables for Cage on Table, Nibbles Thrashes, and RVC Bumps Table that, at a minimum, would revise the conditional probability table for the Nibbles in Cage random variable.

One set of conditional probabilities that would be particularly advantageous for assessing MHC and awareness to support responsible behavior would be the inclusion of states pertinent to human behavior and decision-making. For example, the owner's awareness of whether Nibbles is in the cage or not is conditioned in our model on the Cage Escape sensor and the Owner's Active Status, but might also be conditioned on whether a klaxon for Nibbles' escapes was installed, whether the owner is fatigued or distracted, etc. We've modeled the owner's awareness as affecting whether Nibbles encounters the RVC, but have included no specifics about the owner's distance from the encounter and ability to intervene, the clutter in the apartment, the installation of a remote shut-off switch for the RVC, etc. These sorts of human factors will and should play a critical role in assessing whether conditions permitted a responsible human decision or not. Some of these can be modeled using Bayes Networks, but others may require other modeling techniques, as briefly discussed in the Section below.

We may modify conditional probability tables for reasons beyond new random variables as well. The likelihoods can become more accurate or personalized as information becomes available. If we have to consider a different owner's responsibility for leaving their hamster and RVC alone, then they might have different lifestyles and patterns than Nibbles's owner – we can revise the conditional probability tables to reflect the distributions over their locations, active status, and awareness (via reliance on technology, paranoia that their hamster is out of the cage, trust in the RVC's autonomy, etc.). If we have the resources and permissions to place sensors in the owner's apartment for a while to better understand Nibbles's escape behaviors, then we can also adjust conditional probability tables on the basis of this more realistic information. That means we can use approximations based on general cases (typical hamster sleep behavior) and later repeat the assessment with our specific case when data is available.

Modeling ENT beyond Bayesian Networks

Although the variety of Bayesian Network approaches described above provides insights into the probability of events that imply the ethical responsibility of neglecting or delegating to autonomous systems over time, Bayesian Networks cannot represent everything one might need for probabilistic assessment. In particular, the computational power of conditional probability table composition is a sum of products; this lacks support for other mathematical operators that might describe probabilistic outcomes as a function of other factors. This is different from conditional probability tables that represent continuous functions, which have been studied (Murphy, 1999; Salmerón et al. 2018). For example, if we have some knowledge about how Nibbles and the RVC roam around the apartment, then we can approximate the distance between the two of them while they are both in a shared space: $dist(N,R)$. This distance can be inversely proportional to the probability that they encounter each other with some threshold distance for an impossible (0%) and guaranteed (100%) encounter. N and R do not have to be variables that are assigned values; they can also be functions whose inputs are other measured factors, parameters, or random variables found in a Bayesian Network. The walk pattern that N describes likely depends on Nibbles's Active Status and on Nibbles' average speed, and the cleaning pattern that R describes likely depends on RVC's Active Status and program. In addition to these variables, N and R likely influence each other in some form of a dynamical system. When Nibbles is near the RVC, the loud whirring noise is more likely to send the hamster's walking pattern in the other direction; likewise, with a proper sensor, the RVC is more likely to pause or change direction if sensing Nibbles nearby. Modeling the dependencies between these variables is possible and will capture the ways their behaviors impact one another. One example, in this case, is a random walk Markov Chain (Knill, 2009) where the state definition includes the locations of both Nibbles and the RVC; the transition function represents their joint movements whether independent (far enough away that each simply moves around naturally) or dependent (close enough that each moves around based on the other).

The probability of their encounter might consider yet another distance function $dist(O,A)$ where O is the owner's movement pattern or location in space and A is the apartment's living room's location in space. That is, can the owner rush to where Nibbles and the RVC are expected to collide and intercept them before it is too late? As such, O could be a function that depends on the distribution over the values of the Owner Location random variable, but the probability of her arriving on time likely includes $dist(O,A)$ in addition to the owner's average speed, the Owner's Active Status, and the Owner's Awareness random variables. How do all of these interact with $dist(N,R)$ to compute the queried probability? Modeling this complex system of equations might integrate with information from the Bayesian Network, but a Bayesian Network itself cannot represent this kind of relationship. It is important to recall the story a model needs to tell when considering the factors, their relationships, and what outcomes define the responsibility of one's decision to neglect monitoring an autonomous system for some amount of time.

CONCLUSIONS AND IMPLICATIONS

This chapter introduced the concept of NT analyses as a possible approach to assessing Responsible Human Delegation decisions, along with a methodology for operationalizing the approach based on probability calculations using Bayes Networks. We have worked through a hypothetical scenario revolving around the question of whether, and in what circumstances, it might be "responsible" to decide to allow unmonitored autonomous behavior from a robot when innocents might be harmed. This scenario clearly has analogs in military use of AI technologies in non-lethal support applications that might probabilistically cause unintended harm (e.g., robotic material transport, search and rescue, and mine clearing), defensive applications that might probabilistically target the wrong forces and/or cause unintended collateral damage (e.g., surface to air counter missile and air systems, security patrols, and base defense) and offensive applications which might target the wrong forces, cause collateral damage or produce disproportionate responses (e.g., potentially lethal autonomous robots and other air and ground systems). Indeed, we believe that a similar process of ethical hazard analysis followed by probabilistic modeling of the likelihood of those hazards occurring under various combinations of equipment capabilities, operating procedures and human knowledge and intervention capabilities can be used to compute a quantitative "ethical neglect tolerance interval" in any set of circumstances where reasonable probabilities can be assessed a priori.

This very preliminary analysis provides an initial indication of the viability and utility of the approach. The analysis rapidly adapted the concept of NT to ethical analyses to first identify ethical hazards and then to compute their likelihood under various conditions of "neglect." We could extend these techniques for analyses of information flow and control precision adequacy, etc.

The approach has several limitations, however. Real-world models for ENT assessment might be very large and complex, representing a large data acquisition, representation, and validation problem. This problem is ameliorated to a degree by the fact that one can use multiple versions of such models and the probabilities in them with progressive detail and refinement and accuracy. As new data from more detailed simulations, human-in-the-loop experimentation, and real-world behavior becomes available, one can update models and expand them to reflect these changes.

Methods of intervention, such as human monitoring actions or changes to the autonomous system, are also difficult to realistically implement through evidence of specific random variables. Specifically, the correlation of random variables between the past and present means that setting evidence at a later instance will adjust the distributions of past instances even though they already happened; we presented a workaround for this via copying distributions from prior inference computations into new prior conditional probability tables, but it is still not the most realistic approach.

In addition, as discussed above, our choice of a Bayesian Network implementation and statistical calculation approach is only one possible methodology for computing ENT risk probabilities and intervals. Bayesian Networks make the computation of some properties (e.g., probabilities) easy and others (e.g., functions with non-arithmetic operators) more difficult. We suggested some alternate modeling techniques and adaptations above and other approaches are possible. Whether the set of techniques we can develop is adequate awaits an opportunity to do modeling on a large scale.

Since the ENT analysis is built on an initial Ethical Hazard Analysis, and such analyses will likely always be incomplete, another limitation is that there will be unanticipated hazards and unanticipated pathways to incur known hazards. This incompleteness means the resulting ENT analysis will likely always be an underestimate of the probability of ethical hazards at any point in time or context.

This situation of open-ended hazards that can be only partially known, along with complex modeling approaches, seems not unlike that which prevails for analysis of other forms of system failure – such as Failure Modes Effects Analysis (Ben-Daya, 2009) – which are also costly to perform, subject to the accuracy of initial models and ultimately, generally incomplete. Yet such analyses are an integral part of any complex, high-criticalty system design. While these analyses may not prevent all accidents and system failures, they have proven their worth many times over. ENT as an analytic technique performed before system design or before a mission-specific configuration would seem similarly valuable. It seems plausible that even dynamic, real-time assessment of in-mission conditions could be performed to determine whether some forms of decisions are responsible at the moment – not because the decisions are or are not ethical, but because there was insufficient time or information to make a responsible delegation decision in the current conditions.

A final question remains unanswered, and of concern: is the development of a simple metric for RHD ultimately a good idea? The goal of this RHD approach and of the ENT metric is to retain the ability – through adequate time, training, information flow, control authority, etc., – for a human operator to exert *their* understanding of ethical behavior on the decision to use an automated system, even under non-RT control. Any analysis, especially one that is intrinsically biased toward underestimating the set of ethical hazards, risks being over-generalized and used as a justification that replaces the very human thought processes it is designed to support.

ACKNOWLEDGMENTS

The work and thoughts presented in this paper are entirely the authors'. Some of these concepts were suggested by discussions in the NATO RTO-HFM-322/330 working group on Meaningful Human Control, and we would like to thank them for those discussions, but we do not speak for that group. Additional support and discussion of the concepts was provided by the Dutch Ministry of Defence (MOD) and the Dutch Ministry of Foreign Affairs through their invitation to participate in the expert workshop on Responsible Military Use of Artificial Intelligence held in Amsterdam, The Netherlands, October 31–November 1, 2022.

REFERENCES

Bellman, R. (1957). A Markovian decision process. *Journal of Mathematics and Mechanics, 6* (5), 679–684.

Ben-Daya, M. (2009). Failure mode and effect analysis. In M. Ben-Daya, S. O. Duffuaa, A. Raouf, J. Knezevic, & D. Ait-Kadi (Eds.), *Handbook of maintenance management and engineering* (pp. 75–90). Springer.

Cooper, G. (1990). The computational complexity of probabilistic inference using Bayesian belief networks. *Artificial Intelligence, 42*, 393–405.

Crandall, J. W., & Goodrich, M. A. (2002). Characterizing efficiency on human robot interaction: A case study of shared-control teleoperation. In *Proceedings of the 2002 IEEE/RSJ international conference on intelligent robots and systems* (Vol. 2, pp. 1290–1295).

Crandall, J. W., Goodrich, M. A., Olsen, D. R., & Nielsen, C. W. (2005). Validating human-robot interaction schemes in multitasking environments. *IEEE Transactions on Systems, Man, and Cybernetics-Part A: Systems and Humans, 35*(4), 438–449.

Cummings, M. L. (2019). Lethal autonomous weapons: Meaningful human control or meaningful human certification? [Opinion]. *IEEE Technology and Society Magazine, 38*(4), 20–26. doi: 10.1109/MTS.2019.2948438.

Dagum, P., Galper, A., & Horvitz, E. (1992). Dynamic network models for forecasting. In *Proceedings of the Eighth International Conference on Uncertainty in Artificial Intelligence* (pp. 41–48).

Dean, T., & Kanazawa, K. (1989). A model for reasoning about persistence and causation. *Computational Intelligence, 5*, 142–150.

Department of Defense. (2020). *Human engineering design criteria for military systems, equipment and facilities* (MIL-STD-1472H). Department of Defense. https://everyspec.com/MIL-STD/MIL-STD-1400-1499/MIL-STD-1472G_39997

Department of Defense. (2023). *Department of Defense Directive 3000.09, "Autonomy in Weapon Systems,"* Updated January 25, 2023. https://www.esd.whs.mil/portals/54/documents/dd/issuances/dodd/300009p.pdf

Dillenburger, S. P. (2012). *Minimization of Collateral Damage in Airdrops and Airstrikes.* [Unpublished Dissertation], Air Force Institute of Technology. https://scholar.afit.edu/etd/1204.

Elara, M. R. (2011). Validating extended neglect tolerance model for search & rescue missions involving multi-robot teams. In *Proceedings of International Conference on Intelligent Unmanned Systems* (Vol. 7).

Elara, M. R., Calderon, C. A. A., Zhou, C., & Wijesoma, W.S. (2009). Validating extended neglect tolerance model for humanoid soccer robotic tasks with varying complexities. In *Proceedings of PERMIS* (Vol. 1, pp. 22–29).

Geman, S., & Geman, D. (1984). Stochastic relaxation, Gibbs distributions, and the bayesian restoration of images. *IEEE Transactions on Pattern Analysis and Machine Intelligence, 6*(6), 721–741.

Goodrich, M. A., Olsen, D. R., Crandall, J. W., & Palmer, T. J. (2001). Experiments in adjustable autonomy. In *Proceedings of IJCAI workshop on autonomy, delegation and control: interacting with intelligent agents* (pp. 1624–1629).

Horowitz, M. C. & Scharre, P. (2015). *Meaningful human control in weapon systems: A primer* [Working paper]. Project on Ethical Autonomy, Center for New American Security. https://www.cnas.org/publications/reports/meaningful-human-control-in-weapon-systems-a-primer.

Humphrey, A., See, J., & Faulkner, D. (2008). A methodology to assess lethality and collateral damage for nonfragmenting precision-guided weapons. *International Test and Evaluation Association Journal, 29,* 411–419.

Knill, O. (2009). *Probability and Stochastic Processes with Applications.* Daryaganj, India: Overseas Press India Private Limited.

Memorandum to Convention on Conventional Weapons (CCW) Delegates. (2016). *Killer robots and the concept of meaningful human control.* https://www.hrw.org/news/2016/04/11/killer-robots-and-concept-meaningful-human-control.

Murphy, K. (1999). A variational approximation for Bayesian networks with discrete and continuous latent variables. In *Proceedings of the fifteenth international conference on uncertainty in artificial intelligence* (pp. 457–466).

Miller, C. A., & Parasuraman, R. (2007). Designing for flexible interaction between humans and automation: Delegation interfaces for supervisory control. *Human Factors, 49*(1), 57–75.

Olsen, D. R., & Goodrich, M. A. (2003). Metrics for evaluating human-robot interactions. In *Proceedings of PERMIS* (pp. 4–12).

Roff, H. M. & Moyes, R. (2016). *Meaningful human control, artificial intelligence and autonomous weapons* [Briefing paper]. UN Convention on Certain Conventional Weapons. https://article36.org/wp-content/uploads/2016/04/MHC-AI-and-AWS-FINAL.pdf

Salmerón, A., Rumí, R., Langseth, H., Nielson, T., & Madsen, A. (2018). A review of inference algorithms for hybrid bayesian networks. *Journal of Artificial Intelligence Research, 62,* 799–828.

Sheridan, T. (2002). *Humans and automation: System design and research issues.* Santa Monica, CA: Human Factors and Ergonomics Society/New York: Wiley.

Wang, J., & Lewis, M. (2007). Assessing coordination overhead in control of robot teams. In *Proceedings of 2007 IEEE international conference on systems, man and cybernetics* (pp. 2645–2649).

SECTION IV

Policy Aspects

Strategic Interactions

15

The Economic Complements of AI and the Political Context of War

Jon R. Lindsay

INTRODUCTION

Military futurists have been writing about autonomous systems for over fifty years (Adams, 2001; Deudney, 1983; Dickson, 1976; Singer, 2009; Toffler & Toffler, 1993). But early enthusiasm usually gives way to disappointment. Over this same time period, military organizations have become more dependent on information technologies and experienced more coordination problems (Allard, 1990; Edwards, 1996; Lake, 2019; Lindsay, 2020). The revolution in military affairs seems rather evolutionary in retrospect. Will it be different this time?

Today's machine learning (ML) techniques are more impressive and less brittle than "good old fashioned AI" (GOFAI). Indeed, these are boom times for AI in the commercial economy. AI systems are performing tasks that once seemed to be consummately human. AI is composing orchestral music, writing interesting screenplays, and generating compelling visual art. AI is automating factories, supercharging advertising, and making commercial travel more convenient. AI also excels in video games and competitive strategy games. There is significant potential for disruption and dislocation in the global economy as industries adapt to harness the power of AI.

DOI: 10.1201/9781003410379-19

It is a reasonable assumption that the automation of war is right around the corner. Why shouldn't war also become more efficient and precise, and why shouldn't robotic combatants become even faster and more creative? Now we can ask these questions to AI directly. According to ChatGPT, "AI can enable the development of autonomous weapons systems, such as drones, ground vehicles, and ships. These systems can operate without direct human control, making them faster, more efficient, and potentially capable of executing complex missions with reduced human risk" (OpenAI, 2023). The bot also highlights applications for "Enhanced Situational Awareness…Decision-Making and Command Systems…Cybersecurity and Information Warfare…Logistics and Supply Chain Management…[and] Predictive Maintenance" (OpenAI, 2023).

Major powers are also concerned about the challenges and opportunities of military AI. The United States and China have commissioned numerous studies and developed working prototypes (Kania, 2017; National Security Commission on Artificial Intelligence, 2021). The warfighting advantages of AI, furthermore, seem poised to alter the balance of power and trigger arms races as democracies and autocracies alike seek to substitute autonomous systems for human warriors (Buchanan & Imbrie, 2022; Horowitz, 2018, 2019; Scharre, 2018).

These developments in turn have prompted important conversations about meaningful human control of lethal autonomous weapon systems and the potential for inadvertent escalation (Johnson, 2020, Roff 2014). An even more dire scenario is the rise of AI-enabled systems that transcend human control altogether, leading to worries about the existential implications of so-called artificial general intelligence or Super-AI (Bostrom, 2014; Sears, 2021). Industry leaders have begun calling for more deliberate ethical reflection as well as outright regulations for AI before it is too late. Even ChatGPT hastens to offer reassurance: "while AI has the potential to enhance military capabilities, decisions regarding its use in warfare should be guided by international laws, regulations, and ethical considerations to ensure the protection of civilian lives, compliance with human rights, and prevention of unnecessary suffering" (OpenAI, 2023).

But there is at least one important topic that ChatGPT fails to consider. Does the economic context that created ChatGPT affect the viability of military AI in any way? It is an obvious but underappreciated fact that most of the impressive applications of AI to date are in the commercial world. War, however, is a very different sort of "business". The conditions that make AI economically viable today may not hold in the chaotic and controversial realm of war, or at least not to the same extent (Goldfarb & Lindsay, 2022). For instance, AI depends on the availability of data, but war is full of fog and friction. AI depends on having many opportunities for training, but war is a rare and unpredictable event. AI companies submit to the rule of law, while war is famously anarchic. The success of AI systems in the world of peaceful commerce, therefore, may be a poor guide to the performance of AI in the world of military combat.

Even more fundamentally, the economic conditions that support AI performance may also be associated with important changes in patterns of political conflict. Traditional interstate war, to use the jargon of international relations theory, is a struggle for dominance in a world of anarchy. And yet the modern international system is more globalized, interconnected, interdependent, and institutionalized than ever before.

The so-called liberal order is hardly peaceful, however, as "weaponized interdependence" (Drezner et al., 2021; Farrell & Newman, 2023) enables the proliferation of many other modalities of conflict such as espionage, subversion, covert action, and varieties of "hybrid" or "gray zone" conflict (Gannon et al., 2023; Lanoszka, 2016; Votel et al., 2016). It is no coincidence that limited forms of conflict are prominent in the hyper-globalized 21st century. Shared institutions are a condition for the possibility of subversion and espionage, as well as for their modern manifestation as cyber conflict (Lee, 2020; Lindsay, 2017, 2021; Maschmeyer, 2022; O'Rourke, 2018). With more robust and extensive institutions come more opportunities to subvert them. How, therefore, might we expect people to use AI for conflict *within* social institutions, rather than *between* them?

This chapter explores the unintended consequences of military AI in six parts. First, I discuss assumptions about the substitution of AI for human warriors. Second, I highlight the importance of social complements for technological performance. Third, I briefly summarize the economics of AI, highlighting key complements of data and judgment that depend on social institutions. Fourth, I argue that the political logic of war tends to undermine the institutional complements of AI. Fifth, I argue that military reliance on AI will have unintended consequences including increasing institutional complexity and degraded human security in more protracted, ambiguous conflicts. I conclude with a summary of the argument.

TECHNOLOGICAL SUBSTITUTES

Since its founding, computer science has attempted to automate intellectual processes. Alan Turing imagined his famous universal computing machine as an automated clerk, while Charles Babbage before him imagined the difference engines as an automated parliament (Agar, 2003). Turing's 1950 essay on automating intelligence still resonates, parrying objections from AI skeptics that continue to be voiced today (Turing, 1950). The Macy Conferences on cybernetics, which brought together founding fathers in computing like Claude Shannon and John Von Neumann, were explicitly dedicated to founding a general science of information and control that could be used to build a mechanical brain (Dupuy, 2000; Kline, 2015). The nascent field of computer science aimed to create a new kind of agent, if not a new kind of lifeform.

Since then, AI has gone through several cycles of "AI hype" and "AI winter". GOFAI was great at doing things that seemed hard for humans (like calculating formulae) but stupid at things that were easy (like recognizing images). A common refrain among AI skeptics was that AI lacked common sense or appreciation for why any given computation might be meaningful or useful for human beings (Collins, 1990; Dreyfus, 1992; Rochlin, 1997). The field of computer science continued to grow, nevertheless, but not simply by replacing human beings. Rather, innovation in computing created much more for people to do: designing applications, developing interfaces, building infrastructure, repairing glitches, educating scientists and technicians, developing telecommunications policy, and so on. The economic context of human interaction thus became even more complex (and more lucrative) as computational infrastructure became even more dependent on complementary social activity.

We are now riding the latest wave of AI enthusiasm thanks to connectionist, neural network, deep learning, and ML approaches to AI. These are based on very different principles inspired by the human brain. They are newly feasible thanks to dramatic advances in memory and computing power, together with explosive growth in the "big data" economy. The excitement about AI stems from impressive performance in areas where GOFAI stumbled, such as text translation, image recognition, spatial navigation, etc. Nevertheless, familiar concerns remain that ML has no understanding of why its pattern recognition outputs matter in human context, or why biased data might lead to socially undesirable outcomes (Broussard, 2018; Smith, 2019).

Concerns about the unintended consequences of military AI are also not new. Science fiction movies from the Cold War era continue to inspire our imagination about military AI. A standard assumption is that in the future, robots will replace some human functions, perform some human tasks, and become autonomous characters, which then leads to good things (*Star Trek*) or bad things (*Terminator*). These sci-fi robots usually have something extra (strength, speed, calculating ability) or they are missing something (compassion, insight, understanding, creativity). They may be improved or deficient agents, in other words, but they are fully autonomous agents, nonetheless. The modified capabilities of these human substitutes leads to dangerous or unintended consequences in these stories, which makes it necessary to control, regulate, battle, or banish them.

The basic concerns dramatized in Cold War sci-fi still resonate in modern ethical conversations about AI: we worry that lethal machines will make their own decisions to harm humans without appropriate human control or consent. An important theme that runs through these classic scenarios, in short, is *substitution*. But great entertainment might not necessarily be the best guide to the future. One important reason is that technology is usually not simply a substitute for human labor in economic history. Technological performance depends on social complements.

SOCIAL COMPLEMENTS

Substitutes replace jobs and functions with a cheaper or better improvement. By contrast, complements affect a larger network of jobs and functions throughout society. Often the advent of substitutes will make complements more valuable. If people find a baker who sells cheaper bread, then the market for butter and jam will increase, which means that new shops will open next to the bakery. Thus, the replacement of the horse-drawn carriage with the automobile required a lot of complementary innovation and infrastructure in terms of roads, repair shops, gasoline stations, car dealerships, assembly lines, and so on. One cannot just swap out a horse for a car without considering the profound social changes that make this swap possible.

Likewise for military AI, we need to ask whether the complementary innovations that are unlocking productivity in the AI economy might also be correlated with important changes in the nature or conduct of war. It may be true that an AI drone swarm will be able to defeat a modern company of soldiers in short order, but what are the chances of that company not evolving as well? A machine gun, similarly, would be invaluable when facing an ancient army of hoplite soldiers, but what are the chances that anyone would still fight with spears and swords in the same economic milieu that could produce machine guns? The chances are not strictly zero, as historically lopsided contests between Hernan Cortes and Mesoamericans or the Battle of Omdurman suggest, but these events stand out as exceptional for a reason.

Most of the ethical discussion of AI tends to hold the nature of war constant and consider only the issues around using AI systems to fight the war. Yet, it is further possible that the political context of war itself might change in interesting ways as the use of AI becomes more militarily feasible. At a tactical level, innovations in offensive lethal autonomous weapons may be met with innovations in automated defenses. Very fast AI weapons could end up leading to very protracted wars, as a result. At a more political level, the economic circumstances that make AI feasible may alter the incentives to engage in violent conflict. Somewhere in between, the institutional requirements for designing and using large-scale AI in military scenarios may be simply too hard to meet in real-world scenarios.

In the 1980s, for example, the well-known computer scientist David L. Parnas resigned from the Reagan administration's Strategic Defense Initiative – a.k.a. "Star Wars" – Panel on Computing in Support of Battle Management. Parnas argued that it was impossible to draft precise requirements or optimal system designs for warfighting circumstances that are guaranteed to change: "The military software that we depend on every day is not likely to be correct. The methods that are in use in the industry today are not adequate for building large real-time software systems that must be reliable when first used" (Parnas, 1985, p. 1330).

Parnas was especially pessimistic about AI: "Artificial intelligence has the same relation to intelligence as artificial flowers have to flowers. From a distance they may appear much alike, but when closely examined they are quite different. I don't think we can learn much about one by studying the other. AI offers no magic technology to solve our problem" (Parnas, 1985, pp. 1332–1333). He was obviously talking about a previous AI technology (i.e., GOFAI using theorem proving or expert system databases). But while deep learning technology is different, to be sure, the warfighting problems and warfighting organizations are as complex as ever. Software engineering is always hard, but it is even harder when use cases are infrequent, complex, and unpredictable. Sadly, the uncommon is common in combat.

The concerns raised by Parnas remain relevant because they are ultimately grounded in social conditions rather than just technical considerations. Indeed, engineering computational systems always depend on assumptions about conflict and cooperation (Bowker & Star, 1999; Brown & Duguid, 2000; Ciborra, 2002; DeNardis, 2009). Most successful software engineering is predicated on cooperation among developers and users, to some degree, and everyone who maintains the economic ecosystem in which these systems will be employed. As we shall see, viable AI systems depend on institutional assumptions as well.

THE ECONOMIC COMPLEMENTS OF AI

This section will briefly summarize insights from economics research on AI (Agrawal et al., 2018, 2019, 2022). Economic models of decision-making typically highlight four components: data, prediction, decision, and action. In military command and control these four components are known as the OODA loop, a cybernetic cycle of observing, orienting, deciding, and acting. Information comes in from the world. That information must be processed and combined with information in memory. Decision-makers use this combined information to figure out how to achieve a goal. Finally, they take action to modify the state of the world. Here "prediction" refers to the second step of inferring information that is missing from inputs based on information in memory.

All the forms of AI that are getting so much attention today (i.e., narrow AI or ML) are forms of automated prediction. This statistical notion of prediction applies to actual prediction tasks like forecasting weather or planning navigation routes as well as other forms of filling in missing information in classifying images or translating texts, as well as generative AI applications for producing text copy, software code, and graphical designs. This means that AI automates only part of the decision cycle. Robotics, moreover, may automate aspects of action such as running factory machinery or flying drones. And there are, of course, many automated sources of data available through the internet and remote sensing systems.

Judgment, however, remains a consummately human task. The economic concept of judgment refers to the determination of utility functions. An AI weather forecasting system can tell you whether it is going to rain with some given probability, but it cannot decide whether you should bring an umbrella. That depends on whether you mind getting wet or the hassle of carrying an umbrella whether it is wet or dry. These are value judgments that the AI system cannot make. The concept of judgment can be considered more broadly to encompass all manner of meaning, value, preference, or care.

From an economic perspective, a drop in the price of substitutes makes complements more valuable. AI is a form of prediction, and technical trends in memory, algorithms, and computing power are making prediction both better and cheaper. But this drop in the price of prediction means that its complements of data and judgment become more valuable. To get AI systems to work, it becomes necessary to have a lot of high-quality, unbiased data. And it is necessary to figure out what to predict and how to act on predictions.

The quality of AI-supported decisions, therefore, will be determined by the quality of the data used to train AI and the quality of the judgments that guide them. Conversely, missing or biased data lead to suboptimal system behavior, and decisions about appropriate action become challenging when there is political complexity or controversy in decision-making institutions. All the impressive AI achievements are in areas where companies have figured out how to solve the data and judgment problems, typically in areas where decision problems can be very well constrained and lots of representative data can be collected. For other tasks, such as determining the corporate mission and values of an organization, AI is of little use. Companies that figure out how to reorganize themselves to exploit AI complements and build infrastructures of data and decisions have the potential to gain a competitive advantage. Substitution alone will not provide an advantage, and may even undermine performance.

A very important decision task in this respect is the meta-task of understanding the distribution and flow of decision-making in an organization. Disaggregating decisions makes it possible for administrators to identify decision-making tasks that can be fully or partially automated versus those that must be performed by human beings. If a decision can be fully specified in advance – if X then Y – and if lots of data are available to classify situations – X or not X – then fully automated decision-making may be feasible. Video game-playing AI systems fall into this category, with a clear goal of winning the game by getting the most points and millions of previous games to learn from. But there will be other kinds of difficult decisions about organizational direction as well as implementation at the last mile. Thus, for instance, executives and engineers at a ride-sharing service have created a business model that can automate route-finding and billing in areas where there are standardized geospatial data available and lots of data about previous trips and rider demand patterns. But judgment is still required for passenger safety and navigation in crowded, cluttered environments. Thus, there are many judgment tasks at both the top and the bottom, for high-level organizational design and direction, and for low-level implementation on a case-by-case basis.

The challenge of business leadership lies in determining how and whether to reorganize decision-making to make the most of automation within a given economic niche. Many uses of AI, such as self-driving mining trucks on well-controlled routes, the replacement of taxi drivers, or quality control devices in manufacturing, are still focused on substituting human prediction tasks while providing complementary infrastructure for data and judgment. There are just a handful of industries, most notably in online advertising, that have fundamentally rearranged business processes and the industrial ecosystem to make the most of automated prediction. These two phases of innovation have been described in terms of platform versus systemic innovation, akin to simply replacing steam engines with local dynamos versus inventing assembly lines with distributed energy supplies (Agrawal et al., 2022). We are still largely in the platform substitution phase of the commercial AI revolution, but major realignments may follow from the innovation of systemic complements.

In short, automated prediction depends on the economic complements of data and judgment. These complements, in turn, depend on permissive institutional conditions. Institutions are the human-built "rules of the game" that constrain and enable human beings to solve collective action problems (North, 1990; Ostrom, 1990; Williamson, 1985), and "sociotechnical" institutions include the "tools of the game". Data depend not only on data collection, processing, and communication infrastructure, but also on shared standards and technical protocols as well as access, quality-control, and maintenance agreements. Judgment depends on organizational institutions to solicit opinions, develop ideas, adjudicate disputes, and socialize values. Therefore, AI performance depends on sociotechnical institutions. And the platform innovations of the future that unlock the productive potential of AI will fundamentally depend on complementary innovations in shared sociotechnical institutions.

An underappreciated reason why we are seeing so much dramatic progress in AI is that national and global economies are more complex and institutionalized than ever before. Institutions create reliable conditions for exchange. Institutions stabilize information collection protocols and processes – better data. Institutions create shared expectations about how organizations and contracting partners will behave – better judgment. Each in turn depends on complex systems of shared norms and political monitoring and enforcement mechanisms. The modern concept of "global liberal order" is largely a shorthand for this set of shared expectations, norms, and governance mechanisms. This institutional order contributes to the availability of quality data and collective agreement about judgment, the complements that make AI commercially viable.

THE POLITICAL CONTEXT OF WAR

The political logic of war could not be more different than the global liberal order described above. War, in the realist tradition of international relations, is associated with anarchic political systems. In anarchy, there is no overarching government, so actors must help themselves to survive and thrive. In anarchy, actors will lie, cheat, and steal, and there is no global court or policy to make them behave. War, conquest, and exploitation are always possible in this tragic world. This situation is the exact opposite of the liberal order described above. This means that the conditions that are most likely to lead to war are also the least conducive for the reliable and responsible use of AI systems.

AI performance depends on the institutional complements of data and judgment, but these same conditions are absent or elusive in war (Goldfarb & Lindsay, 2022). War is notoriously uncertain, surprising, and chaotic (Beyerchen, 1992; von Clausewitz, 1976). Modern theories of war stress that uncertainty is a major – if not *the* major – cause of war (Fearon, 1995; Gartzke, 1999; Ramsay, 2017). Actors bluff about their power and may not commit to agreements, both of which can make fighting more attractive than peace. Still, wars are rare events. But this is another way of saying that war is prime evidence that the situation is somehow unpredictable. If there is war, at least one actor and probably both are confused about the true balance of power and interests (otherwise they would find a deal to avoid the terrible costs and risks of war). War is not predictable, which does not bode well for prediction machines.

War is also controversial, obviously. There is not only external combat between adversaries, but also many internal controversies as well. Different components of military organizations will disagree about doctrine or strategy. Different political factions of government will disagree about war aims and the conditions of negotiation. Different interest groups will disagree about what sorts of behavior and targets are legitimate given the stakes of a conflict. Coordination and consensus are always hard in complex distributed organizations, but they may be well-nigh impossible when the goal is the management of violence for politically consequential stakes. This means that the conditions of clear consensual judgment about strategies, missions, rules, limits, and ethics are especially difficult to achieve. This does not bode well for prediction machines, either.

Clearly, there are plenty of military applications of AI already or soon to be deployed. We have seen examples of sophisticated sensors, loitering munitions, and drones already in use on the battlefields of the twenty-first century. Experimental proto-types of swarming drones, uncrewed submarines, and robotic wingmen suggest the art of the possible. These prototypes work in well-constrained problems with institutional scaffolding. The question here is not whether it will be possible to automate numerous tactical functions now performed by human beings or to couple automated classifiers with automated decisions about lethal effects. It will and it is. But there is another ques-tion worth asking as well: how will automated weapons serve the political purposes of war?

For tactical prototypes, combat might be modeled to a game with the goal of win-ning by destroying more enemies while preserving more friendlies. Perhaps modern AI can excel in such games. But at the strategic or political level, war is about solving fundamental disputes. The concern here is not simply that "warbots" will be brilliant tacticians but stupid strategists (Payne, 2021). It is further unclear how and whether the ability of autonomous weapon systems to win set piece battles translates readily into political influence over human societies that hate each other enough to kill and die (Gartzke, 2021).

Put differently, how do the political conditions that give rise to the onset, escala-tion, or duration of war relate to the economic conditions that support AI performance? Should we expect AI-enabled weapons to be more useful in traditional forms of conflict, or for more ambiguous or protracted contests in the "gray zone" between peace and war? It is notable that long-term patterns of political violence appear to be shifting at the same time as the technical means of violence are shifting as well. A reasonable question is whether there is some relationship between these two trends, i.e., the "graying" of conflict and the rise of AI. And if so, what does the concurrent change in the nature or conduct of war mean for widespread worries about using certain weapons in war, which are often predicated on more traditional models of combat?

It is an unappreciated paradox that the same historical trends that are increasing the salience of gray zone conflict, cyber insecurity, terrorism, and subversion, are also increasing the viability of AI. The current Russo-Ukrainian war may be the exception that proves the rule, but even this war escalated from a protracted gray zone conflict and is accompanied by extensive information operations and covert action at the margins. The same economic conditions that make modern AI possible have also resulted in some important changes in war in the modern world. As states become more invested in trade, and as war becomes more destructive, states have reduced incentives to engage in open conquest, or at least they must be highly motivated for other reasons. While this may seem like good news insofar as the risk of total war between nuclear powers becomes less likely, it is still bad news for human security as civil society tends to suffer the most in irregular war, hybrid war, gray zone conflict, and cyber conflict. Yet these are precisely the sorts of conflicts that take place *within* shared institutions rather than across them.

The emergence of viable AI at scale is a product of the global liberal order, which describes a complex constellation of institutions for monetary policy, technical pro-tocols and standards, the rule of law, and so on. The realist tradition of international relations, however, emphasizes that war tends to emerge where institutions are weak or irrelevant, i.e., in a state of political anarchy. Sovereign states without an effective and legitimate overarching authority to adjudicate their disputes help themselves to enhance their own security, relying on their own resources and capabilities rather than common institutions. So, what does it mean for us to imagine an AI-enabled war if large-scale AI is a product of shared liberal order while war is the consummately realist pursuit? The reason that military AI systems are possible at all is that few wars if any are truly anarchic. There are usually some shared constraints, certainly at the level of geography, but also in shared assumptions and perhaps even normative institutions. The more war resembles civil conflict and subversion, the more this is so. Moreover, each combatant is itself an institution or set of institutions. Military doctrine attempts to break things down into reliably repeatable situations. This creates the potential for generating data and training for well-defined combat scenarios.

The scariest scenarios of fully autonomous robot armies may be simply impossible given the severe problems with data and judgment. Yet, there are many other ways in which military organizations are institutionalized. It is the areas of armed conflict that are most bureaucratized that we should expect to see the most promising applications of AI. While lethal drones get all the attention, more promising applications are in the realms of logistics, administration, personnel, recruitment, medicine, civil affairs, intelligence, and any other area of military operations with clear analogs in civilian organizations.

But even highly bureaucratized tasks will probably not be fully automated. Rather we should expect to see instances of human-machine teaming. This includes situations where the human takes the output of prediction systems and then decides, or then feeds back new inputs. Many people are already using generative AI systems in this way to improve their writing, coding, and graphic design. But human-machine teaming must also be understood more broadly to include all the support, maintenance, and repair activity required to keep AI infrastructure up and running. This is certainly true for platform substitution – the use of AI systems to replace human prediction tasks in existing work processes – but it will be even more true for the innovation of system complements – the transformation of military decision processes to better exploit the power of automated prediction.

STRATEGIC IMPLICATIONS

The human support system for institutionalized prediction in military organizations can be expected to become ever more complex. And with more complexity will come more potential for disagreement, bureaucratic politics, and coordination failure, to say nothing of enemy subversion and manipulation. Reliance on AI for almost any military task will require ongoing human intervention, tinkering, and negotiation to modify system functionality and gain access to relevant data, tasks that become all the more difficult in an environment of classified and controlled information.

People like to emphasize the importance of having a "man in the loop" for any AI decision, but this overlooks the fact that any real software system will be a tangled mess of many loops, and loops within loops. This is a long-standing challenge for enterprise software systems (Brooks, Jr., 1995; Ciborra, 2002), and increasing interdependencies in AI systems, data sources, and client organizations will make it worse. More adoption of AI will simply exacerbate a decades-long trend in military organizations of increasing complexity and coordination problems and reliance on human capital. In short, more reliance on AI for even mundane military tasks will make military organizations more reliant on people, not less (Goldfarb & Lindsay, 2022).

We can continue this analysis at the political level. The argument is straightforward: (i) If AI performance depends fundamentally on quality data and clear judgment, and (ii) if military organizations become more dependent on AI and thus data and judgment, then (iii) data and judgment will become critical strategic resources in political conflict, and (iv) adversaries will alter their strategies to complicate data and judgment. The very institutional complements that make it possible to use AI in war change the ways in which that war will be conducted.

What does this mean in practice? It means that cybersecurity and disinformation, which are already prominent and incredibly challenging features of modern war, become even more difficult. Adversaries have incentives to manipulate or poison the data that AI relies on or engage in AI-specific forms of deception. It also means that adversaries have incentives to move conflict in unexpected directions, where AI systems have not been trained and will likely perform in undesired or suboptimal ways. This creates not only data but judgment problems, as combatants will have to reconsider what they want in challenging new situations. As the war spills into new regions or begins harming civilians in new ways, how should AI targeting guidance change, or should AI systems be withheld altogether? We should expect that adversaries facing AI-enabled forces will shift war into ever more controversial and ethically fraught dimensions.

Adversaries facing automated armies may elect to avoid direct engagements altogether. After all, it may be impossible to discriminate whether the enemy is fighting with robots because robots are the most effective means or because the enemy is afraid of losing human lives (Gartzke, 2021; Gartzke & Walsh, 2022). To test the balance of resolve, therefore, the enemy has incentives to alter its behavior, perhaps by targeting civilians, expanding the war to other regions where robots are not used, or protracting the war to test how much the enemy is really willing to suffer. At the end of the day, the politics of violence is not only about the ability to kill – which tactical AI forces can do well – but also about the willingness to die – about which the use of AI forces says less than nothing.

A terrible irony is that the use of AI to fight decisive tactical engagement at reduced risk to military personnel is likely to result in more drawn-out political conflicts that increase the suffering of civilian personnel. This is not simply a problem of bad targeting guidance or failing to incorporate ethical precepts in lethal control systems, the usual focus of conversations about the responsible use of military AI. This is rather a political problem of strategic incentives for war changing as a result of changes in the tactical conduct of war. The core problem here is that AI is a product of stable institutions, but war is a product of anarchy. The conditions that make AI performance better also make traditional war less likely. Conversely, the conditions that allow war to persist or escalate also make it harder to use AI systems in reliable ways. The just-so stories of robots (or even "man in the loop" centaur systems) engaging in decisive set-piece battles are based on a political fantasy. War is more likely to emerge in areas where these systems cannot be used effectively if they can be used at all.

Future research should explore not only the ways in which AI changes the technology and tactics of war but also how it interacts with concurrent changes in the strategy and politics of war as well. This shift of focus may lead to a different set of ethical, operational, and strategic concerns. As military planners and antiwar activists alike focus on applications of AI for high-end conflict, they may be missing some of the most likely and most pernicious applications of AI in political conflict. It would be tragic to succeed in coming to an agreement about the responsible use of AI in war only to fail to consider the ways in which the same use of AI may encourage *humans* to behave less responsibly in war.

CONCLUSION

There is something of a contradiction between the circumstances that enable the development of large-scale AI and the circumstances that promote the onset, duration, and escalation of war. AI, to put it a bit too glibly, is a technology of peace. The most successful applications of AI to date have arisen in the peaceful context of commerce, and it is not clear that AI will work as well in other contests. On the contrary, there are principled reasons to believe it will not.

The economic conditions for the possibility of automated prediction are in tension with the political conditions for the possibility of violent conflict. Most examples of AI success to date are grounded in the pervasive institutionalization of capitalist infrastructure in the international system. This collective information infrastructure is the product of cooperation to a degree unequaled in human history. AI relies on large-scale data and stable collective judgments. But these same conditions are elusive in war. AI, to put it glibly, is an economic product of peace. War destroys the conditions that make AI viable. The conditions that are conducive to AI are not conducive to war, and vice versa.

Reliance on a technology born in peace, such big data-driven AI, for the politics of war, is sure to lead to unintended consequences. At the organizational level, militaries and governments will find themselves struggling with greater institutional complexity to provide the quality data and clear judgment that AI systems require. At the strategic level, AI-enabled combatants may find themselves involved in unexpectedly complex conflicts – not necessarily faster or more decisive but quite likely more protracted and controversial – because adversaries have incentives to contest the data and judgment resources that become sources of strength.

To the extent that AI can be applied in war, therefore, it is most likely in the aspects of war that most resemble peace. These are the "boring" administrative and logistics parts of the military enterprise rather than the "pointy end of the spear" where strategic and ethical considerations become paramount. Or, it is in conflicts in the "gray zone" between peace and war, where adversaries struggle within shared systems and with shared resources and assumptions. The common theme is that institutionalization is a vital complement for military AI. Enhanced institutionalization can be provided by a more complex military and defense establishment, or it can be provided by more robust collective social institutions and shared infrastructures. Yet these scenarios differ considerably from the imagined wars of robotic substitutes for human warriors.

In the final analysis, I expect that the future of military AI will resemble its past in many ways. Great expectations of faster, more decisive, and more automated war will continue to emerge with every new generation of AI, and information technology more broadly. Commercial successes of AI, moreover, will supercharge those expectations. Meanwhile, the problems of implementing AI in complex national security organizations will continue to grow ever more wicked. Military organizations will continue to become more reliant on the civilian economy, civilian technology, and civilian skills. While wars will continue to be as frustrating and full of friction as ever. The only difference is that the increasing complexity of sociotechnical implementations of AI systems will generate even more friction, to include even more opportunities for adversaries to inflict friction.

REFERENCES

Adams, J. (2001). *The next world war: Computers are the weapons and the front line is everywhere*. Simon & Schuster.

Agar, J. (2003). *The government machine: A revolutionary history of the computer*. MIT Press.

Agrawal, A., Gans, J., & Goldfarb, A. (2018). *Prediction machines: The simple economics of artificial intelligence*. Harvard Business Press.

Agrawal, A., Gans, J., & Goldfarb, A. (Eds.). (2019). *The economics of artificial intelligence: An agenda*. University of Chicago Press.

Agrawal, A., Gans, J., & Goldfarb, A. (2022). *Power and prediction: The disruptive economics of artificial intelligence*. Harvard Business Review Press.

Allard, C. K. (1990). *Command, control, and the common defense*. Yale University Press.

Beyerchen, A. (1992). Clausewitz, nonlinearity, and the unpredictability of war. *International Security*, *17*(3), 59–90.

Bostrom, N. (2014). *Superintelligence: Paths, dangers, strategies*. Oxford University Press.

Bowker, G. C., & Star, S. L. (1999). *Sorting things out: Classification and its consequences*. The MIT Press.

Brooks, Jr., F. P. (1995). *The mythical man-month: Essays on software engineering* (Anniversary Edition). Addison-Wesley Longman, Inc.

Broussard, M. (2018). *Artificial unintelligence: How computers misunderstand the world*. MIT Press.

Brown, J. S., & Duguid, P. (2000). *The social life of information*. Harvard Business Press.

Buchanan, B., & Imbrie, A. (2022). *The new fire: War, peace, and democracy in the age of AI*. The MIT Press.

Ciborra, C. (2002). *The labyrinths of information: Challenging the wisdom of systems*. Oxford University Press.

Collins, H. (1990). *Artificial experts: Social knowledge and intelligent machines*. MIT Press.

DeNardis, L. (2009). *Protocol politics: The globalization of Internet governance*. MIT Press.

Deudney, D. (1983). *Whole earth security: A geopolitics of peace*. Worldwatch Institute.

Dickson, P. (1976). *The electronic battlefield*. Indiana University Press.

Dreyfus, H. L. (1992). *What computers still can't do: A critique of artificial reason* (Revised). The MIT Press.

Drezner, D. W., Farrell, H., & Newman, A. L. (Eds.). (2021). *The uses and abuses of weaponized interdependence*. Brookings Institution Press. https://www.brookings.edu/book/the-uses-and-abuses-of-weaponized-interdependence/

Dupuy, J. P. (2000). *The mechanization of the mind: On the origins of cognitive science*. Princeton University Press.

Edwards, P. N. (1996). *The closed world: Computers and the politics of discourse in cold war America*. MIT Press.

Farrell, H., & Newman, A. (2023). *Underground empire: How America weaponized the world economy*. Henry Holt and Co.

Fearon, J. D. (1995). Rationalist explanations for war. *International Organization, 49*(3), 379–414.

Gannon, J. A., Gartzke, E., Lindsay, J. R., & Schram, P. (2023). The shadow of deterrence: Why capable actors engage in contests short of war. *Journal of Conflict Resolution*. https://doi.org/10.1177/00220027231166345

Gartzke, E. (1999). War Is in the error term. *International Organization, 53*(03), 567–587.

Gartzke, E. (2021). Blood and robots: How remotely piloted vehicles and related technologies affect the politics of violence. *Journal of Strategic Studies, 44*(7), 983–1013. https://doi.org/10.1080/01402390.2019.1643329

Gartzke, E., & Walsh, J. I. (2022). The drawbacks of drones: The effects of UAVs on militant violence in Pakistan. *Journal of Peace Research, 59*(4), 463–477.

Goldfarb, A., & Lindsay, J. R. (2022). Prediction and judgment: Why artificial intelligence increases the importance of humans in war. *International Security, 46*(3), 7–50.

Horowitz, M. C. (2018). Artificial intelligence, international competition, and the balance of power. *Texas National Security Review, 1*(3), 37–57. https://doi.org/10.15781/T2639KP49

Horowitz, M. C. (2019). When speed kills: Lethal autonomous weapon systems, deterrence and stability. *Journal of Strategic Studies, 42*(6), 764–788. https://doi.org/10.1080/01402390.2019.1621174

Johnson, J. (2020). Delegating strategic decision-making to machines: Dr. Strangelove Redux? *Journal of Strategic Studies*. https://www.tandfonline.com/doi/abs/10.1080/01402390.2020.1759038

Kania, E. B. (2017). *Battlefield singularity: Artificial intelligence, military revolution, and China's future military power*. Center for a New American Security.

Kline, R. R. (2015). *The cybernetics moment: Or why we call our age the information age*. Johns Hopkins University Press.

Lake, D. R. (2019). *The pursuit of technological superiority and the shrinking American military*. Palgrave Macmillan.

Lanoszka, A. (2016). Russian hybrid warfare and extended deterrence in Eastern Europe. *International Affairs, 92*(1), 175–195.

Lee, M. M. (2020). *Crippling leviathan: How foreign subversion weakens the state*. Cornell University Press.

Lindsay, J. R. (2017). Restrained by design: The political economy of cybersecurity. *Digital Policy, Regulation and Governance, 19*(6), 493–514.

Lindsay, J. R. (2020). *Information Technology and Military Power*. Cornell University Press.

Lindsay, J. R. (2021). Cyber conflict vs. cyber command: Hidden dangers in the American military solution to a large-scale intelligence problem. *Intelligence and National Security, 36*(2), 260–278.

Maschmeyer, L. (2022). Subversion, cyber operations and reverse structural power in world politics. *European Journal of International Relations, 29*(1). https://doi.org/10.1177/13540661221117051

National Security Commission on Artificial Intelligence. (2021). *Final Report*. https://www.nscai.gov/wp-content/uploads/2021/03/Full-Report-Digital-1.pdf

North, D. C. (1990). *Institutions, institutional change, and economic performance*. Cambridge University Press.

OpenAI. (2023). ChatGPT (June 9 version) [Large language model]. https://chat.openai.com/chat

O'Rourke, L. A. (2018). *Covert regime change: America's secret cold war.* Cornell University Press.

Ostrom, E. (1990). *Governing the commons: The evolution of institutions for collective action.* Cambridge University Press.

Parnas, D. L. (1985). Software aspects of strategic defense systems. *Communications of the ACM, 28*(12), 1326–1335. https://doi.org/10.1145/214956.214961

Payne, K. (2021). *I, Warbot: The dawn of artificially intelligent conflict.* Oxford University Press.

Ramsay, K. W. (2017). Information, uncertainty, and war. *Annual Review of Political Science, 20*(1), 505–527.

Rochlin, G. I. (1997). *Trapped in the net: The unanticipated consequences of computerization.* Princeton University Press.

Roff, H. M. (2014). The strategic robot problem: Lethal autonomous weapons in war. *Journal of Military Ethics, 13*(3), 211–227.

Scharre, P. (2018). *Army of none: Autonomous weapons and the future of war.* W. W. Norton & Company.

Sears, N. A. (2021). International politics in the age of existential threats. *Journal of Global Security Studies, 6*(3). https://doi.org/10.1093/jogss/ogaa027

Singer, P. W. (2009). *Wired for war: The robotics revolution and conflict in the twenty-first century.* Penguin.

Smith, B. C. (2019). *The promise of artificial intelligence: Reckoning and judgment.* MIT Press.

Toffler, A., & Toffler, H. (1993). *War and anti-war: Survival at the dawn of the 21st century.* Little, Brown & Co.

Turing, A. M. (1950). I. - COMPUTING MACHINERY AND INTELLIGENCE. *Mind, LIX*(236), 433–460. https://doi.org/10.1093/mind/LIX.236.433

von Clausewitz, C.. (1976). *On war* (M. Howard & P. Paret, Trans.). Princeton University Press.

Votel, J., Cleveland, C., Connett, C., & Irwin, W. (2016). Unconventional warfare in the gray zone. *Joint Force Quarterly, 80*, 101–109.

Williamson, O. E. (1985). *The economic institutions of capitalism: firms, markets, relational contracting.* Free Press.

Promoting Responsible State Behavior on the Use of AI in the Military Domain

16

Lessons Learned from Multilateral Security Negotiations on Digital Technologies

Kerstin Vignard

INTRODUCTION

Following the explosion of interest in AI ethics over the past several years, the concept of "Responsible AI" (RAI) has moved into the spotlight (Jobin et al., 2019; see also NATO, 2021; Anand & Deng, 2023; OECD, n.d.). On the heels of trends in industry

 DOI: 10.1201/9781003410379-20

and academia, the defense domain is beginning to express greater interest in how to adopt RAI as an organizing principle for the integration of AI across their enterprise (U.S. Department of Defense, 2020; United Kingdom Ministry of Defence, 2022).[1] However, as the international community starts to coalesce around the concept of RAI there is a need to develop common understanding as well as concrete actions to operationalize them.

This is not the first time the international security community has faced such a challenge. There are many similarities to multilateral negotiations to develop shared understandings on responsible use of other digital technologies in defense applications: specifically in the fields of cyber and autonomous weapons. In those discussions there was no agreement on the need for new legal instruments; there was a wide disparity of technological literacy among policy makers and of capabilities among countries; and there were divergent perspectives on the military utility, concerns, risks, benefits, and potential responses.

As an international, legally binding instrument regulating military applications of AI is difficult to conceive in the current geopolitical environment, what can be learned from how the international community has approached the development of norms of responsible behavior in the absence of appetite for new treaties? Would a similar approach focusing on reaffirming existing international law, development of norms of responsible behavior, identification of confidence-building measures, and the development of capacity-building initiatives suffice in the field of military applications of AI? Or have these approaches proven too slow to keep pace with the speed of innovation while excluding key stakeholders, such as technologists and the private sector?

This chapter will identify key lessons from the UN negotiations on cyber in the context of international security (from 2004 to 2021) and those on lethal autonomous weapon systems (2014–present) applicable the objective of developing shared understanding of RAI and accelerating international operationalization of RAI practices.

THE EMERGENCE OF THE CONCEPT OF RESPONSIBLE AI

Following the wave of interest in "AI ethics"[2], the more holistic approach of "Responsible AI" (RAI) has gained traction over the past few years, emerging first from industry and then being taken up by governments, international organizations and academia (U.S. Department of Defense, 2022; Accenture, n.d.; Google Cloud, n.d.; Microsoft Azure, 2022; Netherlands Ministry of Foreign Affairs, 2022; Dignum, 2019). And while there is not a single agreed definition of RAI, most descriptions identify as essential components a combination of engineering best practices, safety, trust, and ethics. For example, the UN Institute for Disarmament Research defines RAI as "an emerging AI governance approach that entails different normative tools including principles and ethical risk assessment frameworks to guide lawful, safe, and ethical design, development, and use of AI" (UNIDIR, n.d.).

Importantly, RAI is not simply a rebranding of AI ethics, nor does RAI replace consideration of ethics – rather ethics is the foundation of RAI. It is telling, for example, that the US Department of Defense (DoD) adopted its Ethical Principles for Artificial Intelligence a year before announcing its RAI implementation strategy. RAI helps shift attention from discussions of abstract high-level principles or norms to the action of operationalizing them: essentially moving from principles to practice.

The concept of RAI is gaining currency in the defense domain. The United States has adopted RAI as its strategy for AI in the military domain, with Secretary of Defense Lloyd Austin stating unambiguously that "We have a principled approach to AI that anchors everything that this Department does: we call this Responsible AI, and that is the only kind of AI that we do" (U.S. Department of Defense, 2021). In 2021 Deputy Secretary of Defense Katherine Hicks issued a memorandum on implementing RAI in the DoD, outlining the Department's "holistic, integrated, and disciplined" approach to RAI (Hicks, 2021). The six foundational tenets outlined in that memo were further expanded a year later in the DoD RAI Strategy and Implementation Pathway (U.S. Department of Defense, 2022).

The concept is also spreading among allies, as evidenced by the NATO 2021 Principles of Responsible Use of Artificial Intelligence in Defense, which "help steer our transatlantic efforts in accordance with our values, norms, and international law" (NATO, 2021, para. 7). Regional organizations, such as the European Union, are also producing guidance on the military uses of AI by its members (European Parliament, 2021, para. 94; European Commission AI High-Level Expert Group, 2019).

The Government of the Netherlands organized the first Responsible AI in the Military domain (REAIM) Summit in February 2023 (Netherlands Ministry of General Affairs, 2022; François-Blouin & Vestner, 2023). The Summit, co-hosted with the Government of the Republic of Korea, drew over 2,000 participants from over 100 countries to the Hague. The Summit's ministerial segment culminated with over 60 countries supporting the Summit's Call to Action on RAI in the military domain (Netherlands Ministry of Foreign Affairs & Ministry of Defence, 2023). The Call urges a variety of general actions: interdisciplinary and multi-stakeholder dialogue; sharing good practices and lessons learned; education and training; risk assessments; development and adoption of national strategies, policies and governance mechanisms; and testing and assurance methods. Commonly recognized AI principles, such as responsibility, accountability, and reliability are highlighted within the Call. While only States were invited to "endorse" the Call to Action, the Call is addressed to all stakeholders (governments, industry, knowledge institutions, international organizations, and others). The value of the Call is not so much in its general content, but as a signal of shared international concern and interest.

At the REAIM Summit, the United States launched its Political Declaration on Responsible Use of Artificial Intelligence and Autonomy (U.S. Department of State, Bureau of Arms Control, Verification and Compliance, 2023a), with the objective to develop shared international norms promoting responsible State behavior. The declaration reflects existing DoD policy and practice, including DoD revised Directive 3000.09 on Autonomy in Weapon Systems, the 2022 RAI Strategy and Implementation Pathway, and the Nuclear Posture Review (U.S. Department of State, Bureau of Arms Control, Verification and Compliance, 2023b). The US presented the declaration as a concrete response to the REAIM Call to Action, as well as complementary to, but not replacing, its March 2022 proposal on principles and good practices in the international discussions on autonomous weapons (Australia, Canada, Japan, Republic of Korea, United Kingdom, United States, 2022) as the declaration is not limited to LAWS and includes other concerns, including nuclear weapons systems.

As we look toward the second international summit on RAI in the Military Domain, to be convened by the government of the Republic of Korea (Republic of Korea, Ministry of Foreign Affairs, 2023), what lessons can we learn from the successes and failures of two and a half decades of multilateral negotiations on the international security dimensions of digital technologies? A closer examination of the UN cyber negotiations (2004–2021) and those on Lethal Autonomous Weapon Systems (2014–present) hold useful lessons for developing a shared understanding of responsible State behavior in the area of AI in the military domain.

Building on a brief description of the modalities of these two UN-based negotiations and their key achievements, ten lessons for the nascent international discussions on RAI are presented.

INFORMATION AND COMMUNICATION TECHNOLOGIES AND INTERNATIONAL SECURITY

The international community has over two decades of experience with negotiations on the international security dimensions of digital technologies. The Russian Federation sponsored the first resolution on information and communication technologies (ICTs) in the context of international security in 1998. A few years later, in 2004, the first UN Group of Governmental Experts (GGE) on "developments in the field of information and telecommunications in the context of international security" was convened, comprising 15 experts from a range of countries to consider "existing and potential threats in the sphere of information security and possible cooperative measures to address them" (United Nations General Assembly, 2003). All told, there were six GGE negotiations on ICTs and international security between 2004 and 2021, with four of the six achieving a consensus report transmitted to the General Assembly.[3]

Modality

GGEs are established through a resolution of the UN General Assembly First Committee (Disarmament and International Security) requesting the Secretary-General, with the assistance of a group of governmental experts, to undertake a study and to submit a report of their findings to the General Assembly at a later session. The group's mandate, as well as its size and number of sessions, are crafted in consultations and negotiations in the First Committee. Over the years the GGEs have grown from 15 to 20 to 25 members.

A GGE's composition is determined by the Secretary-General. In practice, the five permanent members of the Security Council are offered a seat on all GGEs and the remaining seats are allocated by UN regional grouping. The Secretary-General takes many factors into consideration when composing the group: geographical and political balance, demonstrated interest in the topic, the number of times that a country's experts have served on other GGEs or currently on a different First Committee GGE, etc. Once the GGE composition has been determined, the selected countries are asked to nominate an expert for the group. A Chair is selected and the group typically meets for four one-week sessions, for a total of 20 days of negotiations. GGEs meet behind closed doors and there are no public records of their meetings. The group must adopt its final report by consensus.

Key Achievements

Following the failure of the first cyber GGE (2004–2005) to reach an agreement, the 2010 GGE (United Nations General Assembly, 2010) produced a short consensus report, outlining the objectives of the cybersecurity negotiations to come: work to develop norms of responsible State behavior, confidence-building measures, and capacity-building measures. The 2013 GGE (United Nations General Assembly, 2013) confirmed an initial shared understanding of the international legal framework applicable to cyberspace, stating that the UN Charter, international law, and the principles of State sovereignty applied in this domain.

The high-water mark of the cyber GGE negotiations was the 2015 agreement on 11 norms of responsible State behavior (United Nations General Assembly, 2015). These voluntary, non-binding norms comprise both positive and negative commitments, including that States should cooperate to prevent harmful ICT practices, respond to requests for assistance, and that States should not conduct or support ICT activity that damages or impairs critical infrastructure (for an overview of the normative framework, see Australian Strategic Policy Institute, n.d.). Although the following GGE in 2017 failed to reach consensus, the 2019–2021 GGE issued a consensus report focused on adding an "additional layer of understanding" on ways to interpret and operationalize the 2015 norms (United Nations General Assembly, 2021a).

EMERGING TECHNOLOGIES IN THE AREA OF LETHAL AUTONOMOUS WEAPON SYSTEMS

A few distinct events heralded the arrival of the issue of autonomous weapons on the international policy making stage in 2012–2013. In late 2012, the US Department of Defense released Directive 3000.09 on Autonomy in Weapon Systems (U.S. Department of Defense, 2012). The directive is a detailed, publicly available document that sets out the US definition of fully and semi-autonomous weapons and details the procedures in place that regulate how increasingly autonomous weapons could be developed and fielded by the United States.[4] In the spring of 2013 the UN Human Rights Council Special Rapporteur on Extrajudicial, Summary, or Arbitrary Executions, Christof Heyns, presented his report to the Council in which he recommended a moratorium on the development of what he called "Lethal Autonomous Robotics" on the grounds that they might pose significant challenges to the right to life and the right to human dignity (United Nations Human Rights Council, 2013). This report built upon and extended many of the concerns that had emerged around the increasing use of armed unmanned systems in conflicts such as Afghanistan and Iraq, including human rights abuses, ethical issues, proliferation risks, and issues of transparency and accountability.[5] And around the same time, an international coalition of influential non-governmental organizations (NGOs) was formed: the Campaign to Stop Killer Robots, which started actively lobbying for a ban on autonomous weapon systems (*Stop Killer Robots*, n.d.; see also *International Committee of Robot Arms Control (ICRAC)*, n.d.).

Modality

At the UN, autonomous weapons (and by extension, artificial intelligence in the military domain) was taken up in an established treaty-specific body – the Convention on Prohibitions or Restrictions on the Use of Certain Conventional Weapons Which May Be Deemed to Be Excessively Injurious or to Have Indiscriminate Effects (known as the Convention on Certain Conventional Weapons or CCW). The CCW's purpose is to ban or restrict the use of specific types of weapons that are considered to cause unnecessary or unjustifiable suffering or affect combatants or civilians indiscriminately (International Committee of the Red Cross, 2014). The CCW is structured as a framework instrument – it comprises a short Convention and then protocols are negotiated as needed and attached to it. The CCW currently has five protocols: on non-detectable fragments, on mines and booby traps, on incendiary weapons, on blinding laser weapons, and on explosive remnants of war.

Between 2014 and 2016 the CCW convened three one-week Informal Meetings of Experts on Lethal Autonomous Weapon Systems (LAWS). At these meetings, international experts and governments exchanged views on technical issues, characteristics, International Humanitarian Law (IHL) and the law of armed conflict, ethics, strategic implications, and possible policy responses. In 2017 these discussions transitioned to a slightly more formal format called a Group of Governmental Experts (GGE). Although they share the same name, GGEs established by the UN General Assembly First Committee as described above in relation to cyber negotiations and those established by the CCW have different structures, formats, and rules of procedures. Some of these differences are touched upon below. Importantly, the GGE on LAWS is open to all CCW High Contracting Parties (HCPs). It is a uniquely transparent international arms control framework, with a high level of civil society access.

Key Achievements

Three key achievements in shared understanding emerged between 2017 and 2019. In 2017, HCP agreed that international law applies to the development and use of autonomous weapons (CCW, 2017). In 2018, the Chairman introduced what became known as the "sunrise diagram" (CCW, 2018, annex III), which illustrated the range of points of human involvement and potential oversight of LAWS – from political decisions pre-development through post-use assessment. Considering issues such as control, responsibility, oversight throughout the whole lifecycle of a weapon system, not just at time of use, enlarged the range of potential actors concerned with governance of autonomous weapons as well as the variety of means of control in addition to internationally negotiated agreements limiting or prohibiting certain types/classes of weapons. Lastly, also in 2018, the HCP agreed ten Guiding Principles, which included, for example, a re-affirmation that IHL applies to LAWS, that human responsibility cannot be transferred to machines, and that risk and mitigation measures should be part of the design, development, testing, and deployment lifecycle. The following year they added an additional principle on human-machine interaction, for a total of 11 (CCW, 2019b, section III, para. 16).

Following the affirmation in 2019 of the 11 Guiding Principles, CCW HCP were invited to submit national commentaries (CCW, 2021) on the principles and how to operationalize them. Considering how few States have articulated national AI principles for the military domain, these submissions were a valuable exercise to prompt internal discussions and to produce artifacts to help understand both national perspectives – as well as commonalities and divergences among them.

Since 2020, the GGE on LAWS has stagnated, with the further entrenchment of divergences about appropriate policy and regulatory responses. A hiatus of physical meetings due to the pandemic slowed movement in all multilateral discussions. The deterioration of geopolitical relations contributed to the substantive failure of the quinquennial CCW Review Conference in December 2021, with negative consequences for renewing the GGE's mandate. Despite HCPs agreeing to "intensify" discussions, the actual number of meeting days mandated for 2022 was reduced.

Although working in the shadow of Russia's invasion of Ukraine and its efforts to block substantive formal discussions, delegations tabled several concrete proposals in 2022, including on IHL, policy considerations, and development and sharing of good practice (CCW, 2022b).

LESSONS

What lessons can be drawn from over two decades of international security and digital technology negotiations that could help the international community address military applications of AI? Would a similar approach as that taken in the GGEs described above (through a process of Member State negotiation attempt to reaffirm existing international law, develop norms of responsible behavior, identify confidence-building measures, and develop capacity-building initiatives) help the international community reach common understandings concerning the military utility, potential risks, and legal and ethical issues raised by AI-enabled military systems? The conversation on RAI has emerged at the national, regional, and international levels, but has yet to be placed on the UN's agenda for consideration by Member States in a standing body. Would it be desirable to focus on developing international consensus on responsible State behavior in the use of AI in the military domain through a UN-based process?

As international engagement on RAI is only beginning, there is still time to shape how the issue is framed, scoped, and structured, as well as what should be prioritized as outcomes. The following ten lessons from previous digital technology negotiations help us consider whether the RAI discussions are being set up for success.

- Limit the scope of the discussions
- Adopt a productive framing
- Take a technology lifecycle approach
- Engage non-State stakeholders from the start
- Keep the focus on responsible State behavior
- Reaffirm that Responsible Innovation is built on an ethical foundation
- Be as transparent as possible, whenever possible
- Prioritize concrete actions to build common understandings before seeking global consensus
- Welcome diverse champions
- Ensure that RAI doesn't have a single point of failure

Limit the Scope of the Discussions

The cyber GGEs benefitted from a clear scope in part due to how they are created. These GGEs were established by the First Committee of the General Assembly, the committee that is mandated to consider items on disarmament and international security. Issues that are not under the purview of the First Committee – such as espionage, human rights, Internet governance, crime, terrorism, development and digital privacy – are outside the purview of the GGE's discussions. This understanding helped to limit the number of potential topics for consideration and thus make the best use of the limited number of days for negotiation.

The GGE on LAWS (and the informal meetings of experts that preceded it) has not benefited from the same clarity of scope. When this topic was opened for discussion in 2014, it created a space for a wide array of important issues to be brought to the multilateral table – these ranged from whether remotely operated systems lower the threshold for the use of force, to what sorts of proxy indicators are acceptable for use in weapon targeting, to whether AI may pose an existential threat to humanity. These and others were valid topics for discussion, but not all of them were directly related to LAWS. In the initial years, for example, some delegations repeatedly raised their concerns about remotely operated systems. This blurry scope of discussion persists today, with some delegations talking about autonomous systems and others about AI.

The problem was initially compounded by the forum in which the issue was considered. The international community decided to discuss what are arguably some of the most intelligent systems imagined in a forum traditionally dedicated to "dumb" conventional weapons such as landmines and booby traps. In the initial years of discussions, this choice had significant implications for the expertise and knowledge resident on delegations, as well as how the discussions were framed. For example, many CCW delegations are staffed with experts knowledgeable about conventional weapons such as mines and cluster munitions. It is also a forum that traditionally has discussed the legality of specific objects and their technical specifications. As a consequence, many delegations came into the room with the mental model that the discussions would be about objects rather than a characteristic or behavior of autonomous systems.

Building on the REAIM 2023 Call for Action, the international community will need to quickly consolidate its understanding of what is in – and out of – scope for these discussions and be disciplined about maintaining that understanding. One approach to an initial triage of issues of focus on, would be to ask two questions:

1. What is unique about, specific to, or amplified by AI use in the military domain?
2. Is the issue already being addressed through another process?

Militaries, like any other sector, are keen to reap the benefits of AI integration across all of the enterprise – including in health care, human resources, logistics, transportation, infrastructure, etc. We need to level set where military uses of AI are similar to other industries/employers and where they are different. For those categories of AI-enabled military activities that are similar to the private sector, collaborating with the private sector on best practices and industry standards may be most appropriate.

The second question helps to address the risk of forum shopping. The norms developing around RAI in the military domain should cover AI-enabled weapon systems, but an international discussion on RAI is not an alternative to existing, dedicated discussions on LAWS. Through its Guiding Principles affirmed in 2019, HCP have agreed that "The CCW offers an appropriate framework for dealing with the issue of emerging technologies in the area of lethal autonomous weapon systems within the context of the objectives and purposes of the Convention, which seeks to strike a balance between military necessity and humanitarian considerations"(CCW, 2019a, Annex III, para (k)). A growing number of States, supported by the Campaign to Stop Killer Robots, have expressed frustration at the slow pace of those discussions and lack of progress toward a ban on autonomous weapons. Pointing to successful precedents of like-minded arms control treaties negotiated outside of the UN framework (in particular the Anti-personnel Landmine Convention and the Cluster Munitions Convention), some would like to see the LAWS debate move outside the constraints of the UN (see for example Human Rights Watch & International Human Rights Clinic, Harvard Law School, 2022). Despite the efforts to frame the first REAIM Summit as a broader discussion of AI in the military domain rather than that of LAWS alone, participants across the board quickly fell into familiar talking points and disagreements well-known to those who have followed the CCW debates.

Without prejudice to the forum in which LAWS should ultimately be addressed, fora dedicated to operationalizing RAI are unique in being able to consider a wider range of AI-enabled military systems that are: (i) of critical importance to international security, stability, and cooperation; (ii) currently underattended in international discussions; and (iii) would most likely fall outside any "arms control" instrument addressing autonomous weapons.

Adopt a Productive Framing

Poor framings in international discussions result in States focusing their limited resources and attention on issues where there is unlikely to be consensus agreement (such as seeking early agreement on definitional issues, particularly when some States have already adopted a definition at the national level) or failing to sequence discussions in an order that identifies first the area(s) of concern before debating specific policy responses. Despite a lack of definitions and disagreement between key actors as to whether existing law is sufficient to address the use of ICTs by States, the cyber negotiations were still able to make relatively steady, cumulative progress on establishing normative frameworks, identifying confidence-building measures, and prioritizing areas for capacity building.

In contrast, from the earliest informal meetings of the LAWS GGE, independent actors such as the UN Institute for Disarmament Research (UNIDIR) warned that poor framing of the area of concern (including adopting a tech-centric or definition-based approach) would risk jeopardizing productive discussions on increasing autonomy in weapon systems (Vignard, 2014b).

Over the past nine years in the GGE on LAWS, States' understanding of technology, characteristics, potential risks and areas of concern has evolved and deepened, yet the mandate to discuss "emerging technology in the area of lethal autonomous weapon systems" remains largely unchanged. Additionally, not only has understanding evolved at different rates, but different States have emphasized different areas of concern and utility – including the international security dimensions of AI writ large, military decision-support tools, or embedding AI in specific domains, such as nuclear command and control. As a result, the discussions have become less focused – but more entrenched – over time.

On the other hand, the GGEs on ICTs repeatedly emphasized that their deliberations and recommendations should remain technology-agnostic. This had a fundamental advantage over the GGE on LAWS: it shifted discussions away from regulating or responding to the "latest-but-soon-to-be-outdated" cyber issue to a discussion of the roles and responsibilities, including legal obligations, of State actors when using ICTs in the domain of international security. Contrast this with the LAWS discussion, which was framed as a discussion of an "object" (a weapon system), which immediately demands consideration of – and, some would argue, agreement on – definitions.

Due to the nature and pace of AI innovation, maintaining a tech-neutral approach to the RAI discussions will be critical to success. Similar to the cyber GGE, focusing on responsible State behavior puts legal obligations at the heart of the discussion. The RAI framing also is an opportunity to focus attention on a range of critical issues that are underattended in the LAWS GGE, such as AI-enabled decision-support tools. It also offers a welcome opportunity to actively consider how to leverage AI for civilian protection and humanitarian missions undertaken by military forces (Beduschi, 2022; Gupta et al., 2021; Lewis & Ilachinski, 2022).

For the international RAI discussion to gather momentum, it is crucial that it not duplicate discussions happening in other international fora, such as CCW. It was disappointing but perhaps not surprising that many of the discussions at the REAIM

Summit quickly turned to autonomous weapons as it is a familiar and contentious subject and a variety of actors are dissatisfied with what they perceive as the slow pace and lack of results in CCW. That said, stakeholders should be cautious about squandering the valuable opportunity to discuss the potential risks and benefits of other AI-enabled systems in the military domain as these are not currently on the multilateral agenda and would unlikely be part of any dedicated instrument on LAWS, whether negotiated within CCW or as part of a like-minded process outside of the UN. Dedicated discussions on LAWS and a broader discussion of RAI can and should exist in parallel.

Take a Technology Lifecycle Approach

The 2018 LAWS GGE sunrise diagram was a valuable contribution to the international discussions on LAWS as it was an attempt to open the aperture of the CCW discussions. However, despite the recognition that the weapon lifecycle as a whole offered many opportunities for different actors to exercise control, after so many years without the engagement of some of the key stakeholders (R&D, tech researchers, industry) who have roles and responsibilities throughout that lifecycle, CCW discussions tend to gravitate back to the familiar terrain of the targeting cycle.

Developing an international understanding of RAI in the military domain is an opportunity to restart a conversation focused on the lifecycle of intelligent systems – including but not limited to weapon systems – used by our militaries, that takes into account the roles and responsibilities of a much larger set of practitioners and actors. This is not a conversation that has a natural "home" in the multilateral system – in part because it isn't limited to weapon systems and requires engagement from non-governmental stakeholders.

Engage Non-State Stakeholders from the Start

As closed-door negotiations, the cyber GGEs had no engagement with civil society and industry. With one exception, industry expressed little interest in the work of the GGE. It was only after the agreement of the 2015 norms of responsible State behavior that we saw wider openness and interest in engagement – on behalf of States toward industry and vice versa. At the same time, for many years, the cyber negotiations attracted minimal interest from civil society: almost unique among the arms control issues on the international agenda, there was no dedicated NGO movement addressing cyber and international security issues at the UN. It was only with the establishment in 2019 of a parallel cyber negotiation with a similar mandate yet meeting in a more transparent and inclusive format (the Open-Ended Working Group or OEWG) that non-State stakeholders were able to observe international cyber negotiations for the first time. The OEWG held a dedicated multistakeholder meeting early in those negotiations, which included advocacy groups, primarily with a human rights focus, industry representatives, and members of technical organizations.

While there has been more active non-governmental engagement from the start in the GGE on LAWS, once the GGE transitioned out of informal mode, it has been a State-led conversation. As previously mentioned, non-governmental experts have participated as part of civil society, through academic work and through organizing activities such as workshops and side events on the margins of CCW meetings. On the whole, civil society engagement in the LAWS GGE has been predominantly from advocacy groups, largely coordinated by the Campaign to Stop Killer Robots (SKR), and think tanks (see for example CCW, 2022a, para 10). A few non-NGO affiliated academics follow the negotiations for research purposes. Absent from the GGE on LAWS is sustained engagement from industry or the scientific/technical research community who are not members of an advocacy group. Over the years, the discussions would have benefitted from much more independent technical expertise to help ground the discussions in technical and operational realities.

As the notion of RAI takes purchase, we will need regular opportunities for multi-stakeholder gatherings to socialize norms, build confidence, and build capacity in RAI. Success in moving from principles to practice in RAI will depend on robust engagement with a broad range of actors from the technical community. State-centric discussions, even with limited civil society participation, will be insufficient. Technologists and the private sector/industry need to be welcome at and encouraged to be engaged in the international policy-shaping discussions. At the same time, governments must find ways to incentivize or compel their engagement. This is the only community with the knowledge and experience to be able to determine good practice, agreed standards, and metrics for testing, evaluation, validation and verification.

Industry also has a role as governments seek partnerships for innovation as well as contractors. Having a national AI strategy, AI principles or guidelines is a prerequisite for a government to be able to set clear expectations for industry. An early example of governmental guidance for industry on how to translate high-level principles into practice is the DoD's Defense Innovation Unit's "Responsible AI Guidelines in Practice" (Dunnmon et al., 2021). The framing of "responsible AI" may help to attract greater industry engagement, as many tech companies have shied away from the weapon-centric discussions on LAWS.

Civil society advocacy organizations also are needed to hold governments accountable for their RAI commitments. Civil society monitoring has proven to be a powerful influence in helping to shape State behavior and draw attention to abuses, failures, or neglected areas. States should welcome and encourage this engagement as a sign of commitment to operationalizing RAI.

Keep the Focus on Responsible State Behavior

Seeking to develop shared, voluntary normative frameworks on "responsible State behavior" in a variety of fields, including digital technologies and outer space, is a growing trend among UN Member States (United Nations General Assembly, 2010, 2013, 2015, 2021b, 2021a, 2022; United Nations Office for Disarmament Affairs, n.d.-a). While some would prefer to see negotiation of consensus-based, legally binding arrangements, fundamental impediments to negotiation of new law include the adversarial geopolitical climate arising from Russia's invasion of Ukraine; and the view of key States expressed in consensus fora such as the Convention on Certain Conventional Weapons that no new law is needed to address AI-enabled weapon systems. Although normative approaches are voluntary, the fact that they too are negotiated by consensus demonstrates a collective understanding of what is expected behavior of responsible State actors. Despite the fact that there are no legal consequences for acting in a manner contrary to these agreed principles, the authority afforded by their consensus support lends legitimacy to a variety of both unilateral and cooperative responses to instances of "bad" behavior.

Focusing on developing a shared understanding of what is responsible State behavior fits with the trends in other multilateral security negotiations. Common among the efforts to develop a shared understanding of responsible State behavior are that they focus on developing voluntary norms, sharing of good practices, building confidence through information sharing, and capacity building. The US Political Declaration announced at REAIM 2023 invites nations to engage in those activities.

The objective of moving from principles and norms to practice is happening much more quickly in RAI than in, for example, the international cyber conversation, where it took 6 years from agreement on norms in 2015 to a more detailed understanding of those norms. And despite discussions and proposals about a need to establish a "Programme of Action" (POA) in support of the international normative cyber framework, a resolution was passed only in 2022 and the POA will be convened only after 2025 – a full ten years following the adoption of the consensus norms (United Nations General Assembly, 2022; Pytlak, 2022). In contrast, moving from "AI principles" to practice is happening much more quickly, with dedicated efforts at the national, regional, and international levels. For example, in addition to the international REAIM summit, NATO's Data and AI Review Board (DARB), has established a subgroup to develop a first draft of an RAI certification standard as well as an inventory of existing assessments and toolkits by the end of 2023 (NATO, n.d.). Notably, however, RAI in the military domain has yet to be placed directly on the UN's agenda by a Member State via a General Assembly resolution. Rather it is through national statements (Netherlands Ministry of Foreign Affairs, 2022), side events (*UNIDIR and UNODA Introduce Delegates to Responsible AI for Peace and Security*, 2022) and conferences (UNIDIR, 2023) that the concept is being socialized among Member States.

As stated previously, the international discussion on RAI in the military domain is not an alternative to or in competition with dedicated discussions, in CCW or elsewhere, on particular applications of AI in the military domain. Rather, by offering an avenue to address some of the broader RAI topics, it may help to concentrate specific efforts such as the LAWS discussion by sharpening its scope and focus.

Reaffirm that Responsible Innovation is Built on an Ethical Foundation

The cyber GGEs, due to their mandate being established by First Committee, have considered that human rights are outside of the groups' mandates. So they have reaffirmed language agreed in other fora on human rights in the digital domain. Ethics has never been mentioned in any of the cyber GGE reports nor has ethics been a point of discussion within the groups themselves. In contrast, the topic of LAWS first arrived on the UN's agenda via the Human Rights Council report by Special Rapporteur Christof Heyns, so it is unsurprising that when discussions started in the CCW framework in 2014, ethical and rights-based concerns were among the first to be articulated (Docherty, 2015; United Nations Human Rights Council, 2013; Vignard, 2015) and remain prominent in its discussions.

The broader lens of responsible innovation may now be turning to computing writ large. In 2022, the United States National Academies study "Fostering Responsible Computing Research: Foundations and Practices" took a broad approach to responsible computing research in general, including AI but also cyber security and software engineering, rather than singling out AI for special ethical or legal consideration (Committee on Responsible Computing Research: Ethics and Governance of Computing Research and Its Applications et al., 2022). This welcome approach will be important when it comes time to deconflict norms and legal interpretations which have developed in the cyber and LAWS negotiating silos (Vignard, 2017). One example of this is the concept of Meaningful Human Control, which has been a focus of discussions within the GGE on LAWS since its origins (see for example Vignard, 2014a; iPRAW, 2019). While there is not a common understanding of the phrase, nor is there an international consensus agreement on its utility, there are groups of States who have embraced the concept as a principle to guide AI development in the military domain. However, the notion of human control has never featured in any of the UN cyber negotiations, even as increasingly autonomous cyber operations have become technologically feasible. In sum, our normative and ethical frameworks for digital technologies are diverging at a time of increasing technological convergence. A more holistic approach to responsible computing will help to bring coherence to our international legal and normative frameworks on responsible use of digital technologies, including AI.

Be as Transparent as Possible, Whenever Possible

Unlike the decade and a half of cyber negotiations behind closed doors, the LAWS discussion has been held from the beginning in a format that permits both attendance and participation of non-governmental stakeholders. While governments remain the decision makers, nearly all the official meetings are open to civil society attendance and they may take the floor to address delegates. The laudable transparency and accessibility of the CCW format have had ripple effects outside the negotiation chamber itself, as civil society dispatches, media engagement and academic research have helped to foster a community of interest and engagement around the world in near real-time. This openness has resulted in valuable external contributions from a variety of actors, including analysis of legal questions, regulatory approaches, and proposals of draft regulation.

Unlike the GGEs on ICTs where negotiations took place behind closed doors and for which there are no public meeting records or summaries, documentation concerning the GGE on LAWS's deliberations, draft reports, national statements, written contributions, and working papers is plentiful and accessible (Reaching Critical Will, n.d.; United Nations Office for Disarmament Affairs, n.d.-b). Through this transparency we can see the evolution of, or shifts in, national positions over time that we might not have had privy to as an internal, domestic policy development process.

As governments take forward their international discussions on RAI in the military domain, being as transparent as possible, whenever possible will help to keep the wide range of stakeholders informed, engaged and actively contributing to the topic.

Prioritize Concrete Actions to Build Common Understandings before Seeking Global Consensus

Most UN international security discussions operate under the rule of consensus. Consensus instruments send a valuable message of unity in purpose. However, consensus outcomes represent the common minimum amount of agreement, not the aspirational maximum. Consensus negotiations deliver the floor, not the ceiling.

Is a negotiated consensus on a political agenda for RAI necessary? In these early days of the international discussion on RAI, there is much more utility in setting the bar high – and as such, demonstrate the actions and goals to reach for, rather than watering down the objectives from the very start. Modeling the principles they have committed to themselves – for example, if government contractors are required to use model cards, the government should also have the same requirement of governmental departments and agencies – can influence both the adoption of those norms as well as the behavior of other actors.

While some high-capacity actors are fairly advanced in their thinking on national AI principles, strategies, policies, or guidelines (and only a few have articulated these specifically for the defense domain), the majority of the world's governments are not. Having a national position is nearly always a prerequisite for agreeing international policy since sovereign States do not want to cut off their options through agreements at the international level before they have worked through their domestic processes. Rather than focusing on seeking consensus agreement, now is the time for capacity building to encourage and support other nations to undertake the necessary reflection, deliberation and legislation at the national level. Higher-capacity States or those who have already done so should be encouraged to share information and lessons learned (as appropriate) about this process in order to support the capability of others to do so.

Those States with high capabilities and standards have the opportunity to influence international discussion in ways that align with their national objectives for innovation, security and stability and offer the chance for raising the bar or socializing good practices on specific issues such as weapons review processes for AI-enabled systems under Article 36 of Additional Protocol 1 of the Geneva Conventions.

Welcome Diverse Champions

The value of having a few States who serve as issue champions within their circles of influence cannot be understated. For example in the cyber negotiations, the Government of Singapore was highly influential in getting all members of the regional organization

Association of Southeast Asian Nations (ASEAN) to commit to the norms of responsible behavior recommended by the GGE, as well as funding numerous capacity-building initiatives, such as establishing the ASEAN-Singapore Cybersecurity Centre of Excellence (Singapore Cyber Security Agency, n.d.). Similarly the Government of the Netherlands established the Global Forum on Cyber Expertise (Global Forum on Cyber Expertise, n.d.) to strengthen international cooperation for capacity building in the cyber domain.

The Dutch have taken an early lead in championing an international dialogue on RAI in the military domain. The Republic of Korea stepped up as a co-sponsor of the first REAIM Summit and has agreed to host the second edition. The diversification of champions will also help RAI to be developed in a more inclusive manner, taking into consideration non-Western perspectives and offering opportunities for ownership and buy-in of a more globally represented set of nations.

Early adoption of a national strategy is insufficient to qualify as a "champion". As previously noted, the United States Department of Defense released Directive 3000.09 on Autonomous Weapon Systems over a decade ago (U.S. Department of Defense, 2012) and although it garnered international attention and criticism, it didn't result in a flood of similar policies from other nations. It remains to be seen whether and how these first national detailed RAI policies like the DoD RAI Strategy and Implementation Pathway will influence other States. It is clear, however, through the emphasis within its RAI strategy on building a global RAI ecosystem, and actions such as inviting other nations to endorse its Political Declaration, that the US is positioning itself as an influential international leader in RAI.

Ensure That RAI Doesn't Have a Single Point of Failure

The cyber GGEs have typically met for 20 days over a year, while the GGE on LAWS has met for as few as five days in a year and for as many as 20. During official meetings, experts and delegates present initial views, respond to the views of others, comment on draft language, offer revisions and negotiate the final report. With those significant constraints, even if consensus can be reached, there is a limit to what can realistically be expected as "progress" each year.

These multilateral processes are also at the mercy of geopolitics, current events, and financial constraints. Using the rules of procedure, the Russian Federation blocked substantive discussions from getting underway for the first two full days of the 2022 discussions,[6] and then only agreed to holding informal discussions for the following three. Meetings have been postponed or canceled due to circumstances that have nothing to do with the substantive work being undertaken: for example, in 2017 some CCW meetings were canceled due to the dire financial situation of the Convention as a result of unpaid contributions by some High Contracting Parties, and the COVID-19 pandemic curtailed nearly all official UN meetings in 2020. At the level of each individual State, a national election, a change in government or an internal policy process timeline may mean that a delegation has not received instructions from its capital at a critical moment on whether to support consensus.

For a field evolving as quickly as AI, a more agile process is required, with different constellations of actors working within and across different communities.

There is a rough framework offered by the REAIM Summit Call to Action, the Political Declaration proposed by the United States, initiatives in security alliances such as NATO, lines of effort detailed in national policy documents, and joint proposals by groups of States in fora such as the CCW. Over the next year, groups of stakeholders will be working on discrete facets of RAI in a variety of fora. This includes the scientific and technical community's invaluable work on topics such as developing test and evaluation tools and working on identifying good practices. Rather than locking the discussions into a single, rigid format, the objective should be to come back together at the following summit to share progress and lessons learned, articulate new challenges, refine the action areas and priorities, and continuously strengthen and expand a community of stakeholders committed to responsible development and use of AI in the military domain. There may be a time when consolidating this objective via a UN framework would be beneficial, but attempting to do it too soon risks hindering the initial necessary momentum needed to set high standards for responsible State behavior.

CONCLUSIONS

International discussions on digital technologies and international security hold many lessons for the nascent international RAI agenda. Traditional arms control approaches – which affirm the applicability of international law or create new law, establish shared norms, build confidence through information sharing and transparency, and build the capacity of others to adhere to their commitments, are applicable to RAI. However, it is too soon to place developing norms of responsible State behavior on AI in the military domain on the UN's agenda for three key reasons.

First, multilateral, consensus-based negotiations are notoriously slow. It took over a decade of cyber GGE negotiations to reach agreement on 11 norms of responsible State behavior, and five years of discussions for the GGE on LAWS to reach agreement on Guiding Principles. The international community cannot afford to approach RAI in the military domain at the same pace. For the most part individual States have not enacted hard stops domestically on AI development. In addition, a technology-centric multilateral process will never get out in front of the tech in order to build speedbumps or barriers that would force development pathways to slow down. In the ever-accelerating field of digital technologies, multilateral processes will always lag behind the speed of innovation: it is not a question of keeping pace: technology developers and policy makers are not even in the same race.

Second, while the United Nations and its Member States have made great strides in creating opportunities for engagement with non-State actors, at the end of the day the UN is not a multi-stakeholder organization. International agreement – although not universal consensus – on a set of high-level AI principles in the military domain is important. But once decision-makers have identified the principles or objectives, the work of operationalization – of turning those principles into practice at the level of design, development, testing, evaluation, validation, and verification – requires a different set of actors, with the scientific and technical community in the lead.

Translating RAI principles to practice must be a multistakeholder endeavor including other essential stakeholders such as industry and the research community.

Third, norms can and do emerge through means other than negotiation among States – they can evolve through practice and, in the case of AI, practice is happening faster. They can also emerge from actors other than governmental security policy makers, such as the technical community, standards bodies or professional societies. Developing and sharing good practice at the technical level will help inform the development of the right norms and the metrics to measure their operationalization, not the other way around.

The international discussion on RAI in the military domain is only beginning, but so far has avoided many of the pitfalls and stumbling blocks of previous international discussions on the use of digital technologies. For this trend to endure, RAI must continue to be considered in flexible arrangements of concerned communities. In civil society, the analytical community has shown early engagement in RAI discussions, while the technical community has been largely absent. For example, the majority of the breakout sessions at REAIM were focused on policy considerations. Only a handful were organized by technical organizations or had a technical focus. Greater engagement with technical actors, and identifying incentives for their participation in informing and shaping how international RAI norms are operationalized should be a priority for governments.

There will always be concern expressed that not every State will engage with a voluntary normative framework and that this may give irresponsible actors an "advantage" over those States who choose to respect existing law and ethical principles. However, the same could be said for international law – the world continues to witness State actors in breach of their most fundamental legal commitments. The conclusion of responsible actors should not be to join them in a "race to the bottom"; rather it must spur urgency in the building and consolidation of shared values, norms and approaches – even if they are not yet universal. The more members of the international community that agree on what "responsible behavior" looks like in this domain, the more sound footing they will be on to address malicious or abusive use of AI.

RAI is in the rational interest of all States and international stability. A government's public commitment to RAI demonstrates its intentions in this space, its priorities for development, and its expectations for providers of defense products and services. It also sets societal expectations for governments to ensure that societal values are reflected in the defense domain – and will help civil society to monitor governmental commitment to RAI.

NOTES

1 At the time of writing (March 2023), only two States, the US and the United Kingdom, have adopted AI principles specific to the defense sector. Additionally, of the 12 intergovernmental organizations identified as having adopted AI principles, only two (NATO and the UN Group of Governmental

Experts on Emerging Technologies in the Area of Lethal Autonomous Weapon Systems) specifically address the military domain (Anand & Deng, 2023).

2 For example, in 2018 Jobin et al. identified 84 different sets of AI ethics principles (Jobin et al., 2019). According to the 2022 AI Index Report, "Research on fairness and transparency in AI has exploded since 2014, with a fivefold increase in related publications at ethics-related conferences. Algorithmic fairness and bias has shifted from being primarily an academic pursuit to becoming firmly embedded as a mainstream research topic with wide-ranging implications." (Zhang et al., 2022).

3 While not treated in depth in this chapter, it should be noted that between 2019 and 2021, a second UN cyber negotiation ran in parallel with the GGE: The Open-Ended Working Group (OEWG). Open to all Member States, it too achieved a consensus final report in March 2021.

4 A revised directive was issued in February 2023, notably with clarification on particular terms and updates reflecting the DoD's ethical principles, RAI strategy, and changed institutional structures (U.S. Department of Defense, 2023; Scharre, 2023).

5 A UN study in 2015 identified five broad categories of international concerns in the use of UAVs, including altering incentives in the use of force, new interpretations of legal frameworks, proliferation potential, lack of accountability or transparency, and automating and compressing the "time to strike" process (United Nations Office for Disarmament Affairs, 2015, pp. 26–32; UNIDIR, 2017, pp. 9–10). These same concerns have been echoed in the CCW LAWS discussion.

6 The absurdity of the situation was described by an influential civil society group as follows: "Arguing that its delegation was being 'discriminated against' – because it was more difficult for its experts to travel from Moscow to Geneva under the sanctions imposed in response to its illegal invasion of and war against Ukraine – the Russian government wasted two of the GGE's five days, and then only permitted substantive discussions to begin in informal mode. This meant there was limited transparency and accessibility for civil society and other delegates that could not travel to Geneva due to the pandemic, war, climate crisis, economic inequalities, or other reasons – resulting in actual discrimination, instead of the perceived "slights" against the Russian delegation." (Acheson, 2021).

REFERENCES

Accenture. (n.d.). *Responsible AI | AI ethics & governance. Accenture.* Retrieved May 14, 2023, from https://www.accenture.com/us-en/services/applied-intelligence/ai-ethics-governance

Acheson, R. (2021). We will not weaponise our way out of horror. *CCW Report, 10*(2), 7.

Anand, A., & Deng, H. (2023). *Towards responsible AI in defence: A mapping and comparative analysis of AI principles adopted by states*. UNIDIR. https://www.unidir.org/sites/default/files/2023-02/Brief-ResponsibleAI-Final.pdf

Australia, Canada, Japan, Republic of Korea, United Kingdom, United States. (2022). *Principles and good practices on emerging technologies in the area of lethal autonomous weapons system*. CCW. https://view.officeapps.live.com/op/view.aspx?src=https%3A%2F%2Fdocuments.unoda.org%2Fwp-content%2Fuploads%2F2022%2F05%2F20220307-US-UK-RoK-JAP-CAN-AUS-Final-proposal-laws-principles-and-good-practices.docx&wdOrigin=BROWSELINK

Australian Strategic Policy Institute (ASPI). (n.d.). *UN cyber norms*. UN norms of responsible state behaviour in cyberspace. Retrieved May 14, 2023, from https://www.aspi.org.au/cybernorms

Beduschi, A. (2022). Harnessing the potential of artificial intelligence for humanitarian action: Opportunities and risks. *International Review of the Red Cross, 104*(919), 1149–1169. https://doi.org/10.1017/S1816383122000261

CCW. (2017). *Report of the 2017 group of governmental experts on lethal autonomous weapons systems (LAWS)*. United Nations. https://docs-library.unoda.org/Convention_on_Certain_Conventional_Weapons_-_Meeting_of_High_Contracting_Parties_(2017)/2017_CCW_GGE.1_2017_CRP.1_Advanced_%2Bcorrected.pdf

CCW. (2018). *Report of the 2018 session of the group of governmental experts on emerging technologies in the area of lethal autonomous weapons systems*. United Nations. https://documents-dds-ny.un.org/doc/UNDOC/GEN/G18/323/29/PDF/G1832329.pdf?OpenElement

CCW. (2019a). *Meeting of the high contracting parties to the convention on prohibitions or restrictions on the use of certain conventional weapons which may be deemed to be excessively injurious or to have indiscriminate effects: Final Report*. United Nations. https://documents-dds-ny.un.org/doc/UNDOC/GEN/G19/343/64/PDF/G1934364.pdf?OpenElement

CCW. (2019b). *Report of the 2019 session of the group of governmental experts on emerging technologies in the area of lethal autonomous weapons systems*. United Nations. https://documents.unoda.org/wp-content/uploads/2020/09/CCW_GGE.1_2019_3_E.pdf

CCW. (2021). *Chairperson's summary 2020: Group of governmental experts on emerging technologies in the area of lethal autonomous weapons system*. United Nations. https://documents.unoda.org/wp-content/uploads/2020/07/CCW_GGE1_2020_WP_7-ADVANCE.pdf

CCW. (2022a). *Report of the 2021 session of the group of governmental experts on emerging technologies in the area of lethal autonomous weapons systems*. United Nations. https://documents-dds-ny.un.org/doc/UNDOC/GEN/G22/264/50/pdf/G2226450.pdf?OpenElement

CCW. (2022b). *Convention on Certain Conventional Weapons - Group of Governmental Experts (2022)* | United Nations. https://meetings.unoda.org/ccw/convention-certain-conventional-weapons-group-governmental-experts-2022

Committee on Responsible Computing Research: Ethics and Governance of Computing Research and Its Applications, Computer Science and Telecommunications Board, Division on Engineering and Physical Sciences, & National Academies of Sciences, Engineering, and Medicine. (2022). *Fostering responsible computing research: foundations and practices*. National Academies Press. https://doi.org/10.17226/26507

Dignum, V. (2019). *Responsible artificial intelligence: How to develop and use AI in a responsible way*. https://doi.org/10.1007/978-3-030-30371-6

Docherty, B. (2015). *Mind the gap*. Human Rights Watch. https://www.hrw.org/report/2015/04/09/mind-gap/lack-accountability-killer-robots

Dunnmon, J., Goodman, B., Kirechu, P., Smith, C., & Van Deusen, A. (2021). *Responsible AI guidelines in practice* (p. 33). Defense Innovation Unit, Department of Defense. https://assets.ctfassets.net/3nanhbfkr0pc/acoo1Fj5uungnGNPJ3QWy/3a1dafd64f22efcf8f27380aafae9789/2021_RAI_Report-v3.pdf

European Commission AI High-Level Expert Group. (2019, April 8). *Ethics guidelines for trustworthy AI.* https://digital-strategy.ec.europa.eu/en/library/ethics-guidelines-trustworthy-ai

European Parliament. (2021*). European Parliament resolution of 20 January 2021 on artificial intelligence: Questions of interpretation and application of international law in so far as the EU is affected in the areas of civil and military uses and of state authority outside the scope of criminal justice (2020/2013(INI)).* European Parliament. https://www.europarl.europa.eu/doceo/document/TA-9-2021-0009_EN.html

François-Blouin, J., & Vestner, T. (2023, March 13). *Globalizing responsible AI in the military domain by the REAIM summit.* Just Security. https://www.justsecurity.org/85440/globalizing-responsible-ai-in-the-military-domain-by-the-reaim-summit/

Global Forum on Cyber Expertise. (n.d.). *About the GFCE.* Retrieved May 14, 2023, from https://thegfce.org/about-the-gfce/

Google Cloud. (n.d.). *Responsible AI.* Retrieved May 14, 2023, from https://cloud.google.com/responsible-ai

Gupta, R., Rolf, E., Murphy, R., & Heim, E. (2021, December 13). *Third workshop on AI for humanitarian assistance and disaster response.* SlidesLive. https://neurips.cc/virtual/2021/workshop/21858

Hicks, K. (2021). *Memorandum: Implementing responsible artificial intelligence in the Department of Defense.* Deputy Secretary of Defense. https://media.defense.gov/2021/May/27/2002730593/-1/-1/0/IMPLEMENTING-RESPONSIBLE-ARTIFICIAL-INTELLIGENCE-IN-THE-DEPARTMENT-OF-DEFENSE.PDF

Human Rights Watch, & International Human Rights Clinic, Harvard Law School. (2022). *An agenda for action: Alternative processes for negotiating a killer robots treaty.* https://www.hrw.org/sites/default/files/media_2022/12/arms1122web.pdf

International Committee of Robot Arms Control (ICRAC). (n.d.). Retrieved May 14, 2023, from https://www.icrac.net/

International Committee of the Red Cross. (2014, September 8). *FACTSHEET: 1980 convention on certain conventional weapons [Legal factsheet].* International Committee of the Red Cross. https://www.icrc.org/en/document/1980-convention-certain-conventional-weapons

iPRAW. (2019). *Focus on Human Control.* International Panel on the Regulation of Autonomous Weapons (iPRAW).

Jobin, A., Ienca, M., & Vayena, E. (2019). The global landscape of AI ethics guidelines. *Nature Machine Intelligence, 1*(9), 9. https://doi.org/10.1038/s42256-019-0088-2

Lewis, L., & Ilachinski, A. (2022). *Leveraging AI to mitigate civilian harm* (p. 82). CNA. https://www.cna.org/reports/2022/02/Leveraging-AI-to-Mitigate-Civilian-Harm.pdf

Microsoft Azure. (2022, November 9). *What is responsible AI - azure machine learning.* https://learn.microsoft.com/en-us/azure/machine-learning/concept-responsible-ai

NATO. (2021). *Summary of the NATO artificial intelligence strategy.* https://www.nato.int/cps/en/natohq/official_texts_187617.htm

NATO. (n.d.). *NATO's data and artificial intelligence review board.* NATO. Retrieved May 25, 2023, from https://www.nato.int/cps/en/natohq/official_texts_208374.htm

Netherlands Ministry of Foreign Affairs, & Ministry of Defence. (2023, February 16). REAIM 2023 Call to Action [Publicatie]. Ministerie van Algemene Zaken. https://www.government.nl/documents/publications/2023/02/16/reaim-2023-call-to-action

Netherlands Ministry of Foreign Affairs. (2022, September 21). The Netherlands to host international summit on artificial intelligence - News item - Government.nl [Nieuwsbericht]. Ministerie van Algemene Zaken. https://www.government.nl/latest/news/2022/09/21/the-netherlands-to-host-international-summit-on-artificial-intelligence

Netherlands Ministry of General Affairs. (2022, September 13). REAIM 2023 - Ministry of Foreign Affairs - Government.nl [Webpagina]. Ministerie van Algemene Zaken. https://www.government.nl/ministries/ministry-of-foreign-affairs/activiteiten/reaim

OECD. (n.d.). The OECD Artificial Intelligence (AI) Principles. Retrieved May 14, 2023, from https://oecd.ai/en/ai-principles

Pytlak, A. (2022). *Advancing a global cyber programme of action: Options and priorities: Options and priorities* (p. 49). WILPF. https://reachingcriticalwill.org/images/documents/Publications/report_cyber-poa_final_May2022.pdf

Reaching Critical Will. (n.d.). *Convention on certain conventional weapons.* Retrieved May 25, 2023, from https://www.reachingcriticalwill.org/disarmament-fora/ccw

Republic of Korea, Ministry of Foreign Affairs. (2023). Closing remarks by H.E. Park Jin Minister of Foreign Affairs of the Republic of Korea Responsible AI in the Military Domain Summit (REAIM) February 16, 2023 View|Minister | Ministry of Foreign Affairs, Republic of Korea. https://www.mofa.go.kr/eng/brd/m_5689/view.do?seq=319597

Scharre, P. (2023, February 6). *NOTEWORTHY: DoD autonomous weapons policy.* CNAS. https://www.cnas.org/press/press-note/noteworthy-dod-autonomous-weapons-policy

Singapore Cyber Security Agency. (n.d.). Cyber Security Agency of Singapore. Default. Retrieved May 17, 2023, from https://www.csa.gov.sg

Stop Killer Robots. (n.d.). *Stop killer robots.* Retrieved May 14, 2023, from https://www.stopkillerrobots.org/

U.S. Department of Defense. (2012). *Department of Defense Directive 3000.09: Autonomy in weapon systems.* U.S. Department of Defense. https://www.hsdl.org/c/abstract/

U.S. Department of Defense. (2020, February 24). *DOD adopts ethical principles for artificial intelligence.* U.S. Department of Defense. https://www.defense.gov/News/Releases/Release/Article/2091996/dod-adopts-ethical-principles-for-artificial-intelligence/https%3A%2F%2Fwww.defense.gov%2FNews%2FReleases%2FRelease%2FArticle%2F2091996%2Fdod-adopts-ethical-principles-for-artificial-intelligence%2F

U.S. Department of Defense. (2021, July 13). *Secretary of defense Austin remarks at the global emerging technology summit of the nation* [Transcript]. U.S. Department of Defense. https://www.defense.gov/News/Transcripts/Transcript/Article/2692943/secretary-of-defense-austin-remarks-at-the-global-emerging-technology-summit-of/https%3A%2F%2Fwww.defense.gov%2FNews%2FTranscripts%2FTranscript%2FArticle%2F2692943%2Fsecretary-of-defense-austin-remarks-at-the-global-emerging-technology-summit-of%2F

U.S. Department of Defense. (2022). *Department of Defense responsible artificial intelligence strategy and implementation pathway.* https://media.defense.gov/2022/Jun/22/2003022604/-1/-1/0/Department-of-Defense-Responsible-Artificial-Intelligence-Strategy-and-Implementation-Pathway.PDF

U.S. Department of Defense. (2023). *Revised Directive 3000.09.* https://www.esd.whs.mil/portals/54/documents/dd/issuances/dodd/300009p.pdf

U.S. Department of State, Bureau of Arms Control, Verification and Compliance. (2023a). *Political declaration on responsible military use of artificial intelligence and autonomy.* U.S. Department of State. https://www.state.gov/political-declaration-on-responsible-military-use-of-artificial-intelligence-and-autonomy/

U.S. Department of State, Bureau of Arms Control, Verification and Compliance, B. J. (2023b). *Keynote remarks by U/S Jenkins (T) to the summit on responsible artificial intelligence in the military domain (REAIM) ministerial segment.* U.S. Department of State. https://www.state.gov/keynote-remarks-by-u-s-jenkins-t-to-the-summit-on-responsible-artificial-intelligence-in-the-military-domain-reaim-ministerial-segment/

UNIDIR and UNODA introduce delegates to responsible AI for peace and security. (2022, November 1). https://disarmament.unoda.org/update/unidir-and-unoda-introduce-delegates-to-responsible-ai-for-peace-and-security/

UNIDIR. (2017). *Increasing transparency, oversight and accountability of armed unmanned aerial vehicles* (p. 44). UNIDIR. https://unidir.org/sites/default/files/publication/pdfs/increasing-transparency-oversight-and-accountability-of-armed-unmanned-aerial-vehicles-en-692.pdf

UNIDIR. (2023). *The 2022 innovations dialogue: AI disruption, peace and security (Conference Report)* | UNIDIR. https://www.unidir.org/publication/2022-innovations-dialogue-ai-disruption-peace-and-security-conference-report

UNIDIR. (n.d.). *How to operationalize responsible AI in the military domain*. REAIM. Retrieved May 14, 2023, from https://reaim2023.org/events/how-to-operationalize-responsible-ai-in-the-military-domain/

United Kingdom Ministry of Defence. (2022). *United Kingdom, MOD, Defence Artificial Intelligence Strategy*. MOD.

United Nations General Assembly. (2003). *Resolution 58/32. Developments in the field of information and telecommunications in the context of international security*. United Nations. https://documents-dds-ny.un.org/doc/UNDOC/GEN/N03/454/83/PDF/N0345483.pdf?OpenElement

United Nations General Assembly. (2010). *Group of governmental experts on developments in the field of information and telecommunications in the context of international security*. United Nations. https://documents-dds-ny.un.org/doc/UNDOC/GEN/N10/469/57/PDF/N1046957.pdf?OpenElement

United Nations General Assembly. (2013). *Group of governmental experts on developments in the field of information and telecommunications in the context of international security*. United Nations. https://daccess-ods.un.org/tmp/9370825.29067993.html

United Nations General Assembly. (2015). *Group of governmental experts on developments in the field of information and telecommunications in the context of international security*. United Nations. https://daccess-ods.un.org/tmp/9373349.54738617.html

United Nations General Assembly. (2021a). *Group of governmental experts on advancing responsible state behaviour in cyberspace in the context of international security*. United Nations. https://documents-dds-ny.un.org/doc/UNDOC/GEN/N21/075/86/PDF/N2107586.pdf?OpenElement

United Nations General Assembly. (2021b). *Resolution 76/231. Reducing space threats through norms, rules and principles of responsible behaviours*. United Nations. https://doi.org/10.18356/9789210021753

United Nations General Assembly. (2022). *Resolution 77/37: Programme of action to advance responsible State behaviour in the use of information and communications technologies in the context of international security*. United Nations. https://digitallibrary.un.org/record/3997617

United Nations Human Rights Council. (2013). *Report of the Special Rapporteur on extrajudicial, summary or arbitrary executions, Christof Heyns*. United Nations. https://doi.org/10.1163/2210-7975_HRD-9970-2016149

United Nations Office for Disarmament Affairs. (2015). *Study on armed unmanned aerial vehicles: Prepared on the recommendation of the advisory board on disarmament matters* (p. 76). United Nations. https://unoda-web.s3-accelerate.amazonaws.com/wp-content/uploads/assets/publications/more/drones-study/drones-study.pdf

United Nations Office for Disarmament Affairs. (n.d.-a). *Open-ended working group on reducing space threats (2022)*. United Nations. Retrieved May 14, 2023, from https://meetings.unoda.org/open-ended-working-group-reducing-space-threats-2022

United Nations Office for Disarmament Affairs. (n.d.-b). *UNODA meetings place*. Retrieved May 25, 2023, from https://meetings.unoda.org/

Vignard, K. (2014a). Considering how meaningful human control might move the discussion forward. In *The weaponization of increasingly autonomous technologies* (p. 12). UNIDIR. https://unidir.org/sites/default/files/publication/pdfs/considering-how-meaningful-human-control-might-move-the-discussion-forward-en-615.pdf

Vignard, K. (2014b). Framing discussions on the weaponization of increasingly autonomous technologies. In *The weaponization of increasingly autonomous technologies* (p. 14). UNIDIR. https://unidir.org/sites/default/files/publication/pdfs/framing-discussions-on-the-weaponization-of-increasingly-autonomous-technologies-en-606.pdf

Vignard, K. (2015). Considering ethics and social values. In *The weaponization of increasingly autonomous technologies* (p. 14). UNIDIR. https://unidir.org/sites/default/files/publication/pdfs/considering-ethics-and-social-values-en-624.pdf

Vignard, K. (2017). Autonomous weapon systems and cyber operations. In *The weaponization of increasingly autonomous technologies*. https://www.unidir.org/sites/default/files/publication/pdfs/autonomous-weapon-systems-and-cyber-operations-en-690.pdf

Zhang, D., Maslej, N., Brynjolfsson, E., Etchemendy, J., Lyons, T., Manyika, J., Ngo, H., Niebles, J. C., Sellitto, M., Sakhaee, E., Shoham, Y., Clark, J., & Perrault, R. (2022). The AI index 2022 annual report (arXiv:2205.03468). arXiv. https://doi.org/10.48550/arXiv.2205.03468

SECTION V

Bounded Autonomy

Bounded Autonomy

17

Jan Maarten Schraagen

INTRODUCTION

The aim of this concluding chapter is not so much to summarize what has been stated so well by the authors of the various chapters, but rather to reflect on some common themes that run throughout this book, as well as to highlight some additional issues and research challenges, particularly in the field of Human Factors and Ergonomics (HFE). First, I will address the use of AI in military systems and how this relates to the heated debate on 'killer robots'. Second, I will discuss the concept of 'autonomy' and its use in 'autonomous weapon systems (AWS)'. I will argue that, just as there is no such thing as 'rationality' in humans, only 'bounded rationality' (Simon, 1955; 1957), there is no such thing as autonomy in systems, only 'bounded autonomy'. This then will lead to a discussion of the third concept, that of 'meaningful human control (MHC)', which, as will be shown, is closely related to the concept of 'bounded autonomy'. I will argue that the existence of bounded autonomy makes MHC over military systems possible. In the final section, I will discuss what it could mean to develop AI in a responsible fashion in military systems, which, after all, is the title and main topic of this book.

DEMYSTIFYING DYSTOPIAN VIEWS ON THE USE OF AI IN MILITARY SYSTEMS

In some dystopian visions, AI is seen as a technology that is beyond our control and that enables particular weapon systems to select and engage targets by themselves. This vision has been reinforced by the Future of Life Institute that has funded a movie

DOI: 10.1201/9781003410379-22   345

showing how swarms of flying robots ('drones') kill large numbers of innocent people (Russell, 2022; Slaughterbots, 2017). Not only is the flying of the robots enabled by AI, but the assumption is that the AI employed in these drones also enables the face-recognition that is required for the targeting process. What is particularly worrisome in this depiction is that, once the AI has been successfully applied in one instance, it can be replicated easily and applied to hundreds of thousands of drones, making targeted mass killings within reach of terrorists or rogue nations or so the proponents of this campaign claim (Russell, 2022). Other arguments of those opposing 'killer robots' are accountability gaps (if no humans are involved, they cannot be held accountable), violation of International Humanitarian Law (IHL; principles of distinction and proportionality), and the dehumanization of warfare.

The general public is mostly concerned with the use of AI in AWS, also framed as 'killer robots'. However, AI – and certainly the recent developments in generative AI – may be applied across the military enterprise, ranging from human resource management systems to maintenance and logistics systems, from cyber defense systems to reconnaissance systems, and from decision support systems to joint protection and warfighting.

Any discussion of the use of AI in military systems, let alone the 'responsible' use of AI in such systems, needs to relate to these concerns and fears. It is very difficult to deal with these emotions on a rational basis, yet this is what those painting a more nuanced picture need to do. Presenting information and myth-busting are required to assuage those fears, without ignoring or diminishing in any way the real concerns many people have with the rapid development of AI.

One of the first arguments against the Slaughterbots movie is that it depicts a fictitious situation and that, as of yet, there are no killer robots enabled by AI. Although this is being debated, with some arguing that killer drones were employed in Libya in 2020 (in effect, the drone being used was the Kargu-2 rotary wing loitering munition, UN Security Council, 2021), the more general line of reasoning is familiar with anyone watching discussions between those who believe some form of autonomous technology is 'just around the corner' versus those who are more skeptical about those claims. The fact that the Slaughterbots movie depicts a fictitious situation is irrelevant to the first group, as they believe that, even though it may not be a reality today, it will be a reality in the foreseeable future (with 'foreseeable' being used as an ever-receding horizon if it does not materialize in time). Those who believe in this technology take the 'fake it till you make it' stance, which they claim is necessary or else there would be no funding for any new technological developments. This particular stance has been so ingrained in Silicon Valley culture that to criticize it is tantamount to criticizing progress in general (though with the funding drying up, faking it may be over, Griffith, 2023). Those warning about killer robots may not be the same as those Silicon Valley entrepreneurs warning about the dangers of AI, but they do share the same belief in technology that is always just around the corner. I am arguing here that this technology may never materialize.

An analogy may be drawn with the prophesies about self-driving ('autonomous') cars. In 2015, Musk predicted 'complete autonomy' by 2017. In 2016, Lyft co-founder and president Zimmer claimed that by 2025 car ownership in US cities would "all but end" (Sipe, 2023). General Motors in 2017 promised mass production of fully autonomous vehicles in 2019. More than $100 billion has been invested in self-driving cars since 2010 (Chafkin, 2022). However, in 2020 and 2021, respectively, Uber and Lyft shut down their efforts. By the end of 2022, Volkswagen and Ford pulled the plug on their self-driving efforts (Sipe, 2023). According to Anthony Levandowski, a fervent believer turned apostate, "You'd be hard-pressed to find another industry that's invested so many dollars in R&D and that has delivered so little." (Chafkin, 2022). Most of the testing of self-driving cars is done in sunny California and Arizona, hardly representative of the rest of the US, let alone the world. What these cars have difficulties with, are what the engineers call 'edge cases'. Edge cases go far beyond the sunny weather these cars usually operate in: from broken traffic lights to a bicyclist crossing a street, the list is endless as the world is a messy place. Human drivers know what to expect of pigeons on the road and how to respond; self-driving cars have no such intelligence and will slam the brakes causing rear-end collisions. This example shows that the issues are not so much with object detection as such, but rather with interpretation of the information obtained in the context of driving.

The more general problem here is what Woods (2016, p.131) has stated as the gap between the demonstration and the real thing:

> Computer-based simulation and rapid prototyping tools are now broadly available and powerful enough that it is relatively easy to demonstrate almost anything, provided that conditions are made sufficiently idealized. However, the real world is typically far from idealized, and thus a system must have enough robustness in order to close the gap between demonstration and the real thing.
>
> *(Doyle/D. Alderson, personal communication, January 4, 2013)*

Demonstrating that one can drive a car autonomously under confined and idealized conditions, for a brief period of time, does not mean the car can be let loose in the real world, under less-than-ideal conditions. Similarly, as Lindsay (this Volume) stated about AI:

> AI relies on large-scale data and stable collective judgments. But these same conditions are elusive in war. AI, to put it glibly, is an economic product of peace. War destroys the conditions that make AI viable. The conditions that are conducive for AI are not conducive for war, and vice versa.

Hence, the second argument against a dystopian view on AI-enabled weapons is that it underestimates the differences between war and peace and that drones that are tested with AI in peacetime conditions are not robust enough to operate in wartime conditions. The difference may actually even be larger than the difference in traffic conditions for self-driving versus human-driven vehicles.

The analogy with self-driving cars may be taken even further to advance a third argument against dystopian views. Our focus on self-driving typically concerns removing the single human driver and replacing them with sensors. Considering the human task of driving a car an instance of perceiving objects in the environment and appropriately acting upon them, is an example of the 'reductive tendency' that humans are prone to (Feltovich, Hoffman, Woods, & Roesler, 2004). It means neglecting the complexity and connectedness of driving and reducing it to a perceptual-motor task. Once we have completed this reduction in our minds, we can then proceed to automate the perceptual-motor task in order to achieve our goal of developing a self-driving car. But this reductive framing of driving misses very important aspects of the real-world driving task that cannot be easily automated. For instance, driving involves the constant awareness of others and their inferred intentions. If I see a bicyclist coming from the left, cycling at speed and avoiding eye contact, I decide to wait even though I have the right of way. If, however, the bicyclist looks at me and slows down, I may either slowly go ahead or wave to let the bicyclist pass anyway. This example may still be an oversimplification of many real-world traffic situations. The coordination between two interacting road users may be shaped by surrounding road users (Renner & Johansson, 2006). Nathanael and Papakostopoulos (2023) describe a range of coordination strategies employed by road users, such as the 'Pittsburgh left' in which a car is allowed to take a left turn at a two-lane intersection immediately after the traffic light turns green (as if there were a left turn signal) provided that the driver of the oncoming car is willing to cooperate. In general, a driver's exploitation of situational opportunities to gain priority is often contrary to regulatory provisions, but favoring overall traffic efficiency. These human coordination strategies pose the following design challenges for self-driving cars in mixed traffic: (i) distinguish these strategies from errant driving, (ii) recognize to whom a 'space-offering' is addressed, and (iii) assess the appropriateness or abusiveness of a particular strategy (Nathanael & Papakostopoulos, 2023). It is clear from actual observations of self-driving cars that these challenges are currently far beyond their capabilities (Brown & Laurier, 2017) and will frequently result in stalled traffic that requires human intervention (Metz, 2023).

Human coordination strategies are obviously of great importance in military operations as well. For instance, in drone warfare operations in Afghanistan, the focus has often been on the 'sharp end' of the drone warriors who operate these drones from a distance. And while these pilots often suffer from high workloads and moral vexations (Philipps, 2022), it is actually the 'blunt end' ('the customer') who designates the targets. Although targeting may be described as the practice of destroying enemy forces and equipment, it is more accurate and contemporary to describe it as a deliberate and methodical decision-making process to achieve the effects needed to meet strategic and operational campaign objectives (Ekelhof, 2018). This decision-making process may involve hundreds of people working over extended periods of time (at least months). Many automated tools already exist to assist in the targeting process and AI may certainly be used to further improve this process. However, this does not imply that AI will completely 'take over' the targeting process. It may change certain tasks, but the system should be designed such that ultimate control is still with the human. This is not to deny that drones could be used in autonomous mode against military objects and that AI increasingly plays a role in identifying these objects. AI-controlled military drones are reportedly being used in the war between Russia and Ukraine. These drones are able to independently identify and attack military objects. According to New Scientist, Ukraine is using the drone in autonomous attack mode: it is the first confirmed use of a "killer robot", the website says (Hambling, 2023). This brings us to the question of what it means for a weapon to be used in 'autonomous' mode. This will be taken up in the next section.

AUTONOMY

The concept of 'autonomy' has recently been used frequently in the discussion on AWS, in particular as a means of banning these weapon systems. It is not my intention in this section to arrive at a definition in order to ban autonomous weapons. Rather, it is my intention to shed light on various definitions of 'autonomy' and its relation with AI. The relation between autonomy and AI has not always been made sufficiently clear. It has frequently and implicitly been assumed that AWS need AI to function properly or even to be able to exist at all, but this largely depends on one's definition of AWS and to a lesser extent on one's definition of AI. For starters, systems such as the Phalanx or Goalkeeper behave more or less 'autonomously' without any use of AI. These systems were developed in the 1970s using linear programming methods. These are so-called close-in weapon systems to defend naval vessels automatically against incoming threats. According to Scharre (2018), these automated defensive systems are autonomous weapons, but they have been used to date in very narrow ways. These systems do need to be activated by humans, but, once activated, will search for, detect, track, and engage targets that match predetermined profiles (based on, for instance, angle of approach and speed – the minimum and maximum limits are set by operators). These systems are used as a last resort for self-defense and as such also destroy friendly targets that match the predetermined criteria. They are used to target objects, not people. Humans supervise these systems in real-time and could physically disable them if they stopped responding to their commands. This category of weapon systems is mentioned here to illustrate some of the conceptual difficulties that arise when using the term 'autonomous weapon system'. If one does not consider these systems to be fully 'autonomous', then what criteria do systems need to fulfill to be considered 'autonomous'? Would it be sufficient to equip these systems with AI, so they can, for instance, distinguish between friend and foe? Or would these systems have to be able to switch themselves on and engage targets without any human intervention? The former addition would make the systems more 'intelligent' perhaps, but no so much more 'autonomous'. The latter addition would make the systems definitely more 'autonomous' (by whatever definition of 'autonomy' one might entertain), but also more unpredictable and less trusted. Hence, it is necessary to discuss the concept of 'autonomy' in some more detail in order to be able to classify particular classes of weapon systems as being 'autonomous' or not.

One definition of 'autonomy' that has been cited frequently in this Volume by various authors is the one provided by the US DoD in its Directive 3000.09 (U.S. Department of Defense, 2023). Although this Directive is called "Autonomy in weapon systems", it does not provide a definition of 'autonomy'. It does, however, provide a definition of an 'autonomous weapon system':

> A weapon system that, once activated, can select and engage targets without further intervention by an operator. This includes, but is not limited to, operator-supervised autonomous weapon systems that are designed to allow operators to override operation of the weapon system, but can select and engage targets without further operator input after activation.

From this definition we can distill a definition of 'autonomy', in the context of weapon systems, as follows: 'the ability to select and engage targets without further intervention by an operator'. This definition is ambiguous as to what 'select target' actually means, but the Directive also provides a definition for target selection, as follows: "the identification of an individual target or a specific group of targets for engagement". What is crucial here, as stressed by Scharre and Horowitz (2015), is that an AWS does not engage a specific target but rather engages targets of a particular class within a broad geographic area without any human involvement. A loitering munition such as the Harpy would be an example of this. According to Scharre and Horowitz (2015, p.13), loitering munitions are "launched into a general area where they will loiter, flying a search pattern looking for targets within a general class, such as enemy radars, ships or tanks. Then, upon finding a target that meets its parameters, the weapon will fly into the target and destroy it". While this is a critical distinction with guided munitions, where the human controller must know the specific target to be engaged, Scharre and Horowitz do not mention the fact that in both cases targets must meet pre-set parameters and that the parameters in loitering munitions are also programmed by human operators. It is clear why Scharre and Horowitz emphasize the difference between specific targets and general classes of targets because otherwise the entire discussion on AWS "would be a lot of fuss for nothing" (Scharre & Horowitz, 2015, p.16), as guided munitions, which have been around for 75 years or more, would then also have to be classified as AWS. This "almost certainly misses the mark about what is novel about potential future autonomy in weapons", according to Scharre and Horowitz (2015, p. 17). It should be noted that the 'general area' and the 'general class of targets' are not without their own problems, such as the increased risk of collateral damage in populated areas and the fact that 'general targets' increase the risk of false positives (e.g., when using a face recognition algorithm).

Two other phrases are also noteworthy. The first is the phrase 'once activated'. The phrase 'after activation' occurs again at the end of the definition. This implies that an AWS first needs to be activated, presumably by a (human) operator, in order for it to identify and engage targets. The second is the phrase 'further intervention' or 'further operator input' which both strongly suggest that the activation of the weapon system is carried out by an operator. An AWS, in this definition, does not switch itself on to go on a killing spree. In this respect, close-in weapon systems fall under the general category of 'autonomous weapon systems', as they also need to be activated and are capable of selecting/identifying and engaging targets without further intervention by an operator. Loitering munitions also need to be activated by human operators, even though the time scale at which close-in weapon systems and loitering munitions are activated may differ.

Finally, it is noteworthy that in this definition of AWS, the operator can override the operation of the weapon system. This does not make the weapon system semi-autonomous. Hence, the defining difference between autonomous and semi-AWS is not in the possibility of humans to override the system, but rather in the capability of autonomous systems to select targets on their own (which should be interpreted as the capability to select classes of targets rather than specific targets). Hence, there is still the possibility of exerting human control over AWS, once they are activated, even though it is not a necessary condition for such systems to be called 'autonomous'. Once again, even close-in weapon systems may conform to this definition as they allow for a mode of control in which the operator may override the operation of the weapon system. This does not make these systems less autonomous, in this definition.

The US DoD Directive does not mention the use of AI in its definition of AWS, and rightly so. AI is merely a technology to accomplish certain functions, in this case, the ability to select and engage targets. Given that close-in weapon systems do not use AI and conform to the definition of AWS, one may conclude that AI is not required for AWS to exist or even function properly. However, neither does the definition exclude AI for future use in AWS. It is foreseeable that AWS may function better using AI, but this does not make these systems more autonomous than they would have been without AI.

NATO has also provided a definition of autonomy that is noticeably different from the one provided by the US DoD. The Official NATO Terminology Database (NATOTerm, 2023) provides the following definition of 'autonomy':

> A system's ability to function, within parameters established by programming and without outside intervention, in accordance with desired goals, based on acquired knowledge and an evolving situational awareness.

If we parse this, it first and foremost states that autonomy is a system's ability to function in accordance with desired goals. This ability to function is bounded by parameters established by programming, hence the system is not capable of setting its own parameters. These boundaries are presumably set by humans, although the definition does not make this clear (the parameters could also be set by software outside of the system, but this would lead to an infinite regress). Furthermore, the system functions without outside intervention. This is a difference with the US DoD definition, where operators could override the operation of the system, and the system would still be called 'autonomous.' Finally, the definition says something about how the system is capable of functioning in accordance with desired goals: this is based on 'acquired knowledge' and an 'evolving situational awareness (SA)'.

This definition does not make clear who sets the desired goals, who acquires the knowledge, and how SA can evolve, or even what it means for a system to have SA. It is interesting, furthermore, that the NATO Database of terminology does not contain a definition of 'autonomous weapon system'. If it would have contained such a definition, and by applying the definition of 'autonomy' to a 'weapon system', we would have to arrive at the conclusion that such weapon systems would be able to operate without outside intervention, within parameters established by programming, that they would achieve desired goals based on acquired knowledge, and that they would possess an evolving SA. It is clear that such a definition would bring us much closer to AI than the definition stated in the US DoD Directive 3000.09. The NATO terminology is largely derived from the fields of cognitive science, cognitive engineering, and artificial intelligence. It sets a rather high bar for systems to be called 'autonomous', or so it seems. At first sight, one would have to exclude close-in weapon systems. However, if one equates 'acquired knowledge' with 'pre-programmed specifications' and 'evolving situational awareness' with 'dynamic model of target features', then these systems could still be considered 'autonomous' under this NATO definition.

Taddeo and Blanchard (2022, p. 15) have critically reviewed 12 definitions of 'autonomous weapon systems' and, based on cognitive systems engineering, arrived at the following definition themselves:

> An artificial agent which, at the very minimum, is able to change its own internal states to achieve a given goal, or set of goals, within its dynamic operating environment and without the direct intervention of another agent and may also be endowed with some abilities for changing its own transition rules without the intervention of another agent, and which is deployed with the purpose of exerting kinetic force against a physical entity (whether an object or a human being) and to this end is able to identify, select or attack the target without the intervention of another agent is an AWS.

According to Taddeo and Blanchard (2022), once deployed, an AWS can be operated with or without some forms of human control (in, on or out the loop). In this regard, they are in agreement with the US DoD Directive 3000.09. A lethal AWS is specific subset of an AWS with the goal of exerting kinetic force against human beings.

Although one may critique the term 'artificial agent' for legal purposes (Seixas Nunes, SJ, this Volume), given that agency is inherently linked to the notion of liability, I take it that Taddeo and Blanchard are referring to software agents, or software in brief. This allows the system to identify, select, or attack a target, which again is in agreement with the US DoD. The definition mentions no less than three times 'without the intervention of another agent', which is in agreement with the NATO definition that states 'without outside intervention'. However, Taddeo and Blanchard go further than either the US DoD or NATO, particularly when they stress that an AWS is capable of 'changing its own internal states' and 'changing its own transition rules' in order to achieve a set of goals in a dynamic operating environment. Regardless of whether this can be accomplished with rule-based AI or with machine learning using neural networks, this requirement specifies some kind of learning system. As such, it would in all likelihood be unacceptable for military commanders, as it would render the system practically unpredictable and therefore untrustworthy. What makes it all the more questionable is that this kind of self-learning is accomplished without the intervention of another agent. Hence, there is no level of human control whatsoever over this kind of AWS. Although Taddeo and Blanchard claim that their definition is 'value-neutral', it sets the bar so high that any system that potentially meets their criteria will in all likelihood be unacceptable for military commanders. It should also be clear from this definition that close-in weapon systems do not fall within this categorization of AWS, as close-in weapon systems are unable to change their own internal states or transition rules.

Finally, Kaber (2018) presented a conceptual framework of autonomous and automated agents. Although not geared to weapon systems, it is nevertheless an interesting perspective on autonomy, as it contrasts this concept with the concept of automation. Kaber makes clear that the Levels of Automation approach should not be evaluated from a 'lens' of autonomy, as the concepts are quite distinct. Kaber conceptualizes autonomy as a multifaceted construct including: (1) viability of an agent in an environment; (2) agent independence or capacity for function/performance without assistance from other agents; and (3) agent 'self-governance' or freedom to define goals and formulate an operational strategy. An agent could be a software agent, but also a 'thing' in an environment with sensing and effecting capabilities. Viability is the capability of an agent/human to sustain the basic functions necessary for survival in context. Self-governance requires cognitive abilities such as learning and strategizing so the agent can formulate goals and initiate tasks. According to Kaber, this is beyond the capabilities of present advanced computerized and mechanical systems (although AI is capable of formulating subgoals, it is not capable of formulating the top-level goal).

The facet of self-governance is critical to differentiating autonomy from automation. Loss of capacity for self-governance relegates an autonomous agent to an automated agent. According to Kaber (2018, p. 413): "In general, when autonomous agents are pushed past the boundaries of their intended design context, they become forms of automated or functional agents.". In Kaber's conceptual framework of autonomy, one can only speak of autonomy when an agent scores high on all three facets of autonomy (i.e., viability, independence, self-governance) for a particular context. There are no levels of autonomy and therefore it is a 'misnomer' to speak of 'semi-autonomous systems' (Kaber, 2018, p. 417).

Using this framework, we can establish whether a system is autonomous as a result of the absence of specific characteristics. For instance, close-in weapon systems are hardened to their environment (i.e., they are able to operate in the specific naval context for which they were designed), they do not require monitoring or intervention by humans in a defined operating context (even though they are designed for, and might require, monitoring in different contexts), but they are not 'self-governing' (i.e., they are not responsible for mission goals or control of resources as they do not have the capacity to learn or strategize). The absence of the latter facet of autonomy means these systems are not autonomous, according to Kaber's (2018) framework. In fact, according to Kaber, there are currently no autonomous systems beyond known and static environments. This reinforces what was stated above when the difference between the simulation and the 'real thing' was discussed in the context of self-driving cars, or when the context of the use of AI (peace versus wartime) was discussed.

It is important to note that while autonomous agents pose low demands on humans for supervision or management (whereas automation requires human supervision), this does not imply that they cannot serve as partners for humans in achieving a broader mission. Also, many application environments or work systems require humans to support autonomous agents and vice versa. This dictates additional agent design requirements, particularly from a coordination perspective, as already stated above when discussing the targeting process or the sophisticated coordination strategies employed by humans in traffic. Finally, there may also be a dynamic shifting of functions back and forth between humans and autonomous agents, particularly when environmental conditions change beyond the capacities of an autonomous agent. This may be the case, for instance, when road and weather conditions force self-driving cars to enlist the driver's assistance, fully recognizing that human drivers may also experience difficulties under these circumstances. It is well-known in the human factors literature that such sudden transitions of control may lead to 'automation surprises' (Sarter et al., 1997). It is not sufficient to state that humans should be able to exercise 'appropriate levels of human judgment' (U.S. Department of Defense, 2023) or 'meaningful human control' (Ekelhof, 2019) in these cases of shifting control. Even if, in the far future, there will be AWS that are able to deal with unknown and dynamic environments, dynamic shifts in control will occur and humans will have been out of the loop for so long that they either lack the skills to regain control ('deskilling'; Bainbridge, 1983) or are confronted with an 'automation conundrum' (Endsley, 2017). The latter reflects the fact that the better the automation, the less likely humans are able to take over manual control when needed (Endsley deliberately uses automation and autonomy interchangeably).

Comparing Kaber's (2018) definition of 'autonomy' with the other definitions discussed, we may note some similarities with Taddeo and Blanchard's (2022) definition. Both definitions stress agent independence and self-governance. However, Kaber additionally stresses the viability of an agent in an environment, an aspect that other definitions have overlooked or have ignored. Viability is not absolute, obviously. Just as humans, who are generally considered to be 'autonomous' creatures, display limits to their viability across different contexts (Kaber suggests relocating a human outside the Earth's surface atmosphere to reveal the limits of any autonomy), so other agents' viability is always relative to a particular context. Self-driving cars may be viable under Sunny State contexts, but not viable in harsh winter weather. What this means for AWS is not immediately clear. At the very least, one would, when accepting Kaber's framework, have to add viability to the definition provided by Taddeo and Blanchard. This would imply that AWS would have to be able to sustain their operations across at least a range of contexts (imposed by, e.g., weather, terrain, enemy operations, available time) and be able to adapt themselves, through rule modification, to these various contexts. It may be that this is what Taddeo and Blanchard meant by achieving goals 'within its dynamic operating environment'. In that case, their definition of autonomy meets Kaber's viability characteristic.

In summary, we have seen a wide variety of definitions of 'autonomy', as used in 'autonomous weapon systems'. We are left with a choice between definitions that set a high bar for AWS, insofar they are required to possess learning (Taddeo & Blanchard, 2022) or self-governing capabilities (Kaber, 2018), versus definitions that set a lower bar and that include close-in weapon systems and loitering munitions as a class of AWS (US DoD Directive 3000.09). Systems that learn without outside intervention and that are therefore 'self-governing' may not be acceptable to military commanders as they are essentially unpredictable, may not conform to Rules of Engagement, and can therefore not be trusted. Accepting this definition would in effect mean that any use of the term 'autonomous weapon system' would be inappropriate, at least when describing current weapon systems and possible weapon systems for the foreseeable future. It does not preclude that there will ever be weapon systems that conform to this definition, yet, if they are developed, they will in all likelihood be unacceptable for responsible military use. Finally, there is also a very pragmatic reason not to adopt this 'high bar' definition, which is that the current usage of the term 'autonomous weapon system' is much more in line with the US DoD Directive 3000.09 definition. From an academic point of view, the high bar definitions might be preferable, but they leave us mostly empty-handed: we would have to exclude all current weapon systems from this definition, as well as most to come, and, in the unlikely case there will be a future weapon system conforming to these definitions, we would have to ban it from being used, as there will be no guarantee that it will conform to Rules of Engagement. From an ethical point of view, this could be precisely what is desirable, and it could be the entire point of advancing this definition. But then we are left with countless current weapon systems that are 'highly automated' rather than 'autonomous' and whose effects are just as lethal.

In the end, what is important is the level of human control that can be exerted over the weapon systems. This brings us to the discussion of 'meaningful human control', which will be taken up in the next section. For now, it suffices to say that accepting the US DoD Directive 3000.09 definition of 'autonomous weapon system' explicitly includes that level of human control. First, by requiring that the system has to be activated by a human operator. Second, by noting that an operator can override the operation of the weapon system, although not necessarily so. Third, by making clear that although the weapon system may 'select' targets on its own, this in fact boils down to selecting classes of targets rather than specific targets. The latter point is not immediately obvious when first reading this definition. The discussion is also muddled by the US DoD's use of the term 'semi-autonomous', which is reserved for systems that only engage targets but do not select them. However, there are two uses of the term 'target selection'. One is target selection once the system has been activated. This is the sense in which the US DoD uses the term. If, once activated, an operator does the target selection, the system is called 'semi-autonomous'; if the system does the target selection (meaning it does not select a specific target but rather a class of targets over a wide geographic area), it is called 'autonomous'. However, a second use of the term 'target selection' applies to the programming and development phase of the weapon system: 'target selection' here means the specification, frequently in software, of the target parameters. This is clearly under human control and no AWS that meets the definition of the US DoD Directive 3000.09 can do without such target parameters programmed into the weapon system. Parameters in loitering munitions are also programmed by human programmers. This leaves us with the uneasy conclusion that even loitering munitions, considered by Scharre and Horowitz (2015) to be the only examples of autonomous systems (apart from close-in weapon systems), are in fact under human control and cannot be considered 'autonomous'. This would mean that the distinction between a specific target (chosen by a human in, for instance, guided munitions) and a general class of targets (programmed by a human in loitering munitions) would not be as large as Scharre and Horowitz (2015) claimed, making the discussion on the definition of AWS indeed a lot of fuss for nothing.

Accepting the high bar definitions leads to the conclusion that there are no AWS yet, and they are not likely to be developed in the near future. Accepting what Scharre and Horowitz (2015) called a 'common sense definition' yields a superficial distinction between widely used guided missile systems that engage specific targets versus AWS that engage a class of targets on their own. This distinction is superficial in that both classes of weapon systems are ultimately under human control. We therefore seem to be caught between a rock and a hard place and will need to expand our view on autonomy and AWS.

BOUNDED AUTONOMY

In this section, I will introduce the concept of 'bounded autonomy' in order to arrive at a broader and more acceptable definition of autonomy. Introducing this concept will also serve the purpose to gain more clarity on the issue of 'meaningful human control' or being able to 'exercise appropriate levels of human judgment', as the US DoD states.

The concept of 'bounded autonomy' is introduced here by analogy to the concept of 'bounded rationality', as introduced by Simon (1947, 1955). According to Simon (1957, p. 198):

> The capacity of the human mind for formulating and solving complex problems is very small compared with the size of the problems whose solution is required for objectively rational behavior in the real world – or even for a reasonable approximation to such objective rationality.

As a result, the human actor must "construct a simplified model of the real situation in order to deal with it" (Simon, 1957, p. 199). Humans behave rationally with regard to these simplified models, but such behavior does not even approximate objective rationality. Rational choice exists and is meaningful, but it is severely bounded. Our knowledge is necessarily always imperfect, because of fundamental limitations to our information-processing systems (e.g., limits on working memory capacity) and because of fundamental limitations to the attention we can pay to the external world. When confronted with choices of any complexity, we necessarily have to 'satisfice' rather than optimize. Rational behavior is as much determined by the "inner environments" of people's minds, both their memory contents and their processes, as by the "outer environment" of the world on which they act (cf. Simon, 2000).

In terms of economics theory, we could reformulate this as a relaxation of one or more of the assumptions underlying Subjective Expected Utility (SEU) theory underlying neoclassical economics:

> Instead of assuming a fixed set of alternatives among which the decision-maker chooses, we may postulate a process for generating alternatives. Instead of assuming known probability distributions of outcomes, we may introduce estimating procedures for them, or we may look for strategies for dealing with uncertainty that do not assume knowledge of probabilities. Instead of assuming the maximization of a utility function, we may postulate a satisfying strategy. The particular deviations from the SEU assumptions of global maximization introduced by behaviorally oriented economists are derived from what is known, empirically, about human thought and choice processes, and especially what is known about the limits of human cognitive capacity for discovering alternatives, computing their consequences under certainty or uncertainty, and making comparisons among them (Simon, 1990, p. 15).

If we take Kaber's (2018) framework for autonomy as the equivalent of SEU theory, we can reformulate our proposed concept of 'bounded autonomy' as a relaxation of one or more of Kaber's three facets of autonomy (i.e., viability, independence, and self-governance). Rather than relegating the agent to the domain of automation, when the facet of self-governance is lacking, we may view the agent as displaying bounded autonomy. Instead of assuming the viability of basic functions in particular contexts, we may postulate the viability of a limited set of functions for a shorter duration. Instead of assuming independence, we may postulate dependence on parameters established during a preceding targeting process or dependence on mission and task constraints. Instead of assuming self-governance, we may postulate performance in accordance with desired mission goals, based on knowledge acquired during controlled training sessions and a continuously updated model of the environment.

By analogy to bounded rationality, I will now put forth the following description of bounded autonomy:

> The capacity of a system to display viability, independence and self-governance (i.e. 'autonomy') is very limited compared with the variety of the environments to which adaptation is required for objectively autonomous behavior in the real world – or even for a reasonable approximation to such objective autonomy.

This is in accordance with Kaber (2018) who notes that autonomous systems are currently restricted to known and static environments. As a result, the system is dependent on a simplified model of the real situation in order to deal with it (cf. Woods, 2016). Systems behave autonomously with regard to these simplified models, but such behavior does not even approximate objective autonomy, that is, autonomy that fully meets all three facets of viability, independence, and self-governance.

Applying this general concept of bounded autonomy to AWS, brings us to the following maxim:

> The capacity of an autonomous weapon system to select and engage targets on its own ('platform autonomy') is highly dependent on parameters established during the preceding targeting process as well as constraints imposed by legal, ethical, and societal considerations ('mission autonomy').

According to this maxim, a distinction has to be made between platform autonomy and mission autonomy. Mission autonomy concerns what an autonomous system should do and within which constraints. This is the domain of the human. This might be restricted to a single commander, but this is frequently an oversimplification and usually involves hundreds of people (cf. Ekelhof, 2018), not merely in the targeting process, but more generally in the governance and design loops (Heijnen et al., this Volume). This involves the entire design and development process preceding the deployment of an AWS, including testing, evaluation, validation and verification, training with humans in the loop, as well as post-deployment evaluation processes. Mission autonomy is what makes the platform boundedly autonomous. Focusing exclusively on the selection and engagement of targets ('platform autonomy') misses the point. Humans are in control of assessing the necessity and applicability of autonomy. They set the boundaries within which a platform can then operate.

Autonomy is the possibility of an unmanned system to execute an ordered task. The military commander, assisted by countless others, orders the task for the unmanned systems to execute, the AI can help in developing the plan, and the plan is presented to the military commander, he or she can adjust it or he or she can approve it, and then the plan is transferred to the platform (e.g., robot or drone). Hence, the platform is specifically ordered what to do in terms of tasks, conditions, and constraints. Only then do we have controllable, that is, bounded, autonomy.

In the previous section, we were caught between a rock and a hard place in terms of what definition to choose for 'autonomous weapon system' and how to apply that definition to a range of existing and future weapon systems. Given the discussion above, it is now clear that there currently are only 'boundedly autonomous' weapon systems. Moreover, each of these weapon systems contains varying degrees of platform and mission autonomy. There are no hard and clear-cut distinctions to be made between various weapon systems in terms of their 'autonomy'. Loitering munitions may be considered to exhibit less bounded autonomy than guided missiles, as their mission autonomy allows them to select and engage a wider range of targets than guided missiles. The mission autonomy for guided missiles is fairly restrictive, in that these weapon systems have generally been programmed to attack a single target. That this constraint has been relaxed somewhat in the case of loitering munitions (as well as in the case of close-in weapon systems), does not make these systems qualitatively different from guided missile systems. All current systems display bounded autonomy and the discussion would be more fruitful if we focused on the various ways platform autonomy and mission autonomy are achieved than on whether these weapon systems belong to qualitatively different categories.

One could, of course, deliberately restrict one's definition to the phase after activation, but this would be an oversimplification of a complicated process of decision-making finally leading to weapons release. It is akin to blaming a nurse for a medication error that clearly is the result of an entire work organization or design issue. 'Human error', then, is a symptom of a system that needs to be redesigned, not a symptom of a human that needs to be retrained or fired. The emphasis, in system safety, has changed from preventing failures to enforcing constraints on system behavior (Leveson, 2011). In the same way, the emphasis in discussions on AWS needs to change from preventing failures that occur after such weapon systems have been activated to enforcing constraints, through mission autonomy, on such weapon systems. Eggert (this Volume) counters this conclusion by stating: "The fact that humans are in control *before* a weapon system is activated hardly shows that no human control is required *after* a weapon system is activated". This is true and is called 'human on the loop' or the ability to intervene if the weapon system fails or malfunctions. There are many systems, particularly defensive systems that target objects, not people, where such control is possible after the weapon system is activated. Systems where there is no human control after they are activated are rare, however (Scharre, 2018). Loitering munitions are the primary example and the Harpy (and possibly the Saker Scout drone; Hambling, 2023) are currently likely the only operational examples where the human is 'out of the loop' (other circumstances can be imagined in which autonomy without real-time human control could be desirable, for instance in underwater or silent operations). The Harpy dive-bombs into a radar and self-destructs (as well as destroying the radar) after it detects any radar that meets its criteria. The Saker Scout drone is said to be able to identify 64 types of military objects. It is not clear why Eggert would want human control over the Harpy *after* activation, unless she is suspicious about the criteria that the Harpy has been provided with *before* it is activated, and she is distrustful of the entire testing, evaluation, validation, and verification cycle that the Harpy presumably has undergone. Granted, there could be collateral damage if the Harpy destroys a radar, just as there will be collateral damage if the Saker Scout drone destroys a tank. However, human operators can also make mistakes in drone warfare and cause collateral damage (Philipps, 2022). Moreover, in these cases the Harpy and Saker Scout would still comply with key principles of IHL, notably those of distinction and proportionality (Eggert would counter this by stating that complying with IHL is not the same as complying with morality). The real fear, of course, is with loitering munitions that are programmed with face-recognition capabilities and dive-bomb unto humans meeting the face-recognition criteria. But again, this is an issue of human control *before* a weapon system is activated. This brings us to a further discussion of the limits of human control in the next section.

MEANINGFUL HUMAN CONTROL

As noted by Eggert (this Volume), a common claim is that AWS must remain under 'meaningful human control' (see also Ekelhof, 2019, and NATO STO, 2023, for a critical discussion). As noted by Ekelhof (2019, p. 344): "the concept of MHC or a similar concept, is one of the few things that states agree could be part of such a CCW outcome". And Vignard (this Volume) notes: "While there is not a common understanding of the phrase, nor is there international consensus agreement on its utility, there are groups of States who have embraced the concept as a principle to guide AI development in the military domain". Numerous states have explicitly declared their support for the idea that all weaponry should be subject to MHC (Crootof, 2016).

There are several conceptual distinctions to be made when defining 'meaningful human control'. The first is to ask: "control over what"? Do we mean informed human approval of each possible action of a weapon system ('human in the loop'), the ability to intervene if a weapon system fails or malfunctions ('human on the loop'), or do we mean control over the entire distributed targeting process (Cummings, 2019; Ekelhof, 2019)? Given the increased speed of modern warfare, the need for rapid self-defense in some situations, inherent human limitations in time-critical targeting scenarios, and the effects of high workload, fatigue, and stress on human decision-making, it is a far cry from 'meaningful human control' to put a human 'in' or even 'on the loop'. In this sense, it could be more in accordance with IHL to use a precision-strike Tomahawk guided missile that is not under direct operator control, but is under control by the programmers who enter targeting information or the hundreds of people involved in the targeting process. Obviously, there is no guarantee that programmers or others involved in the targeting process do not make any mistakes, but at least the likelihood of such mistakes is much smaller compared to "putting military operators into a crucible of time pressure, overwhelming volumes of information, and life and death decisions in the fog of war (…)" (Cummings, 2019, p. 24).

A second question to ask about MHC is: "control by whom"? This is usually interpreted as control exercised by an individual operator or commander. However, as argued by Ekelhof (2018; 2019) and Cummings (2019), targeting is frequently, though by no means always, a distributed process involving many people (exceptions may be found in urban warfare where individuals, assisted by robots, need to make split-second decisions). This does not make the line of accountability less clear, it just means we need to shift our attention away from the individual operator at the sharp end to the organization at the blunt end. At the very least we should make a distinction between what Cummings (2019) refers to as the 'strategic layer' and the 'design layer', and what I have referred to above as the distinction between 'mission autonomy' and 'platform autonomy'. Human control is exercised at the strategic layer when targets are designated in accordance with mission objectives and abiding by principles of proportion and distinction within the Law of Armed Conflict framework. At the design layer, it depends on whether the target is static, well-mapped, and within the rules of engagement, whether human control can or needs to be exercised. When, in the minority of cases, targets meet these criteria, particular weapon systems may be used without human control at the design layer (e.g., the deployment of a Tomahawk missile against a particular building). When targets do not meet these criteria, and are dynamic and emerging, Cummings (2019, p. 25) argues that there needs to be human certification rather than human control. By this, she means that the use of weapon systems for such targets "should be proven through objective and rigorous testing, and should demonstrate an ability to perform better than humans would in similar circumstances, with safeguards against cybersecurity attacks".

A third question to ask is: "meaningful human control to what end"? Do we want to involve human beings in the decision-making process regarding the use of force? Do we want to ensure compliance with existing legal obligations? Do we want to establish a higher legal or ethical standard? Do we want to improve military effectiveness? As argued by Crootof (2016), MHC will usefully augment the humanitarian norms of proportionality and distinction. However, an overly strict interpretation of what constitutes MHC may actually undermine fundamental humanitarian norms governing targeting. For instance, if MHC is taken to mean that the human commander or operator has full contextual or SA of the target area and the means for the rapid suspension or abortion of the attack, then this would rule out the use of entire classes of precision-guided munitions or close-in defensive weapon systems. This may actually increase collateral damage and the killing of non-combatants. Crootof (2016) argues that the distinction, proportionality, and feasible precaution requirements should serve as an interpretive floor for a definition of MHC. This means that if we augment existing humanitarian norms governing targeting, for instance by strengthening the certification process as suggested by Cummings (2019), or by paying more attention to the distributed nature of the targeting process as suggested by Ekelhof (2019), then the notion of MHC, however imprecise, can fruitfully advance the conversation regarding the appropriate regulation of AWS.

Eggert (this Volume) challenges this widespread faith in MHC and discusses three main problems with AWS: the compliance problem, the dignity problem, and the responsibility problem. She argues that MHC does considerably less to address these problems than typically assumed. The compliance problem was already discussed above, in that legal compliance is not the same as moral compliance. Making AWS comply with moral principles would make them effectively not autonomous. Eggert raises the important question of whether control over autonomous weapons is not just an *apparent* but a *real* contradiction, meaning that we must ultimately choose between autonomy in machines and control in human hands. Can we argue for MHC while at the same time arguing for AWS? I think the distinction introduced above between platform autonomy and mission autonomy largely answers this question. We can both have platform autonomy while at the same time having mission autonomy as well: we have human control via mission autonomy *before* a weapon system is activated, while we have platform autonomy *within the constraints set by mission autonomy after* a weapon system is activated. Platform autonomy is never absolute and is always bounded by higher-order constraints. Advocates of MHC do not need to call for a ban on AWS, provided they make the distinction between mission autonomy and platform autonomy and hence adopt the concept of bounded autonomy.

The second problem, the dignity problem, essentially states that MHC is no guarantee that human dignity is respected. Humans are capable of extreme cruelty and they violate human dignity all the time, according to Eggert. MHC does nothing to prevent this. Indeed, the moral constraints that we impose upon a physical platform may, in the hands of a dictator, have the opposite effect of what we intended with MHC. The dictator will have a different set of moral values and the AWS may be programmed in such a way as to kill many innocent civilians. There will be human control, but it will not be 'meaningful' in the sense in which democratic countries use this word. This argument does not imply, however, that democratic countries would have to abstain from MHC altogether. Quite the opposite, particularly given that MHC is a distributed process that avoids concentration of power in any one malevolent individual. MHC is thus a form of democratic control and this leads to questions on who should be involved and why. Even though Eggert claims that militaries are hardly democratic institutions, at least in some countries, they do act in accordance with democratically elected governments and can be held accountable (at least in democracies). Furthermore, there are currently efforts underway to establish Ethics Advisory Boards and processes to consult with all kinds of stakeholders. This makes MHC even more of a distributed process, as well as a more ethically inspired process.

Finally, the responsibility problem states that if humans are in control, they can be held accountable if something goes wrong. According to Eggert, the most promising sense in which MHC addresses concerns about responsibility is by allowing us to categorize harms wrongfully inflicted by AWS as *human omissions to prevent such harms*. These are cases in which humans omit to act; instances of allowing harm to be caused by an AWS which human agents should prevent. However, the kind of control necessary for ascribing responsibility may altogether rule out autonomy in weapon systems, as humans should always be able to intervene in order to prevent harms wrongfully inflicted by AWS. The conclusion should be that MHC is incompatible with AWS. Again, I think this conclusion is incorrect. First, as noted by Eggert herself, when time is of the essence, for instance in self-defense, humans cannot reasonably be asked to intervene. AWS, such as close-in weapon systems, are used against objects, not people, so the risk of harm wrongfully inflicted by an AWS is minimal anyway. Secondly, in footnote 8 Eggert states:

> Humans might program an AWS to target enemy combatants at t1, but it is still up to the machine, at t2, to select individual targets, and it may not always be clear why it targeted one person rather than another. The fact that a human person, at t1, programmed an AWS to behave a certain way doesn't mean that the AWS is under human control at t2.

If this would be the case, then such an AWS would not have been tested properly and it would not be trusted and accepted by commanders. If there is a disconnect between mission autonomy and platform autonomy, then something is wrong with the constraints put upon the platform (or weapon system) and this should have become apparent during the testing and evaluation phase (I will deal with some of the challenges with

Testing, Evaluation, Validation, Verification (TEVV) later). If the AWS displays signs of self-governance, setting its own goals, and behaving in ways it was not programmed to do, then this is a reason not to use the AWS.

THE RESPONSIBLE USE OF AI IN MILITARY SYSTEMS

So far, little discussion has been devoted to AI, and even less to the responsible use of AI in military systems. This is partly due to the fact that military systems are broader than weapon systems, but even when we restrict our discussion to weapon systems, we have seen that AI is not always necessary (e.g., close-in weapon systems and loitering munitions can fulfill their intended functions perfectly without AI). The notion of developing AI in a responsible fashion has not emerged primarily within the context of AWS. It was primarily driven by concerns over the use of AI in general commercial applications, and efforts to formulate policies and guidelines for the responsible use of AI have emerged first of all in the commercial sector. Nearly every respectable company these days has a responsible AI (RAI) framework, approach, or toolkit. It is beyond the scope of this chapter and this book to discuss these. Instead, I will be focusing on the responsible use of AI in (autonomous) weapon systems.

Any standard for RAI needs to cover the entire AI system lifecycle, from initial design to the decommissioning of the system. This is because decisions are made along the entire lifecycle that can and will impact later decisions. RAI may then be described as a dynamic approach to the design, development, deployment, and use of AI-enabled capabilities that ensure the safety of these capabilities and their ethical employment. By 'ethical employment', I mean conformance to AI principles and guidelines, such as bias mitigation, explainability, traceability, governability, and reliability. Exactly how the process of developing AI-enabled capabilities should conform to these principles is a matter of ongoing research.

In the following sections, three important challenges will be discussed: (i) testing, evaluation, validation, and verification; (ii) human–AI teaming; (iii) transparency and explainability. Although the challenges may seem to arise primarily from an AI perspective, there are important HFE aspects to these challenges as well. These will be explicitly discussed at the end of each section, particularly since these HFE aspects have received insufficient attention in the other chapters in this Volume.

Testing, Evaluation, Validation, and Verification

Reliability is an important ethical value in AI development. Reliability means that a system will provide the same output given the same input. With self-learning AI systems, that could be an issue, as these systems constantly improve themselves, in ways not always transparent to human users. But it is also an issue with AI systems that encounter slightly different situations than they have been trained for (e.g., the camera angle is slightly different; the object is slightly different than the one encountered during training). In practice, this means that a lot of effort needs to go in the validation and verification process in all imaginable contexts. Hence, there should be more focus on TEVV rather than on regulation afterward. A comprehensive approach is crucial during the entire lifecycle: from design to maintenance, from training to doctrine development, and ethics (see Dunnmon, Goodman, Kirechu, Smith, & Van Deusen, 2021 for a first attempt at operationalizing ethical principles for AI within the lifecycle process). Algorithm auditing to ensure explainability, robustness, fairness, and privacy will become an important part of the technical lifecycle of an AI system (Koshiyama et al., 2022). Validating, verifying, and testing AI in all kinds of scenarios and use cases is very important because context is an important factor, also with regard to values such as responsibility and accountability.

All of this is easier said than done. Consider the following example. In a military conflict, a coalition commander has to decide whether or not to employ an AI-enabled system that was developed in a particular country within the coalition. The commander asks whether the system has been thoroughly tested and evaluated (ideally, of course, he or she should be aware of that, but in this case the system was developed in a different country), and receives a confirmatory response by AI experts. There can be two problems here. One is the classical issue that the conditions under which a system has been tested are not representative of the conditions under which the system will be employed. Surprise is continuous and ever-present. There is always the need to close the gap between the demonstration and the real thing (Woods, 2016). These are called 'AI blind spots', or conditions for which the system is not robust. The other issue is that the system may not be sufficiently explainable at this point, and under these time-pressed circumstances, for the commander to have sufficient trust in this system. Given that the commander is currently ultimately responsible and accountable for the use of this system on the battlefield, should the commander trust his or her experts and their V&V process? There is no answer to this question without considering the exact context in which the system will be deployed. If situations are not completely routine, the commander will have to make a decision based on fundamentally incomplete knowledge and taking into account principles of IHL. For this reason, an 'ethical risk analysis' should be developed integrally, together with the technical and human factors aspects of system development. In addition to Technology Readiness Levels, a Moral Organization AI Readiness Level (MORAL) could be developed. MORAL would describe how well AI has been evaluated and tested in terms of ethical risks. Yet it could also be argued that the burden of proof of performance and safety should fall on the shoulders of the industry as well as the military branches who buy their weapons (Cummings, 2019), rather than on the individual commander.

For the HFE community, there are important knowledge gaps and future developments to address the emerging challenges in the TEVV field. Validating, verifying, and testing AI in all kinds of scenarios and use cases leads to the questions "how representative are the scenarios and use cases" and "how many are sufficient"? Given that most AI applications will always work together with humans (see the Human–AI teaming section below), we need to ask whether the AI is intended to support only routine cases or also 'edge cases' where the AI is likely to fail. Thus, HFE professionals need to look at the cognitive support objectives to understand the range of situations where the AI is intended to provide effective support and sample broadly from that range of situations (e.g., Roth et al., 2021). The 'how many are sufficient' question is a familiar one in the usability evaluation literature (e.g., Hwang & Salvendy, 2010), yet the answer may be quite different depending on the criticality of the device to be tested (Schmettow et al., 2013). Measuring and controlling the effectiveness of formative evaluation, usability testing, in particular, is crucial for risk reduction in the development of AI. The number of scenarios that need to be taken into account should not be set to a fixed (or even 'magical') number, but rather should be dependent on the nature of the risk involved when applying the AI (see also Panwar, this Volume). In general, the higher the risk, the larger and the more representative the scenarios to be included need to be. This 'late control strategy' (Schmettow et al., 2013) was developed to determine the number of users required to test a particular device. This strategy may be useful to determine the number of scenarios required as well.

A second emerging challenge for the HFE community lies in the role it may play in the determination of the Moral Organization AI Readiness Level (MORAL). Currently, it is difficult to identify relevant moral values (especially considering that they may change over time) and to ensure that the human–AI system continues to operate in accordance with these moral values and their context-dependencies. There is a role for the HFE community in applying requirements analysis to ethical, legal, and societal aspects of AI, as well as applying Value Sensitive Design methods (Friedman & Hendry, 2019). A necessary discussion is on who to involve with what responsibility to derive requirements from identified relevant moral values, ethics, and laws given the application that is considered. Methods are required that can facilitate this, on top of (existing) methods that shape the more general design process (see also Heijnen et al., this Volume). The IEEE 7000™-2021 standard is an important first step toward value-based engineering, as it is the only value-based engineering standard worldwide so far. The Value Lead, a new profession introduced by IEEE 7000, is trained in ethical and value theories, yet also fits into the system engineering process. HFE professionals are well-suited for fulfilling this profession of Value Lead, as they are used to working together with both end users and system engineers.

Human–AI Teaming

A second HFE challenge that has received a lot of attention recently is 'human–AI teaming', 'human-autonomy teaming', or 'human-machine teaming'. A comprehensive state-of-the-art report on human–AI teaming, including research needs, was published by the National Academies of Sciences, Engineering and Medicine (NAS) in December 2021 (National Academies of Sciences, 2021). Several journals have devoted special issues to this topic, reflecting the burgeoning field. However, neither the NAS report nor recent empirical work in this field is concerned with the moral, ethical, or legal aspects of military decision-making using AI (a notable exception is the NATO STO, 2023, report).

Embedding AI in a Human–AI Team, taken in its broadest organizational context, is an essential part of achieving responsible military AI, both for ethical and legal reasons and to realize the 'multiplier effect' that comes from combining human cognition and inventiveness with machine-speed analytical capabilities. Research gaps are in what the human needs to know about the AI, but also what the AI needs to know about the world and the human. What human needs to know about AI has implications for the training and education of military personnel, and also for human–AI interfaces (see the section on display transparency below). What the AI needs to know about the world and the human has implications for real-time model updating as well as for human enhancement questions (e.g., operator state monitoring). Yet, the responsible use of AI in military systems goes beyond these well-known HFE challenges. An AI system could critique a human's moral reasoning in terms of presenting the moral acceptability of what-if scenarios. Research gaps are in what information is needed for moral SA and how to augment and support the commander's moral model. From an HFE perspective, the use of virtual or augmented reality technology could have the potential to improve moral SA. Sushereba et al. (2021) developed a framework for using augmented reality to train sensemaking skills in combat medics and civilian emergency management personnel. The four key elements of sensemaking that they list – perceptual skills, assessment skills, mental models, and generating/evaluating hypotheses – also appear relevant for training and supporting a commander's moral model.

A second venue for future research lies in the area of trust repair in human-autonomy teams. As artificial teammates may increasingly behave like human teammates, the question arises how human teammates respond when an artificial teammate violates their trust. Trust violations are an inevitable aspect of the cycle of trust and since repairing damaged trust proves to be more difficult than building trust initially, effective trust repair strategies are needed to ensure durable and successful team performance (Kox et al., 2021). A trust repair strategy could be an expression of regret that accompanies the apology, providing promises or explanations, or delaying the repair strategy until the next trust opportunity. Research shows that a single unethical behavior immediately worsened participants' perceptions of an autonomous teammate and that apologies and denials following unethical behaviors were insufficient in rebuilding trust in a military-based human–AI teaming context (Textor et al., 2022). Other research has shown that the intelligent agent was the most effective in its attempt of rebuilding trust when it provided an apology that was both affective and informational (Kox et al., 2021). Hence, future research should evaluate the efficacy of alternative trust repair strategies.

Transparency and Explainability

Transparency (also referred to as observability, and sometimes as traceability) can refer to many objects and processes, for instance, data, algorithms, decisions, or organizations. In the context of this chapter, I follow the definition used in the NAS report (2021, p. 33): "(…) 'the understandability and predictability of the system' (Endsley, Bolte, and Jones, 2003, p. 146), including the AI system's 'abilities to afford an operator's comprehension about an intelligent agent's intent, performance, future plans, and reasoning process' (Chen et al., 2014, p. 2)". Display transparency provides a real-time understanding of the actions of the AI system, whereas explainability provides information in a backward-looking manner. AI explainability is the "ability to provide satisfactory, accurate, and efficient explanations of the results of an AI system" (National Academies of Sciences, 2021, p. 38).

Transparency is important because AI systems learn and change over time and may be applied in contexts and situations they were not initially trained for. In order for humans to keep up with these changes and maintain adequate situation awareness, system changes need to be presented in a transparent manner. Explainability is important due to the black-box nature of machine learning AI. One kind of explainability is post-hoc explanations that make a non-interpretable model understandable after an action has been executed. This is considered more active than it is in transparency because the system only gets understandable with the provided explanation (Arrieta et al., 2020).

As far as empirical evidence is concerned for the effects of transparency and explainability, a recent systematic literature review covering 17 experimental studies found a promising effect of automation transparency on situation awareness and operator performance, without the cost of added mental workload (Van de Merwe, Mallam, & Nazir, 2024). According to Endsley (2023), SA is best supported by display transparency that is current and prospective, whereas explainable AI is primarily retrospective and directed at building mental models. In a study directly comparing the effects of transparency and explainability on trust, situation awareness, and satisfaction in the context of an automated car, Schraagen et al. (2021) showed that transparency resulted in higher trust, higher satisfaction, and higher level 2 SA than explainability. Transparency also resulted in a higher level 2 SA than the combined (transparency + explainability) condition, but did not differ in terms of trust or satisfaction.

These results are promising and show how abstract ethical principles such as transparency and explainability can be operationalized in AI systems. According to Endsley (2023), future research is needed to design effective AI transparency techniques and to demonstrate that needed SA and trust at manageable workload levels are achieved in the real-world conditions where these systems will be used. It is particularly important to keep in mind the time-constrained conditions under which most military commanders operate. Care must be taken not to overload commanders with information. In some cases, there just is not enough time for a lengthy explanation. To safeguard these situations, we need to focus more on the development process up-front and make sure that things go right there. AI should be able to explain itself during this development process, not merely to system designers but also to end users. And once end users have obtained sufficient trust in these systems, they may be fielded in operational contexts, with high time pressure and high stakes. And even then, the military leader should always be aware of contextual limitations.

CONCLUSIONS

The responsible use of AI in military systems is a relatively recent development. Operationalizing ethical principles in the design and development process is an ongoing challenge. There are also many research challenges, not only in the software engineering field but also in the human factors engineering field. The discussion on the interconnections between AI, AWS, and MHC is a complicated and politically charged one. This has sometimes led to oversimplifications and a tendency to reduce the inherent complexity and ambiguity of the subject matter.

The central thesis in this chapter has been that there is no such thing as absolute autonomy, only various gradations of 'bounded' autonomy. Autonomy is bounded by the mission control, or strategic control, exercised by humans over the platforms or weapon systems that they develop. This type of mission control should lead to extensive testing and evaluation, in order to verify that the systems being developed do what they are supposed to do. No military commander would want this otherwise, or else they would not trust the systems they are in control of. We should therefore shift some of our attention from the 'sharp end' of weapon system impact or use, to 'blunt end' weapon system design and development, without neglecting the human-machine interaction that is crucial for the soldier on the ground to interact effectively with AI. This is not to say that this is easy or without its own challenges. There are still, and for the foreseeable future, huge challenges in the certification process, the requirement for AI to be able to explain itself, certainly during the development process, and the way humans and AI need to be able to work together. Over the past couple of years, research has made considerable progress in these areas, yet, there is still a lot to learn.

At the same time, current developments on the battlefield impose their own dynamics on the weapon systems being developed. Systems are being deployed that use AI to recognize targets. Some of these systems are said to operate in 'autonomous mode'. It is my hope that the reader of this book will be able to look critically at these developments and discussions. While the term 'autonomy' is often used loosely, such usage does conjure up images of 'killer robots' that are out on a killing spree on innocent citizens. This is not what modern warfare, with its highly distributed and deliberate targeting process is about. In democratic societies, we should hold on to accountability and conformance with IHL. The distinction between mission autonomy and platform autonomy makes it clear that weapon systems need to conform to ethical and legal standards, and that MHC is a way to impose those standards on the design, development, and deployment process (rather than on 'controlling' the weapon systems after activation). The fact that terrorists, rogue nations, or even democratic states, may use MHC for their own, undemocratic, authoritarian, or immoral ends, and may deploy weapon systems to such ends, violating humanitarian norms of proportionality and distinction, should not dissuade us from using AI responsibly in military systems.

REFERENCES

Arrieta, A. B., Díaz-Rodríguez, N., Del Ser, J., Bennetot, A., Tabik, S., Barbado, A., García, S., Gil-López, S., Molina, D., Benjamins, R., Chatila, R., & Herrera, F. (2020). Explainable Artificial Intelligence (XAI): Concepts, taxonomies, opportunities and challenges toward responsible AI. *Information Fusion, 58*, 82–115.
Bainbridge, L. (1983). Ironies of automation. *Automatica, 19*(6), 775–779.

Brown, B., & Laurier, E. (2017). The trouble with autopilots: Assisted and autonomous driving on the social road. In *Proceedings of the CHI conference on human factors in computing systems* (pp. 416–429). Denver, CO.

Chafkin, M. (October 6, 2022). *Even after $100 billion, self-driving cars are going nowhere.* Bloomberg. Retrieved September 26, 2023.

Chen, J. Y. C., Procci, K., Boyce, M., Wright, J., Garcia, A., & Barnes, M. (2014). *Situation awareness-based agent transparency.* Aberdeen Proving Ground, MD: Army Research Laboratory. https://apps.dtic.mil/sti/pdfs/ADA600351.pdf

Crootof, R. (2016). A meaningful floor for 'meaningful human control'. *Temple International and Comparative Law Journal, 30*(1), 53–62.

Cummings, M. L. (2019). Lethal autonomous weapons: Meaningful human control or meaningful human certification? *IEEE Technology and Society Magazine, 38*(4), 20–26.

Dunnmon, J., Goodman, B., Kirechu, P., Smith, C., & Van Deusen, A. (2021). *Responsible AI guidelines in practice: Lessons learned from the DIU portfolio.* Washington, DC: Defense Innovation Unit.

Ekelhof, M. A. C. (2018). Lifting the fog of targeting: "autonomous weapons" and human control through the lens of military targeting. *Naval War College Review, 71*(3), 61–94.

Ekelhof, M. A. C. (2019). Moving beyond semantics on autonomous weapons: Meaningful human control in operation. *Global Policy, 10*(3), 343–348.

Endsley, M. R. (2017). From here to autonomy: Lessons learned from human-automation research. *Human Factors, 59*(1), 5–27.

Endsley, M. R. (2023). Supporting human-AI teams: Transparency, explainability, and situation awareness. *Computers in Human Behavior, 140.* https://doi.org/10.1016/j.chb.2022.107574

Endsley, M. R., Bolte, B., & Jones, D. G. (2003). *Designing for situation awareness: An approach to human-centered design.* London: Taylor and Francis.

Feltovich, P. J., Hoffman, R. R., Woods, D. D., & Roesler, A. (2004). Keeping it too simple: How the reductive tendency affects cognitive engineering. *IEEE Intelligent Systems, 19*(3), 90–94.

Friedman, B., & Hendry, D. G., (2019). *Value sensitive design: Shaping technology with moral imagination.* MIT Press.

Griffith, E. (April 15, 2023). The end of faking it in Silicon Valley. *The New York Times.* Retrieved November 1, 2023.

Hambling, D. (2023). Ukrainian AI attack drones may be killing without human oversight. *New Scientist.* Retrieved October 16, 2023.

Hwang, W., & Salvendy, G. (2010). Number of people required for usability evaluation: The 10+/-2 rule. *Communications of the ACM, 53*(5), 130–133.

Kaber, D. B. (2018). A conceptual framework of autonomous and automated agents. *Theoretical Issues in Ergonomics Science, 19*(4), 406–430.

Koshiyama, A., Kazim, E., & Treleaven, P. (2022). Algorithm auditing: Managing the legal, ethical, and technological risks of artificial intelligence, machine learning, and associated algorithms. *Computer, 55*(4), 40–50.

Kox, E. S., Kerstholt, J. H., Hueting, T. F., & De Vries, P. W. (2021). Trust repair in human-agent teams: The effectiveness of explanations and expressing regret. *Autonomous Agents and Multi-Agent Systems, 35*(2), 1–20.

Leveson, N. G. (2011). *Engineering a safer world: Systems thinking applied to safety.* Cambridge, MA: The MIT Press.

Metz, C. (February 1, 2023). Self-driving car services want to expand in San Francisco despite recent hiccups. *The New York Times* (nytimes.com). Retrieved September 26, 2023.

Nathanael, D., & Papakostopoulos, V. (2023). Three player interactions in urban settings: design challenges for autonomous vehicles. *Journal of Cognitive Engineering and Decision Making, 17*(3), 236–255. https://doi.org/10.1177/15553434231155032

National Academies of Sciences, Engineering, and Medicine (2021). *Human-AI teaming: State of the art and research needs*. Washington, DC: The National Academies Press.

NATOTerm (2023). NATOTerm: The Official NATO Terminology Database. NATOTermOTAN - Home

NATO Science and Technology Organization (STO) (2023). Meaningful human control of AI-based systems workshop: Technical evaluation report, thematic perspectives and associated scenarios. In *STO Meeting Proceedings* (MP-HFM-322). Neuilly-sur-Seine Cedex: NATO STO.

Philipps, D. (April 15, 2022). The unseen scars of those who kill via remote control. *The New York Times* (nytimes.com)

Renner, L., & Johansson, B. (2006). Driver coordination in complex traffic environment. In *Proceedings of the 13th European conference on cognitive ergonomics (ECCE 2006): Trust and control in complex socio-technical system* (pp. 35–40). Zurich, Switzerland.

Roth, E. M., Bisantz, A. M., Wang, X., Kim, T., & Hettinger, A. Z. (2021). A work-centered approach to system user-evaluation. *Journal of Cognitive Engineering and Decision Making, 15*(4), 155–174.

Russell, S. (2022). Banning lethal autonomous weapons: An education. *Issues in Science and Technology, 38*(3), 60–65.

Sarter, N. B., Woods, D. D., & Billings, C. E. (1997). Automation surprises. In G. Salvendy (Ed.), *Handbook of human factors and ergonomics* (2nd ed.) (pp. 1926–1943). New York: Wiley.

Scharre, P. (2018). *Army of none: Autonomous weapons and the future of war*. New York: W.W. Norton & Company.

Scharre, P., & Horowitz, M. C. (2015). *An introduction to autonomy in weapon systems*. Center for a New American Security Working Paper.

Schmettow, M., Vos, W., & Schraagen, J.M. (2013). With how many users should you test a medical infusion pump? Sampling strategies for usability tests on high-risk systems. *Journal of Biomedical Informatics, 46*, 626–641.

Schraagen, J. M. C., Kerwien Lopez, S., Schneider, C., Schneider, V., Tonjes, S., & Wiechmann, E. (2021). The role of transparency and explainability in automated systems. In *Proceedings of the 2021 HFES 65th international annual meeting* (pp. 27–31). Santa Monica, CA: Human Factors and Ergonomics Society.

Simon, H. A. (1947). *Administrative behavior*. New York: Palgrave Macmillan.

Simon, H. A. (1955). A behavioral model of rational choice. *Quarterly Journal of Economics, 69*, 99–118.

Simon, H. A. (1957). *Models of man: Social and rational*. New York: John Wiley.

Simon, H. A. (1990). Bounded rationality. In J. Eatwell, M. Milgate, & P. Newman, (Eds). *Utility and probability* (pp. 15–18). London: Palgrave Macmillan.

Simon, H. A. (2000). Bounded rationality in social science: Today and tomorrow. *Mind & Society, 1*(1), 25–39.

Sipe, N. (March 23, 2023). We were told we'd be riding in self-driving cars by now. What happened to the promised revolution? (theconversation.com). Retrieved September 26, 2023.

Slaughterbots (2017). Future of Life Institute. Retrieved October 13, 2023, from https://youtu.be/9CO6M2HsoIA?feature=shared

Sushereba, C. E., Militello, L. G., Wolf, S., & Patterson, E. S. (2021). Use of augmented reality to train sensemaking in high-stakes medical environments. *Journal of Cognitive Engineering and Decision Making, 15*(2–3), 55–65.

Taddeo, M., & Blanchard, A. (2022). A comparative analysis of the definitions of autonomous weapons systems. *Science and Engineering Ethics, 28*, 37–59.

Textor, C., Zhang, R., Lopez, J., Schelble, B. G., McNeese, N. J., Freeman, G., Pak, R., Tossell, C., & de Visser, E. J. (2022). Exploring the relationship between ethics and trust in human-artificial intelligence teaming: A mixed methods approach. *Journal of Cognitive Engineering and Decision Making, 16*(4), 252–281.

United Nations Security Council (2021). Letter dated 8 March 2021 from the Panel of Experts on Libya established pursuant to resolution 1973 (2011) addressed to the President of the Security Council. N2103772.pdf (un.org). Retrieved October 13, 2023.

U.S. Department of Defense (2023). DoD Directive 3000.09. Autonomy in weapon systems. Washington, DC: Department of Defense. DoD Directive 3000.09, "Autonomy in Weapon Systems," January 25, 2023 (whs.mil).

Van de Merwe, K., Mallam, S., & Nazir, S. (2024). Agent transparency, situation awareness, mental workload, and operator performance: A systematic literature review. *Human Factors, 66*(1), 180–208

Woods, D. D. (2016). The risks of autonomy: Doyle's catch. *Journal of Cognitive Engineering and Decision Making, 10*(2), 131–133.

Index

Printed and bound by CPI Group (UK) Ltd, Croydon, CR0 4YY

17/10/2024

01775656-0017